Springer Series in Design and Innovation

Volume 1

Springer Series in Design and Innovation (SSDI) publishes books on innovation and the latest developments in the fields of Product Design, Interior Design and Communication Design, with particular emphasis on technological and formal innovation, and on the application of digital technologies and new materials. The series explores all aspects of design, e.g. Human-Centered Design/User Experience, Service Design, and Design Thinking, which provide transversal and innovative approaches oriented on the involvement of people throughout the design development process. In addition, it covers emerging areas of research that may represent essential opportunities for economic and social development.

In fields ranging from the humanities to engineering and architecture, design is increasingly being recognized as a key means of bringing ideas to the market by transforming them into user-friendly and appealing products or services. Moreover, it provides a variety of methodologies, tools and techniques that can be used at different stages of the innovation process to enhance the value of new products and services.

The series' scope includes monographs, professional books, advanced textbooks, selected contributions from specialized conferences and workshops, and outstanding Ph.D. theses.

Keywords: Product and System Innovation; Product design; Interior design; Communication Design; Human-Centered Design/User Experience; Service Design; Design Thinking; Digital Innovation; Innovation of Materials.

How to submit proposals

Proposals must include: title, keywords, presentation (max 10,000 characters), table of contents, chapter abstracts, editors'/authors' CV.

In case of proceedings, chairmen/editors are requested to submit the link to conference website (incl. relevant information such as committee members, topics, key dates, keynote speakers, information about the reviewing process, etc.), and approx. number of papers.

Proposals must be sent to: series editor Prof. Francesca Tosi (francesca.tosi@unifi.it) and/or publishing editor Mr. Pierpaolo Riva (pierpaolo.riva@springer.com).

More information about this series at http://www.springer.com/series/16270

Daniel Raposo · João Neves · José Silva
Editors

Perspective on Design

Research, Education and Practice

Springer

Editors
Daniel Raposo
Campus da Talagueira
Escola Superior de Artes Aplicadas do
Instituto Politécnico de Castelo Branco
Castelo Branco, Portugal

João Neves
Campus da Talagueira
Escola Superior de Artes Aplicadas do
Instituto Politécnico de Castelo Branco
Castelo Branco, Portugal

José Silva
Campus da Talagueira
Escola Superior de Artes Aplicadas do
Instituto Politécnico de Castelo Branco
Castelo Branco, Portugal

ISSN 2661-8184 ISSN 2661-8192 (electronic)
Springer Series in Design and Innovation
ISBN 978-3-030-32414-8 ISBN 978-3-030-32415-5 (eBook)
https://doi.org/10.1007/978-3-030-32415-5

This Springer imprint is published by the registered company Springer Nature Switzerland AG
The registered company address is: Gewerbestrasse 11, 6330 Cham, Switzerland

Preface

Design processes have been modified over time, not only to adjust to each problem but above all to respond to systemic changes in new cultural, social, environmental, and economic contexts that impact on artifacts, theories, meanings, and behaviors. Also, design research processes have accompanied these emerging changes, focusing on issues inherent to the design project (procedural, methodological, plastic, symbolic, and efficient) and understanding the variables that influence the context. This book provides a spectrum of valuable contributions to better understand the challenges posed to design research by contemporary reality, but also possible avenues of research development proposed by a wide variety of practitioners and researchers. Therefore, the book can be considered from a global perspective of design research as well as observed from specific subjects on design research.

This book presents preliminary and final results of research experiments and research processes, which represent new contributions and an advance in scientific, academic and practical knowledge in the areas of communication design, fashion design, interior design, product design, and intersection areas. It is intended especially for the design community (graphic, fashion, product and interior design) and for people working in the communication fields, such as marketers, journalists, publicists, artists, entrepreneurs, brand or corporate communication managers. Teachers and students in these fields will also benefit of this book; graphic artists, the printing press, researchers, and all those who practice an activity related to the design and communication may find useful information and a source of inspiration in this book.

All in all, this book opens up new perspectives in design policy, design teaching, and design research. It presents studies specifically related to design methods, cultural aspects, symbolic components, drawing and visual expression, visual, and territorial language systems. It covers brands, information and interaction design, digital media design, character design, color perception, and design teaching methods. It deals with fashion trends, technologies in design, furniture design, interior design, and car design. It is organized in four parts:

First Part: Design Education, Research and Society;
Second Part: Communication Design;
Third Part: Drawing and Image in Design;
Fourth Part: Fashion Design, Interior and Product Design.

The chapters are the result of an invitation, addressed by the EIMAD Executive Committee, to the authors of the best articles approved by blind peer review and presented at the 6th EIMAD in February 2018. The EIMAD—Music, Arts, and Design Research Meeting, organized by the Higher School of Applied Arts of the Polytechnic Institute of Castelo Branco, Portugal, aims to be a space for scientific discussion in the areas of design and music.

This book gathers together 29 chapters written by researchers and teachers in design. They show design is a transdisciplinary practice that cuts across various fields within the design discipline and beyond a singular academy perspective.

The models and practices described in various chapters question specific learning, complementing designers' professional profiles with "connective" competencies—competencies that allow designers to transition and lead design processes across different professional sectors collaboratively.

The new mental models and professional profiles of future designers require several didactic changes throughout the curriculum. The chapters present best teaching practices aimed at fostering collaboration and collective learning.

Challenges presented in the publication relate to cross-cultural analysis in the areas of branding, graphic design, and fashion, also different perspective approaches, including multi-disciplinary methodologies benefiting more effective design results.

The chapters describe the qualitative methodology, some of which are exploratory, based on case studies or learning experiences, complemented by a thorough literary review aimed at developing project design practices.

The chapters aim to contribute to the discussion in design teaching and practice and its consequent research context, considering participatory practices and methodologies and differentiating in the design object the cultural artifact of the discursive object.

The spectrum of themes brought together expresses the intention of design practices in the organization of discursive organization and knowledge. As mentioned by Findeli et al. (2008), the set of methods, modes, and means that belong to the material culture that when applied or appropriate within other methodologies, remains the right syntax and grammar of the design project. The design project "organizes" knowledge to create a whole. It is in the hermeneutic transformation of knowledge into action, in the pragmatic approach of knowledge for design, that the transdisciplinarity resides.

Although the variety of topics the breadth of focused themes reveals a unique ability in the design project to extract information. According to Findeli and Russell (2010) being able to use the design as a way to extract and generate knowledge requires a "project-oriented perspective," i.e., the use of designed things and their possibilities as the primary research tool in research.

Castelo Branco, Portugal

Daniel Raposo
João Neves
José Silva

References

Findeli A, Brouillet D, Martin S, Moineau C, Tarrago R (2008) Research through design and transdisciplinarity: a tentative contribution to the methodology of design research

Findeli A, Russell K (2010) Searching for design research questions: some conceptual clarifications. In: Chow R, Jonas W, Joost G (eds.) «FOCUSED»—Current design research projects and methods Swiss design network symposium 2008, Berlin, pp 278–293

Contents

Design Education, Research and Society

Why Design Schools Should Take the Lead in Design Education

Jan Eckert

Abstract Over the past decades, we have gained more design literacy in both the educational landscape and in the professional field. As a consequence, more and more actors have become involved in design and amongst them many non-designers. While in business this change has brought up a number of new role models in design, in education instead many programmes such as "Design Thinking", "Design Management", "Strategic Design" or "Design Engineering" emerged at schools of management, departments of computer sciences or engineering schools. This change has led to a situation where educating designers it not limited to design schools and designing is a profession not exclusively limited to designers anymore. Within this wide range of actors, it becomes hard to understand who is able to lead the design conversation across different sectors and accordingly who is willing and able to train these leaders. This chapter gives an insight into the development of our new MA curriculum in Design. The main objective of this curriculum is stepping out of the mental model of designers as problem-solvers or authors and shifting the attention towards problem identification and design leadership. As part of the research done during the development of our curriculum a new model, called the Y-shaped-Designer has emerged. The model questions discipline-based learning by complementing the professional profile of designers with what we call "connective competences"—competences that enable designers to transition in a collaborative mode and lead design processes across different professional sectors. The new mental model and professional profile of our future graduates required a series of didactical changes along the curriculum. These changes will be laid out in this chapter by illustrating a couple of didactical best practices that aim at fostering collaboration and collective learning. As a conclusion two concepts emerge as key to how design schools could take back the lead in design education: the fundamental shift from authorship towards leadership in design and the ability to connect and collaborate beyond the design domain.

J. Eckert (✉)
Lucerne School of Arts and Design, Lucerne University of Applied Sciences and Arts, Lucerne, Switzerland
e-mail: jan.eckert@hslu.ch

D. Raposo et al. (eds.), *Perspective on Design*, Springer Series in Design and Innovation 1,
https://doi.org/10.1007/978-3-030-32415-5_1

Keywords Design education · Y-shaped designer · Design leadership · Issue-based learning

1 In Reality Design Is not that Important Anymore …

During a recent talk at the Design-In-Tech 2019 conference, interaction designer and former associate director at MIT Media Lab, John Maeda discussed the topic of design advocacy in (tech) business. After years of pushing for design leadership in companies, now, he apparently comes to the conclusion that "in reality, design is not that important" (Schwab 2019). Hearing someone like Maeda making such a statement raises the question about what reason had led to this rather drastic change of someone's opinion. The reasoning behind Maeda's critique is his observation that many designers develop bossy attitudes when assigned leadership roles. And apparently rather than teaming up with non-designers these new born design bosses tend to look for allies coming from a design background. In Maeda's words a "microworld of aesthetic high-fives" (Schwab 2019) gets created where designers cultivate their "invisible language" and decide about what good design is and what isn't. In his recent talk, Maeda even emphasizes his critique and complains that "Over half the designers still want to make things beautiful and can't help it. That's a built-in competency" (Schwab 2019).

When in 2018, I held a keynote about the development of our new MA curriculum in Design, I supported the long quest designers have taken to gain more decisional power in the professional field. I also cited Kate Aronowitz, partner at Google Ventures, who stated, that on the one hand, nowadays, designers have the opportunity to have "a seat at the table" (Aronowitz 2018), on the other instead many designers are falling short when taking bigger leadership roles. According to Aronowitz as design leaders "we are paying too much attention to the 'design' part of the role and not enough to 'leadership'" (Aronowitz 2018). Both Maeda and Aronowitz seem to confirm that the change designers have been struggling for such a long time comes with new challenges and responsibilities that designers were not equipped for. The question raises where this shortfall in leadership competencies originates from? Why can't so many designers help it to focus exclusively on making things beautiful instead of paying attention to other aspects, too?

As a design educator my guess is: because no one really taught them to. The change design is currently undergoing apparently has not been foreseen by many design schools and, therefore, students were not able to build up the necessary competences to face their new roles in design and design leadership.

At the same time, other educational facilities such as management schools, engineering schools, faculties in computer science and even medicine came up with programmes such as *Strategic Design, Design Management, Design Thinking* or similar. A recent study at our University of Applied Sciences and Arts in Lucerne has shown that out of 123 listed programmes using keywords such as *Design Management, Design Thinking, Design & Innovation* or *Strategic Design* over 30% are being offered at non-design departments or schools.

2 Towards a New MA Curriculum in Design

When in 2016, I took over the lead of our MA programme in Design at Lucerne University of Applied Sciences and Arts, we instantly started reviewing the curriculum at that time in order to understand how to develop a new one that was ready to meet the change in the design profession. We did so by a two-folded approach: understanding where and how Swiss designers work and where the people we educate (will) come from? Furthermore, we transferred this double point of view to both our curriculum at the time and the hypothetical one that we were about to develop (Fig. 1).

In order to get an accurate description of the current status, we conducted a series of surveys, interviews and in-classroom evaluations with our students, followed up by evaluation workshops with both students and faculty members. At the same time, an alumni survey reached out to gain insights over the challenges that graduates might have encountered when transitioning into the professional field. To incorporate also the professional's perspective, a series of interviews with representatives of the Swiss professional associations helped to complete this image.

During that same period, Weckerle et al. (2016) published the 2016 Swiss Creative Economies Report: a biannual study that depicts the current state of creative occupations in Switzerland by presenting a number of future prospects for the field. One of these perspectives happened to stand in contradiction with one of our findings when talking to our alumni: according to the Swiss Creative Economies report around 50% of Swiss creatives work in non-creative contexts whereas only 14% of our alumni reported to work outside of the traditional design domain. This first finding raised the question whether there was an ongoing change that, so far, has not been

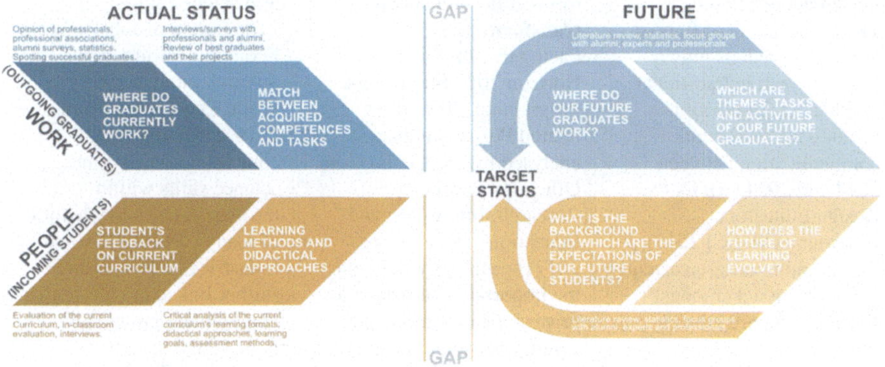

Fig. 1 Developing a new MA curriculum in Design (Eckert 2018)

noticed. Based upon this insight we conducted an extended literature review on the creative economies, the future development of work as a whole and the significance of creative occupations in that future.

As a result of our review, we found that most studies were stating that traditional design occupations were assigned to be less future-proof than those that actively extend into non-design contexts (Table 1). This insight led to the conclusion that according to most studies the current occupational field of our graduates corresponded to a rather outdated status and it was highly probable that this status is about to change in the upcoming years. As a successive insight, we found that the relevance of design skills and topics in non-design sectors was regarded as one of the fundamental drivers for innovation and prosperity by most of the reports (Table 1). It appeared that not only the design sector was reaching out to other sectors but it happened to occur also the other way around—an assumption that is backed by

Table 1 Key findings from our literature review on the future of design occupations

Designers embedding in non-design sectors In their 2016 report on the Swiss Creative Economy, Weckerle et al. (Weckerle et al. 2016) describe the "Embedded Creative" as a figure, who primarily works in a non-design context. According to the same report already today, 50% of Swiss creatives work as so-called "Embedded Creatives".	**Cross-sector collaborations** In its 2016 report on Creative Economies and Innovation (Bundesministerium für Wirtschaft und Energie 2017), the German Federal Ministry for Economic Affairs and Energy predicts Design the highest growth potential in cross-sector innovation and names collaboration and cooperation as two main factors, which might foster this sort of innovation in the Creative Industries.	**Interdisciplinary expertise is key to designers** The 2017 "AIGA Designers 2025 "report (AIGA Design Educators Community (DEC) 2017) discusses today's design problems that "are increasingly situated within larger systems characterised by interdependent relationships" These Relationships "are physical, psychological, social, cultural, technological, and economic in their effects, (and) require interdisciplinary expertise" (ibid.).
Less growth in traditional design occupations The U.S. Bureau of Labour Statistics, predicts only a 0–1% growth in traditional graphic design positions versus 27% in network communication between 2014 and 2024 (AIGA Design Educators Community (DEC) 2017). Similar numbers are currently being discussed amongst European reports, too.	**Design as driver for innovation and social transformation** Facing the 9th European Horizon 2020 Framework Programme, 20% of the related Position Papers analysed by the "Swiss Core Office for European Research, Innovation and Education" (Swiss Core 2017) specifically mention the importance of design as driver for innovation and catalyst for social innovation.	**Combination of creative and science skills** "The Future of Jobs" (World Economic Forum 2016)—a research published by the World Economic Forum in 2016 states that "the combination of arts and science skills within businesses (is) a key feature of many parts of the Creative Industries". A fact which is further linked to "6% higher employment growth and 8% higher sales growth." (Balzagette 2017).

the increasing number of non-designers applying to our MA programme in design. While evaluating all of our findings and according to our point of departure (Fig. 1) we came to the following two conclusions:

(a) The working context of our future graduates is extending into other sectors than design.
(b) At MA level we will educate an increasing number of non-designers that want to transition into the design domain.

Consequently, our hitherto approach of providing designers the opportunity to continue their studies MA programme by specialising in their original discipline appeared to be obsolete and the question rose: what could be an alternative model to the disciplinary approach?

3 Y-Shaped-Designers and Connective Competences

As discussed in section two of this chapter. Our preliminary research on design curricula led to the conclusion that our former approach of our MA curriculum to focus on the same design disciplines that were being taught on BA level was very limiting when compared to the cross-sectoral extension that is happening in the professional world. Especially, if considering the fact that we might need to address more and more non-designers with our programme. So, the question arose how to meet these new requirements with our re-visited curriculum.

For decades, the disciplinary approach in both education and the professional field has been described by using the so-called "T-Shape" (Guest 1991; Brown 2010) representing a vertical *root* that represents the disciplinary base of a person's knowledge and skills and a horizontal dimension standing for the ability to collaborate across other disciplines than the own one (Fig. 2).

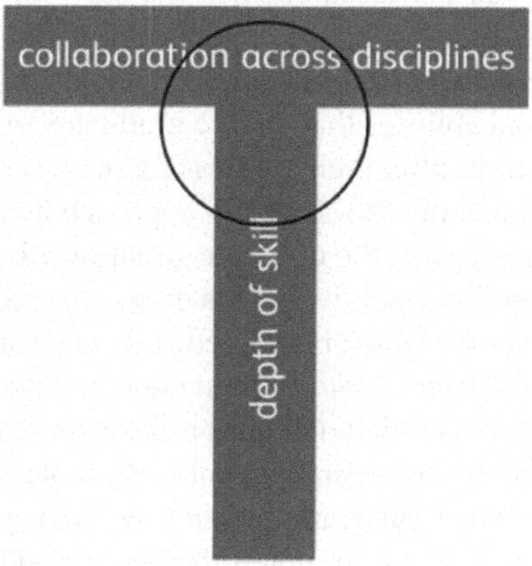

Fig. 2 T-Shaped model (Guest 1991; Brown 2010)

In the post-Bolognese landscape of educational programmes, this model often has been interpreted in two different ways. In some cases, the programmes provide a sort of preliminary and interdisciplinary foundation course (in the german-speaking design world also known as *Gestalterischer Grundkurs*, transl.: design foundation course) that would be followed-up by a more disciplinary and specific programme (e.g. a Master's programme). In other cases, students might first enrol a rather disciplinary programme such as e.g. Graphic Design in order to extend their studies into interdisciplinary fields after their first degree.

In our case we were facing a mix of both: all of our BA programmes at Lucerne University of Applied Sciences and Arts would be subdivided by design disciplines such as e.g. Graphic Design, Illustration, Textiles, etc. In our Master's we would then gather students from different design backgrounds and group them by so-called "specialisations" again. So, in many cases, someone who studied Graphic Design in his or her BA would continue the studies in the same discipline again, even if the programme was meant to support multi-disciplinary learning. This observation inspired us to question the T-Model and reflect on how we could meet the confusion between disciplinary and multi-disciplinary design education.

According to our research regarding the development of creative professions, the idea of collaborating across disciplines was not wrong in the first place. It was rather the self-evident way that the T-Shape implies the ability to transition from a disciplinary to a collaborative working mode. If we think back of the new design leaders criticised by John Maeda and Kate Aronowitz in section one of this chapter it becomes quite evident that nobody told those designers how to transition from a designerly way of working (= disciplinary) to a truly collaborative way by leading interdisciplinary or cross-sectoral collaborations. And if considering that the ebility to collaborate and lead collaborative processes doesn't come natural T-Shape obviously lacks some *connecting piece* between the vertical and the horizontal dimension of the shape. We therefore decided to *zoom into* the junction between the vertical and horizontal part of the T-model in order to understand which could be those *connective competences* that link the disciplinary root to a collaborative working mode (Fig. 3).

In his research, Swiss didact and learning researcher Hansruedi Kaiser (2005, 2011) established the concept of *situative competences* (formerly known as *concrete competences).* Kaiser's situative competences represent a combination of knowledge and abilities that enable graduates to manage concrete practical or professional situations after their studies (e.g. negotiating skills while working with a large number of stakeholders). Kaiser's approach helped us to align our quest for connective competence with the concrete situations that our graduates might encounter in their future professional lives. This quest also required a revisited mental model of the designers that we were about to educate and train. Since the traditional self-contained designer or *design-author* would not correspond with the situation of the cross-sectoral design leader depicted in our preliminary research, there was a need of changing that mental model by defining a more suitable one. Together with faculty members from both our BA programmes and our MA programme in design, we conducted a series of workshops aiming at the better understanding of the Y-Shape Model and it's implications onto the designers mental model or role. During these workshops, we started

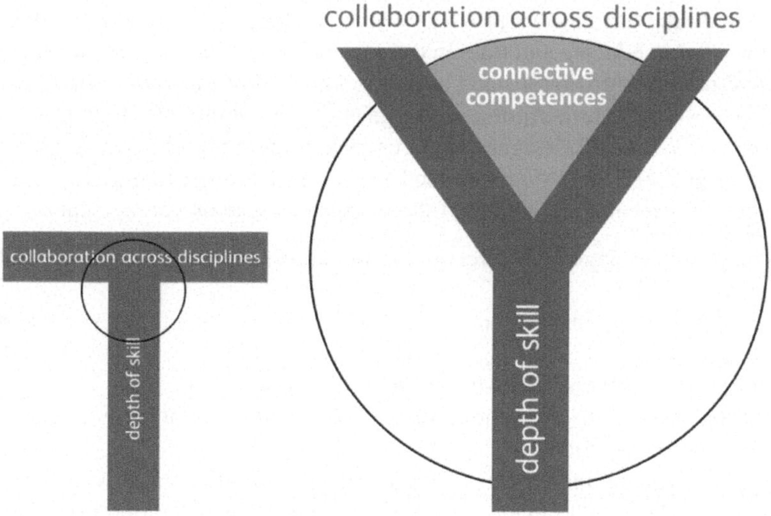

Fig. 3 T-Shaped model and connecting Y-Shape model (Eckert 2018, 2017a, b, 2018)

mapping properties and characteristics that would either be associated with our BA graduates in design and those graduating from our MA programme in design. The Result was a chart showing that while most characteristics and learning goals associated with our BA education would correspond to a rather disciplinary approach, those associated with the MA level, all together were located in the upper *connective* part of the Y-Shape (Fig. 4).

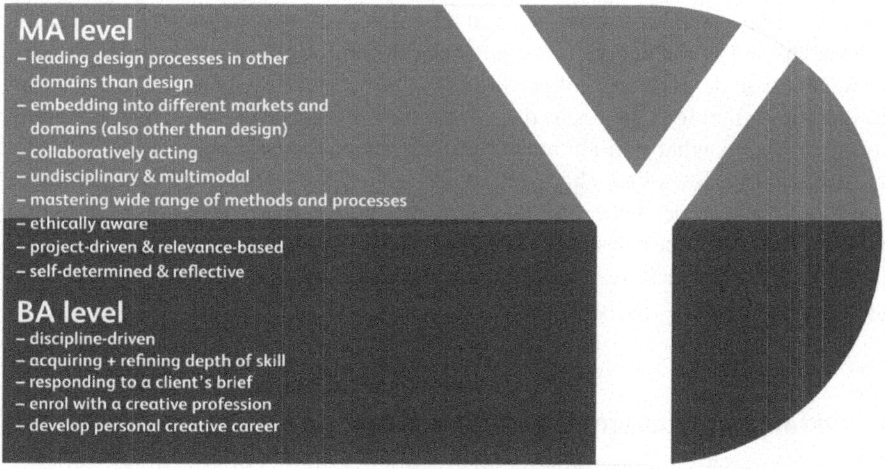

Fig. 4 Y-Shaped designers graduating from BA and MA programmes (Eckert 2017a)

The same diagram gave us also a better understanding of the type of designers that we wanted to educate and train in our master's programme or in other words of the mental model that should stand at the background of our new curriculum—the Y-Shaped Designer. Furthermore, it also gave us some hints of which professional situations this Y-Shaped Designer might encounter and thus which situative competences (Kaiser 2005, 2011) we will need to train and support in our programme. In a first attempt, we came up with the following list of situative competences:

- Handling a wide range of different opinions expressed by both designers and non-designers.
- Evaluating these opinions and creating a common ground and vocabulary for discussion.
- Leading and facilitating cross-disciplinary discussions.
- Identifying areas of friction and leverage points which might be the starting point for a design intervention.
- Critically analysing these areas of friction from multiple point of views.
- Choosing from a wide range of methods in order to lead an appropriate and participatory process.
- Foreseeing, evaluating and controlling the impact of the design process and intervention in ethical, social, economic, ecological and technological terms.

Summarising this section, it became clear to us that the former models used to develop design curricula demonstrated a major gap when compared to the predictions made about the future job market and future design occupations. Especially the latest "Future of the Jobs Report" published by the World Economic Forum in 2019 (World Economic Forum 2019) predicts an increasing need of cross-sectoral job profiles such as "Human-Machine Interaction Designers", "Service and Solutions Designers" or "User Experience Designers". It becomes quite evident that such profiles need interdisciplinary and even cross-sectoral training covering a wide spectrum of skills, knowledge and competences—competences that in our Y-Shaped model we term *connective competences.* Caroline Sinders, who is working as a machine learning designer and therefore relates to the job profiles predicted by the World Economic Forum describes what it might mean to be a Y-shaped Designer:

"The future is less of a T-shaped designer, but a designer that's a Jill-of-all-trades. You should have an aesthetic eye, but you need to be a systems designer when dealing with AI, you need to be a designer that focuses on ethics or (as) an ethicist, you need to have a little bit of a technical understanding. And you need to be highly data-minded and data-questioning" (Schwab 2017).

4 Didactical Changes—Best Practices

After having developed the Y-shaped approach and a first set of competences that our future students were meant to acquire, between fall 2016 and spring 2018, we started a series of didactical pilot schemes and together with our university's didactics we elaborated a list of didactical aspects that were meant to support students when building up what we call connective competences (Table 2). Aim of testing these first didactical schemes was to gather experiences that would allow us establish the right combination of didactical approaches meant for our new curriculum starting in fall 2018. The present section reports on two of these pilot schemes.

4.1 Collective Learning and Conversation Training

As discussed in this chapter before, our former MA curriculum in design would split up students into disciplinary groups called specialisations (e.g. Graphic Design, Illustration, …). According to their specialisation, students would be supervised by one or two teachers having the same disciplinary background as their students. Compared to our Y-Shaped model and our move towards collective learning this separation into disciplinary groups did not make sense anymore. Especially, the fact that formerly, each group would meet on a different day made it impossible for students and faculty members to meet and have a common discussion.

We, therefore, decided to find a way to gather all students and faculty members together in order to collaboratively focus on project-based questions and problems that might be relevant to all of our students. Furthermore the aim was to give them the opportunity to learn from each other instead of being exclusively supervised by teachers. To render this collective way of learning possible, we decided to introduce a weekly *Atelier Day*. A day, where all students and teachers from our programme would meet together in the morning and decide based upon subjects and matters which groups would work together for the rest of the day.

The range of possibilities we explored to train the connective competences in this setting, was surprisingly wide and convincing. After the first week, students took over most of the conversation by bringing in different subjects, organising micro-learning sessions (Hattie 2013), workshops or even lunch cooking sessions. Especially, the high level of self-management showed by the students came as an evidence that it was worth opening up the studio in order to turn it into an open platform for discussion between learners and educators coming from different backgrounds.

In order to even emphasize this openness, together with our students we started inviting external guests such as e.g. experts from a specific field that we were discussing in a project. This way, the conversation got enriched with an external point of view and besides reflecting their work from multiple design angles, students now also had the opportunity to blend in perspectives from disciplines outside of the design domain. A fact that relates a lot to the described shift in the professional field, where designers more and more embed into sectors that don't necessarily belong to the design domain itself.

Table 2 Didactical implications pre- and post-curriculum (Eckert 2018)

Didactical aspect	PRE-curriculum	POST-curriculum
Project- and Issue-based learning	Disciplinary learning-track (e.g. graphic design) and disciplinary studio-work	Project- and issue-based design studio. Taught by multidisciplinary faculty
Collaborative learning	Individual project-work and studio-coaching	Collaborative project-work and group-coaching in the design studio
Context-based learning, scaffolding	One disciplinary project; often approached from a design-only point of view	An initial series (1st semester) of brief interdisciplinary projects in collaboration with external partners shifts the attention to the real-world context
Self-determined learning and micro teaching	Students can choose between a variety of courses and pick one personal subject for their project	Students get involved into the organization of student-led courses, alumni lessons and focus groups. Subjects and projects get discussed collectively by both, learners and educators
Competence-oriented learning and self-evaluation	Students get evaluated and graded by a set of criteria	Students start evaluating themselves with a competence-matrix. Based upon this evaluation a *learning-agreement* is made after the 2nd semester. This agreement is part of the final evaluation and grading as well as the base for an individual coaching to achieve the established learning-goals
External referencing systems	Students develop their MA project in the design studio and mainly get taught and evaluated by the program's own faculty	During the 1st semester, short-projects are held together with external partners and companies. During semester 2 and 3, students set up partnerships for their own project and work at the partner's or company's site
Learning that traverses institutional boundaries across different sites of expression	Students mainly work in the studio and the university's different labs	The studio and the lab (newly collocated on the same floor) merge together. Project-weeks held at companies' sites and collaborations with external partners extend the studio into a *real-world lab*

4.2 Re-Briefing the Brief And Scaffolding

A second format that we introduced in fall 2016 is called CONNECT While in our former curriculum students would work for their entire MA studies on their own project, we realised that this led to a lack of opportunities to test, experience and learn a variety of design processes and methods. For instance, there was little room for students to explore different methods they would learn in our methods class without taking a risk at their own project. Consequently, CONNECT *Projects* were meant to provide that room for exploration and learning by limiting the project's time to 2 months. Starting point for these projects is a primary brief by an external partner (e.g. company or organisation), which then gets re-briefed and critically analysed by a group of students. This not only allows students to engage with the project in a personal way but also pushes them into a leading role, when convincing the external partner why and how the project might be re-framed. At the same time students need to accept and facilitate a large number of different opinions expressed by their fellow group members and the partners (who usually come from a non-design context)—another connective competence that we started training.

From a didactical point of view the CONNECT projects not only turned out to be excellent platforms to train some of the connective competences we wanted to support, they also related to the concept of *scaffolding*. By giving our students the opportunity to test and experience different approaches and methods, together, we build a so-called *scaffold* that enables them to transfer the experience and knowledge to their own projects, too.

5 Conclusions—Taking the Lead by Unlearning Design

This chapter departed from a controversial critique of designers pushing into leading roles in non-design contexts without bringing the necessary requisites. It also started with the hypothesis that the gap might be caused by a flaw in design education. A flaw that pushed other actors such as non-design schools into the domain of design and as a consequence, the question arose of how design schools could gain back the lead in design education. The report on the development of our new MA curriculum in design at Lucerne University of Applied Sciences and Arts has shown that taking this lead back might imply change on multiple levels: not only the structure and content of a programme needs to change, also the mental model standing behind that programme and the didactical approaches meant to transfer it require a lot of research and testing. Especially, the constant monitoring of the professional field's development happened to be one of the main drivers of our research and development of the new curriculum and should be considered in any similar development as well.

Since we introduced our new MA curriculum in fall 2018, we were able to gather experiences with our first group of students. So far, two main observations came as evidence that changing the perspective from a disciplinary and design-focused curriculum to one that focuses collaboration was the right direction to go:

- Due to the weekly collective coaching all of our current student's project proposals include multi-disciplinary perspectives—before, most of the proposals were driven by personal and disciplinary arguments.
- At the end of semester one, more than 75% of our students presented project proposals focusing topics originating from a non-design context with regards to social, economic or ecological issues.

Especially, the second observation comes as proof for a radical shift towards an issue-based design approach that allows our learners identifying problems outside of the design domain. With this new generation of our students John Maeda's notion on designers, who "still want to make things beautiful and can't help it." (Schwab 2019) becomes disproven and we are confident that this upcoming generation of designers has the potential to become design leaders that truly engage with responsibilities within and beyond the design domain.

Coming back to the question of how design schools might take back the lead in design education, our case study of developing a new MA curriculum in design shows that taking this lead comes with similar challenges and responsibilities as the ones in the professional world described by Maeda and Aronowitz: designers need to step back from an author-driven approach that is focused exclusively on aesthetics and design language. At the same time, they need to acquire connective competences enabling them to extend that language to a conversation beyond the design domain and leading that conversation by acting as leaders in cross-sectoral design processes. This conclusion scrutinises the way we kept teaching and learning design so far. And inevitably, it leads to another question: does former design education provide even the right base to transition into the new way we propose? Or: do we first need to *unlearn* part of our design understanding in order to clear the way for this transformation? When we started some of our classes based upon the Y-Model, some of our students first were disappointed that it appeared not being that much about themselves and their own creative work anymore. Some of them even got afraid they might need to forget about all the precious skills and competences they had acquired during their undergraduate studies. And probably the same goes for many of us being faculty members and design educators, too. As an answer to this uncertainty there is probably no better response, than Mark Bonchek's understanding of the concept of unlearning:

"Unlearning is not about forgetting. It's about the ability to choose an alternative mental model or paradigm. When we learn, we add new skills or knowledge to what we already know. When we unlearn, we step outside the mental model in order to choose a different one" (Bonchek 2016).

References

AIGA Design Educators Community (DEC) (2017) Aiga designer 2015—why design education should pay attention to trends. Retrieved 13 Jan 2018 from: https://educators.aiga.org/wp-content/uploads/2017/08/DESIGNER-2025-SUMMARY.pdf

Aronowitz K (2018) Designers finally have a seat at the table. Now what? Retrieved 13 Jan 2018 from: https://www.fastcodesign.com/90156186/designers-finally-have-a-seat-at-the-table-now-what

Balzagette P (2017) Independent review of the Creative Industries

Bonchek M (2016) Why the problem with learning is unlearning. Harvard Bus Rev. Retrieved 26 Nov 2017 from: https://hbr.org/2016/11/why-the-problem-with-learning-is-unlearning

Brown T (2010) T-Shaped stars: the backbone of IDEO's collaborative culture. An interview with IDEO CEO Tim Brown, By Morten T. Hansen. Retrieved 19 Dec 2017, from https://chiefexecutive.net/ideo-ceo-tim-brown-t-shaped-stars-the-backbone-of-ideoaes-collaborative-culture__trashed/

Bundesministerium für Wirtschaft und Energie (2017) Kreativwirtschaftliche Innovationen für mehr Wachstum und Wertschöpfung im Mittelstand – Begleitanalyse von Fördermodellen in Sachsen-Anhalt und Thüringen

Eckert J (2017a) RE-shaping the master curriculum in design. In: REDO - CUMULUS Conference 2017, Kolding, Denmark

Eckert J (2017b) The Y-Shaped designer: connective competences as key to collaboration across disciplines. J Educ Learn 6(4)

Eckert J (2018). O ENSINO DO DESIGN E O DESIGNER DE ACORDO COM O MODELO Y – DESIGN EDUCATION AND THE "Y-SHAPED DESIGNER". EIMAD Encontro de Investigação em Música, Artes e Design, Castelo Branco, Portugal

Eckert J (2018). Leading the conversation—why design education should care more about leadership and stewardship in design. Cumulus Paris 2018—TO GET THERE: DESIGNING TOGETHER, Paris

Guest D (1991) The hunt is on for the Renaissance Man of computing. The Independent

Hattie JAC (2013) Lernen sichtbar machen. German Edition of "Visible learning". Schneider, Hohengehren, Baltmannsweiler

Kaiser H (2005) Wirksame Ausbildungen entwerfen—Das Modell der Konkreten Kompetenzen. h.e.p. verlag, Bern

Kaiser H (2011) Situative Anforderungen und individuelles Können. Retrieved Jan 2016 from http://www.hrkll.ch/typo/fileadmin/Texte/SiKo/Anforderung_und_Koennen.pdf

Schwab K (2017) AI dominated 2017. Here's what's coming next. Fastcodesign. Retrieved 19 Dec 2017 from https://www.fastcodesign.com/90154248/ai-dominated-2017-heres-whats-coming-next?utm_source=postup&utm_medium=email&utm_campaign=Co.Design%20Daily&position=3&partner=newsletter&campaign_date=12182017

Schwab K (2019) John Maeda: "In reality, design is not that important', FastCompany.com [Web Magazine], Retrieved 19 Mar 2019 from: https://www.fastcompany.com/90320120/john-maeda-in-reality-design-is-not-that-important

Swiss Core (2017) How to connect Switzerland and Europe in the field of the arts. Art Driven Innovation in Horizon 2020 Euresearch. Retrieved 13 Jan 2018 from: https://www.euresearch.ch/index.php?id=443&tx_seminars_pi1%5bshowUid%5d=709&no_cache=1&user_id=46336&job_id=3227

Weckerle C, Page R, Grand S (2016) Von der Kreativwirtschaft zu den Creative Economies—Kreativwirtschaftsbericht Schweiz 2016. Zürcher Hochschule der Künste, Departement Kulturanalysen und Vermittlung

World Economic Forum (2016) The future of jobs—employment, skills and workforce strategy for the Fourth Industrial Revolution

World Economic Forum (2019) The future of jobs report

Rethinking the Role of the Contemporary Designer: Is There a Mismatch Between Theory and Practice in Design Education?

Suzana Dias and Ana Baptista

Abstract Nowadays, factors such as the accelerated cultural dematerialization, the unprecedented participatory culture, and a new kind of social values have become crucial for understanding the human experience. This scenario implies an inevitable revision of design, and its educational approaches so it is adaptable to new circumstances. The relevance of this study lies in the critical approach used to review the current design models, and the discrepancy that has been observed between research processes and practices, and the effective practice and application of design. We will thus be discussing the (underdeveloped) competences and skills of today's designers, namely relating to the participatory approach and the decentralization of their role. These latter concepts are of fundamental importance to reflect on the articulation with the immaterial and relational culture of current days. We argue that what is new in the concept of design is its understanding as a process. It should thus be considered a means to achieve an end, and not an end in itself. As Dunne (2008) states, the design has the potential of being 'a tool of doing'. Therefore, design is seen by the latest research within the field as a means to drive flexible solutions, to produce continuous adaptations in order to contribute to a constantly changing context. In this book chapter, we will thus underline that the education of the future designer should (i) incorporate an articulated understanding of the problems, which are considered within a system or web of relationships, and (ii) promote the development of competences and skills that are essential to manage a consciously decentralized role by using open and flexible methodologies applied to different actors.

Keywords Designer's role · Socially responsible design · Participatory design · Design education

S. Dias (✉)
IPCA ID+/Instituto Politécnico do Cávado e do Ave, Instituto de Investigação em Design, Media e Cultura, Barcelos / Aveiro, Portugal
e-mail: cdias@ipca.pt

A. Baptista
Queen Mary University of London, London, UK
e-mail: a.baptista@qmul.ac.uk

D. Raposo et al. (eds.), *Perspective on Design*, Springer Series in Design and Innovation 1,
https://doi.org/10.1007/978-3-030-32415-5_2

1 Introduction

In recent times, we have been living in a deeply unstable environment that has been demanding a growing adaptation from individuals. We can thus find ideas and projects that seem to react to the lack of unity and meaning caused by current historical circumstances. Some of those aspects were already dealt with and further developed in previous publications (Dias 2015, 2017, 2018; Dias and Moura 2017). Within this background there is a clear perception that we are living in a cultural turning point where the discourse of design is transforming itself.

The need to understand the impact of the current historical context in the area of design and its teaching emerges from the motivation of the authors of this article. As researchers and higher education (HE) educators, we seek to establish a conscious and dialogic reflection about the changes that may change pedagogical processes within a specific disciplinary area: design. This is particularly due to shared concerns and deep interest in tuning an interpretative and critical reasoning about the contemporary world with the new expectations and characteristics of the current design student.

The relevance of this article lies in the critical approach used to review the current design models, and the discrepancy that has been observed between research processes and practices, and the effective practice and application of design. We will thus be discussing the (underdeveloped) competences and skills of today's designers, namely relating to the participative approach and the decentralization of their role. These latter concepts are of fundamental importance to reflect on the articulation with the immaterial and relational culture of current days. In this chapter, we thus intend to highlight a new set of questions related to the context of education and design that, in recent years, have been seen as a concern and thus debated in important conferences (e.g. The 3rd International Symposium for Design Education Researchers in 2015) and publications (e.g. *Design Integrations—research and collaboration*, by Poggenpohl and Sato 2009).

In this chapter, we aim to stress that the education of the future designer should (i) incorporate an articulated understanding of current problems, which are considered within a system or web of relationships, and (ii) promote the development of competences and skills that are essential to manage a consciously decentralized role by using open and flexible methodologies applied to different actors. At the end of their path in HE, students should understand their professional activity as an area of interactions, where their role cannot be individualised or authoritarian. Instead, they should be moderators or coaches, and should be able to determine the structures and rules of a game where one will participate in order to improve the final solution.

Notwithstanding the importance of new labels by which new methodologies and practices in the current design approaches begin to be known, it seems to us that what is new is its understanding as a process. It should thus be considered a means to achieve an end, and not as an end in itself where it is created something that is fixed and physical in order to produce continuous adaptations to a constantly changing context. Or, as stated by Anthony Dunne, 'what is new is trying to use design as a tool of doing' (2008).

2 Aiming for Socially Responsible Design

Throughout the 20th century, we have noticed changes in the approach to design in particular in relation to the role of the user and his/her needs. There has been a shift from an industrial context, which was centred on production, to a post-industrial context, which was focused on services. As Freire points out this transition can be noticed by a move from a total indifference to an actual attempt to understand users' needs, by simulating their role, and including them in the design process (Freire 2009). Nowadays, it is undeniable the importance of understanding each individual as a whole when considering the design project.

The continuous immaterial nature of today's culture has been giving rise to new products and social requirements that are highly different from previous demands. This has profound consequences in the economy, and therefore in the field of design. Norman (2011) stresses that current design practices rather than having a concern about appearance focus primarily on interaction. This is even more necessary in an economy based on services and on the experience of citizens, where technology permeates current daily lives. Also, this encompasses immaterial situations where design is called to change social behaviours.

Since the Industrial Revolution, thinking about product design directed to a market has been the dominant paradigm of the modern and industrial society. However, the model that we have been following, which is product centred design, needs to be restructured and adaptable to the new reality. These changes, namely within economy, demand (re)new(ed) design skills and models that need to be suitable to a changing environment. Despite having been widely discussed, these issues seem to reinforce concerns that emerged in 1960s and 1970s when it was debated whether the post-industrial society should match a post-industrial design process (Cross 1975, quoted by Lawson 2005).

In the early 1970s, Victor Papanek's book—Design for the Real World (1971)—was considered one of the first alternative approaches to the 'market model'. Also, it was a key landmark on reflections about the role of the designer. Other discussions followed Papanek's initial theory that demonstrates the feasibility of thinking about design models for the development of social practices. This expands and reinforces the discussion around the concept of socially responsible design.

Within graphic design, Frascara (2000, 2006) points out that no significant examples and attention had been given to the real needs of the citizen before the 60s. Following the same author, we are able to find moral imperatives in the past linked to the ideologies of William Morris or the Bauhaus—particularly when we think about objects that should be part of ordinary citizens' life (and not only affordable to elite groups), such as beautiful and high quality furniture, books, and architecture. However, this ideology was mainly focused on the formal aspects of objects.

In the 1960s and 1970s, a shift towards social concerns was initiated, and we witnessed a fight for freedom. Social actions with claims expressed in the popular culture triggered social movements. In this historical moment, the awareness of different discourses and the need to communicate with specific groups of society were intensified.

Communication ceased to be perceived as an exclusive domain of the Government or political parties. It started to be possible to hear the voice of anonymous citizens spreading new ideas about the social order for instance. Simultaneously, other discourses and voices addressed many other topics, such as the rejection of the Vietnam War, warnings reflecting a mistrust about quality of the air, unethical approaches of corporations or individuals, and injustice or frauds on different areas.

In 1964, the First Things First manifesto became a milestone within the context we have been describing. In addition to continue to raise awareness, it inspired some practitioners to claim a different positioning for the field of Design: practitioners' skills should be used more meaningfully, and should not serve instrumental, mercantilist or market needs. This manifesto was published again in 1999 in Emigrè n. 51, and was updated at this point: First Things First Manifesto 2000. It stressed that its original message had become even more crucial, particularly due to the explosive growth of globalised commercial culture (Blauvelt 1999).

3 Rethinking the Role of the Designer and the Participatory Approach

The importance of citizens' experience and their active participation in society leads Schalk (2011)—in a lecture entitled Relational aesthetics and institutional change—to argue for a critical review of the role of architecture and design. From her perspective, citizens are increasingly prepared to take initiative by their own. She observes the increasing capabilities on the growing demand of the users or communities for better information, transparency, and willingness in being actively involved in local issues. Therefore, it is not enough that experts 'search for purely technical solutions, simply addicted to the status quo without questioning professional protocols and identities', as Schalk pointed out in a lecture in 2011. This critical perspective asks from specialists and institutions, which have been based on an authoritarian modernist model, to be able to adopt a relational approach to society. As it has been done in the contemporary artistic universe, explained by the relational aesthetics theory (Bourriaud 2002), the approach is underlined by 'their values, systems, their material cultures, social structures and institutionalised practices of their disciplines' (Shalk 2011). If citizens are more enlightened, their demand for quality will be greater, thus becoming more useful collaborators in the design process.

The design researcher Bryan Lawson considers that a modernist approach has kept students' or professionals' actions disconnected from the reality and fundamental cultural elements. Within this model, History is reduced to facts learned academically without connection with the current reality. He points out that, fortunately, these notions of modernism have been overcome by creating the opportunity for this knowledge to influence and guide contemporary design projects (Lawson 2005).

Lawson further considers that the mismatch of existing models in design is justified by a political dimension, due to the willingness to understand the decentralisation of power centres within our society. Focusing on theoretical approaches, Markus points out that there are three possible frameworks for the designer's role in society: conservative, participatory and revolutionary. This follows the designer's willingness in decentralising his/her domain (Markus 1973, quoted by Lawson 2005). In fact, Lawson highlights the reflections of Christopher Jones in the 1970s about the openness of the designer's activity to 'collective control' as a turning point in thinking about the role of the designer. This reflection was taken on by designers to explore new models in design processes. The architect and designer Thomas Markus was one of them (Lawson 2005).

Lawson also argues that the participatory approach is the most appropriate methodology to our reality. Though the designer is at the centre of the design process, we cannot neglect the importance of the role of the client and other people involved in the solution. In fact, contrary to a linear logic within the modernist ideology, the reality increasingly tends to demonstrate that 'problems and design solutions tend to emerge when working together' (Lawson 2005, p. 153).

We can understand the participatory approach within a continuum. On one end, we find a conservative approach, where the role of the designer remains separate from his/her client and markets, and constitutes a professional body not concerned with the public good. On the other end, we find the revolutionary approach. In this case, the designer does not see him/herself as a leader but as a partner and citizen, who aspires to change social structures. However, the designer may fail to prove him/herself as a leader since his/her role is merged with that of the user.

Within this theory, the participatory approach necessarily implies a conscious rejection of the dominant role of the designer, who usually acts individually. Nevertheless, designers need to continue to demonstrate specialised skills within the decision making process. In other words, in this approach the designer is the one who manages the rules of the game, which is not linear nor made entirely of individual decisions. It is rather done by a team throughout the process. By adopting different techniques (such as surveys, games, and simulations), the designer tries to identify the crucial aspects of the problem, make them explicit, and suggest alternative actions, but always by involving non-designer participants (Lawson 2005).

Lawson concludes that the professional area of a designer is very distinctive from other activities. The decentralisation of power proposed by the participatory approach cannot represent a reason for weakening the field's authority. It should rather represent a revitalisation of the area, a new way of looking at the context, and the new role of the designer. Lawson has argued for the importance of the mental processes of designers, in line with research carried out by Nigel Cross (2001)—Designerly ways of knowing. This strand of work highlights that, in addition to 'traditional' human abilities, 'designerly' is perceived as a particular form of knowing that relates to the nature and characteristics found within the mental process of design.

Although the participatory approach implies that we all are potential participants in the design process, Buxton (2007) emphasises that we are not all designers. The non-specialist individual can play a critical role during the design process. Also, creative activities from other fields are critical in the design process. Still following Buxton (2007), the role of the professional designer is to work with users (or clients) within a trainer/trainee dynamics, in order to help them to find an appropriate design solution. Therefore, a distinction of roles becomes necessary to maximise the roles of all those involved in the design process.

4 Impact on Design Education

The growing need for the designer's social involvement and the recognition of the individual experience are key elements that lead us to question the (mis)match of educational models in relation to the development of appropriate competences, skills and attitudes of future designers. Back in the 1930s, Bauhaus professor and designer Moholy-Nagy acknowledged how crucial it was for the designer to develop the ability to see everything in relation and intertwined, having applied this ideology to the teaching and learning model he proposed for his New Bauhaus in Chicago. The researcher Alain Findeli reinforces the importance of Moholy-Nagy idea – 'seeing everything in relationship' (2001, p. 10)—recognising it as a key feature of our era, since it requires a strong adapting to the dynamic reality, and the need to deal with relationships that are invisible in nature.

Although these issues are not recent, and many authors have contributed greatly to rethinking the role and social importance of the designer, this awareness and consequences are not very visible in practice. We can thus observe a delay in the adjustment between what we argue it is today's role of the designer and the education models that should be put into practice in design. This raises questions about the quality of the education and training that a future designer will experience and its impact on his/her future professional performance. Norman (2011) strongly suggests that if the situation does not change, it will impact on the value designers will have for society. If future designers are not properly educated for the real world, then they will not be able to fulfil their professional and social value, as they will not be able to critically participate in the dynamics of the world.

Towards the end of the 1980s, Margolin—the first America PhD holder in History of Design and co-founder of Design Issues in 1984—identified a need for changing the design educational landscape. He clearly pointed out the importance of creating 'a new discipline in design studies' (Margolin 1989, p. 5). Despite the urgency of discussing this issue, Margolin and Margolin (2002) reinforce a lack of attention paid to social design, and to the design model that is centred in well-being. They blame the lack of research approaches as well as the lack of proper advanced training tailored for future designers. In particular, the same authors strongly criticise the unpreparedness especially from product design students—and professionals—as they were not able to respond to the needs of the populations with a similar socially responsible approach that was already being seen in other areas. Margolin and Margolin (2002) highlight enriching collaborations being carried out between architects, psychologists, therapists and other professionals in order to explore the intersection of multiple human needs with landscapes, neighbourhoods or interior spaces to increase positive emotions and well-being. However, these developments were not being observed in the field of product design (Margolin and Margolin 2002).

Niedderer (2013), who was focusing on design for behavioural change in the social context, also demonstrates concerns with the education of future designers. This researcher acknowledges the social model of design practice, which had been proposed by Margolin and Margolin in 2002, and highlights the slow progress of educational issues and designers' overall preparation for what is required from them professionally and socially. The impetus given by social innovation, essentially in design contexts for sustainability, is highly recognised. Simultaneously, Niedderer (2013) stresses the insufficient discussion about ethical implications of design projects for social and/or individual behavioural change, despite their relevance. She thus argues that, within the academic community, there is a lack of interest in researching the social potential of design for behavioural change (Niedderer 2013).

In 2010, Norman stresses that, despite the fact that designers are increasingly asked to deal with behavioural and experience-based issues, schools have not been training future professionals for complex subjects such as human and social behaviour, behavioural sciences, technology, and business. Norman actually states: 'There is little or no training in science, scientific method and experimental design' (2010). In 2011, the same author presents strong criticisms about the traditional approach to design education as it remains authoritarian and closed. Davies (2008) had already reinforced this idea by considering that the priority had been given to the designer who defines 'meaning, value, and good form', and to a design education where 'individual performance and results' control' is (wrongly) valued (Davis 2008, p. 5).

It is urgent to recognise that social design practices are based on the user experience. Therefore, design education is seen as based on integrated and collaborative knowledge, which also come from other areas. Research carried out by Sharon Poggenpohl demonstrate the relevance in using methods of interdisciplinary collaborations within the academic environment, and between the academic and business contexts (Poggenpohl 2009). This perspective thus blurs the boundaries that, from the 1990s, seem to differentiate practices, such as process and project, testing and project, workshop and project. The Dutch Masters' project Werkplaats Typografie, founded in 1998, is one of many examples.

Recently, IDEO has been applying design thinking in schools. This promotes experimental education that encourages and enhances the natural tendency of children to create and experiment things (Brown and Katz 2009). The 'd-School' project at Stanford University, within the Hasso Plattner Design Institute or with a public school, is one example. Brown and Katz (2009) further suggest that the design thinking methodology could help reinventing education in order to 'unleash the vast reservoir of human creative potential' (p. 222).

All that has been stated, mainly regarding the changes of design objectives in the new knowledge society, demand us to pose a new set of questions related to the context of education and design. In recent years, there have been requests for an urgent re-evaluation of current standards within design education. These have been presented at conferences and publications. We may identify, for example, the e-design Conference Visions for teaching in Europe, held in Lisbon in 2009; the 3rd Latin-Design Progress Forum Innovation in Design Education, held Italy in 2011; the International Symposium for Design Education Researchers, Cumulus Association (1st edition in France in 2011, 2nd edition in Norway in 2013); and the VII Ibero-American Conference of University Teaching—Innovation and Quality in Teaching, held in Porto in 2012.

5 Conclusions

The designer's activity and attention have, for a long time, been focused on the object (product-centred) as an end in itself or on the person as a mere user (user-centred). The integration of the design project within a system of relationships has not been considered, and therefore many questions were ignored throughout the design process. Nowadays, the designer cannot ignore that s/he is working for and within complex systems that involve unfamiliar values and identities. Additionally, quite frequently the design project will integrate and coexist with several components of this complex system.

Therefore, we cannot continue to ignore concerns related to the education and training of future design professionals as they need to master new competences and skills. They, for instance, will need to solve technology, innovation, competence, readiness, and generality tensions, also bringing together different design specialisms (Keinonen 2009). Furthermore, they need to be aware of distinctions within 'the increasingly complex domain of contemporary design practice': 'multidisciplinary design', crossdisciplinary design', 'interdisciplinary design', and 'transdisciplinary design' (Dykes et al. 2009).

It is thus central that the training and overall HE experience of future designers combine the understanding of the need of seeing connections and inter-relations, and manage a decentralised role within their own responsibilities. Design education that is based on open and collaborative methodologies, allow the creation of flexible structures, while also being a powerful tool that generates involvement, meanings, and well-being. Therefore, the creation of learning experiences through innovative solutions is needed and should be continuously (re)created. Rather than putting education at the service of design, design should be put to the service of education. We think that this is a challenge for today's (and future) designers and researchers.

In practice, we need to reflect, as a community, on the implications—of current and future, better improved HE models—for students' development (and engagement throughout the degree), curriculum development (actual content and nature of knowledge, and overall pedagogical approaches and decisions), and trained, reflective HE teachers (through continuous professional development opportunities and pedagogical experiences). In fact, the role of HE teachers is undeniably relevant, as stated by several experts:

- 'Teachers need to reflect on the art of teaching and learning as there are no one formula, particularly in a creative, future-oriented discipline' (Poggenpohl 2012, p. 14);
- '(…) we need to develop and teach appropriate conceptual and practical tools, and educate students in those tools so that human needs and wishes—and even emotional, cultural, cognitive or other flaws—could be addressed' (Frascara and Noel 2012, p. 40);
- 'Indeed, educators need to develop pedagogies that meet students where they are, and provide the right conditions for scaffolding their knowledge and helping them to reflect on what they learn (or do not learn)' (Frascara and Noel 2012, p. 50).

In addition with the previous ideas, we conclude this chapter with a question towards the future. The designer Sandy Speicher (2011) mentions that everything can be designed, even the learning experiences. So, how can collaborative research between design and diverse scientific areas (such as anthropology, sociology or behavioural sciences) contribute to the development of truly engaging and meaningful teaching and learning experiences?

References

Blauvelt A (1999) First things first manifesto 2000. Eye Mag (33). http://www.eyemagazine.com/feature/article/first-things-first-manifesto-2000. Cited 10 Nov 2010

Bourriaud N (2002) Relational aesthetics. Les Presses du Réel, Dijon

Brown T, Katz B (2009) Change by design: how design thinking transforms organizations and inspires innovation. Harper Collins, New York

Buxton B (2007) Sketching user experiences: getting the design right and the right design. Morgan Kaufmann, San Francisco

Cross N (2001) Designerly ways of knowing: design discipline versus design science. Des Issues 17(3):49–55. http://doi.org/10.1162/074793601750357196

Davis M (2008) Toto, I've got a feeling we're not in Kansas anymore… In: AIGA Boston Presentation. http://www.aiga.org/resources/content/4/8/5/7/documents/davis_keynote_paper_and_images.pdf. Cited 10 Dec 2011

Dias S (2018) Reflexão sobre o papel do designer contemporâneo – o desajuste entre a teoria e a prática no Ensino do Design. Convergências - Revista de Investigação e Ensino das Artes XI(21). http://convergencias.ipcb.pt

Dias S (2017) O Momento Presente e os Novos Contornos Culturais – reflexões para a área do design. Convergências - Revista de Investigação e Ensino das Artes X(20). http://convergencias.ipcb.pt

Dias S, Moura M (2017) Aspetos do mundo contemporâneo e a tendência relacional no design: uma síntese. In: UD17 Noisewise – Design Research in face of current challenges to knowledge. Sixth Annual Forum on Doctoral Design Research proceedings. Universidade do Porto, Portugal

Dias S (2015) Design Como Processo: Uma reflexão sobre a dimensão lúdica, participativa e relacional (Tese de Doutoramento). Faculdade de Belas Artes, Porto, Portugal

Dunne A (2008) Interpretation, collaboration and critique: interview with Anthony Dunne by Raoul Rickenberg. J Des Manag 3(1):22–28. http://www.dunneandraby.co.uk/content/bydandr/465/0.

Dykes TH, Rodgers PA, Smyth M (2009) Towards a new disciplinary framework for contemporary creative design practice. CoDesign 5(2):99–116
Findeli A (2001) Rethinking design education for the 21st century: theoretical, methodological, and ethical discussion. Des Issues 17(1):5–17
Frascara J, Noel G (2012) What's missing in design education today? Visible Lang 46(1–2):36–53
Frascara J (2006) Designing effective communications: creating contexts for clarity and meaning. Allworth, New York
Frascara J (2000) Diseño grafico para la gente: Comunicaciones de masa y cambio social, 2ª edn. Ediciones Infinito, Buenos Aires
Freire K (2009) Reflexões sobre o conceito de design de experiência. Strateg Des Res J 2(1):37–44
Keinonen T (2009) Immediate and remote design of complex environments. Des Issues 25(2):62–74
Lawson B (2005) How designers think: the design process demystified, 4ª ed. Elsevier
Margolin V, Margolin S (2002) A social model of design. Issues Pract Res 18(4):24–30
Margolin V (ed) (1989) Design discourse. University of Chicago Press, Chicago
Moholy-Nagy L (1969) Vision in motion. Paul Theobald and Company, Chicago
Niedderer K (2013) Mindful design as a driver for social behaviour change. In: Consilience and Innovation in Design. Proceedings of the 5th International IASDR Conference, Tokyo, Japan. http://niedderer.org/po.html. Cited June 2014
Norman D (2011) Design education: brilliance without substance. Core77, Design Magazine & Resource. http://www.core77.com/blog/columns/design_education_brilliance_without_substance_2036.asp. Cited Jan 2011
Norman D (2010) Why design education must change. Core77, Design Magazine & Resource. http://core77.com/blog/columns/why_design_education_must_change_17993.asp. Cited 14 Jan 2011
Papanek V (1971) Design for the real world: human ecology and social change. Pantheon Books, New York
Poggenpohl S (2012) Envisioning a future design education: an introduction. Visible Lang 46(1–2):8–19
Poggenpohl S (2009) Time for change: building a design discipline. In: Poggenpohl S, Sato K (eds) Design integration, research and collaboration. Intellect Ltd., The University of Chicago Press, Chicago, pp 3–34
Schalk M (2011) Relational aesthetics and institutional change. Royal Institute of Technology of Stockholm (KTH School of Architecture). https://www.youtube.com/watch?v=CVggSyV8t8o. Cited 12 Jun 2013
Speicher S (2011) Everything is designed, even learning experiences. https://beplayfuldesign.tumblr.com/tagged/IDEO. Cited 26 Jan 2018

Learning to Learn: Lessons from 25 Years Managing a Design Company Abbreviated in 3 Ideas by the Secretary General of the United Nations

Gonçalo Falcão

Abstract This chapter compiles some key thoughts from my 25-year experience in leading a graphic design company. The chapter was developed with the intention of building a reflective writing process, and specifically to consider how my education and training background shaped my practice, and how my professional experience can be used in a transformative way, back again, in education. I summarize the key lessons I gathered, and articulate them with some relevant points from the speech of António Guterres, the United Nations Secretary General, at the award ceremony of his Honoris Causa degree at the University of Lisbon. Guterres's reflection about the role of his own college education helped transform small design issues into a broader picture about the superiority of academic studies.

Keywords Professional design practice · Reflective writing · Design education · Design ethnography

1 Introduction

Having an experience is not enough to learn. We experience many more things than those we understand; however, it is experience—more than understanding—that influences behaviour (McLuhan [1964] 1994). As such, this chapter tries to understand this or, at least, provide some insightful thoughts about experiences:

> Without reflecting upon this experience, it may quickly be forgotten, or its learning potential lost. It is from the feelings and thoughts emerging from this reflection that generalisations or concepts can be generated. And it is generalisations that allow new situations to be tackled effectively. (Gibbs etal [1988] 1994).

Accordingly, in order to develop this text I will use the Reflective Writing lessons (Watton et al. 2001; Moon 2006). Reflective Writing provides an opportunity for people to gain further insights from their own work through a deep reflection on

G. Falcão (✉)
L Universidade de Lisboa, Faculdade de Arquitectura, CIAUD, Lisbon, Portugal
e-mail: gfalcao@fa.ulisboa.pt

D. Raposo et al. (eds.), *Perspective on Design*,
Springer Series in Design and Innovation 1,
https://doi.org/10.1007/978-3-030-32415-5_3

their experiences, and through further consideration of other perspectives from people and theory. Through reflection, we can deepen our learning from work. Using this method I could better understand how my background training and education molded my practice, and how my professional experience can be used in a transformative way, back again, in education. That is, how to build "a form of mental processing – like a form of thinking – that we may use to fulfil a purpose or to achieve some anticipated outcome (…) [that is] applied to relatively complicated, ill-structured ideas for which there is not an obvious solution" (Moon 2006, p. 37).

Having an experience is not enough to learn. We experience many more things than those we understand; however, it is experience—more than understanding—that influences behavior (McLuhan [1964] 1994). As such, this chapter tries to understand this or, at least, provide some insightful thoughts about experiences: without reflecting upon an experience, it may quickly be forgotten, or its learning potential lost. It is from the feelings and thoughts emerging from this reflection that generalizations or concepts can be generated. It is generalizations that allow new situations to be tackled effectively (Gibbs et al [1988] 1994).

I was an entrepreneur for 25 years. In December 1992 I've founded my own company with two other colleagues from my University. It closed in December 2017 because we couldn't work together anymore.

It is a *cliché* to say that most successful magnates never went to university or are dropouts— the most used examples of this are Bill Gates, Steve Jobs, Mark Zuckerberg, Amancio Ortega, and Richard Branson. However, it is also true that some other very successful businesspersons did finish university studies: Elon Musk, Jeff Bezos, Warren Buffett, Bernard Arnault, or Carlos Slim Helu. Universities are neither a curse to entrepreneurs, nor a road to become a billionaire. In my case, university—higher education—gave me a lot, so the main reason I am writing this chapter using personal experience is to reflect about some of the things given to me as a student, and to try to give them back to new students. Before arriving in what is the "superiority" of university studies, I will address other issues. Recognizing that this personal experience might not be of great value per se, I hope to widen it, going beyond descriptive writing, and with the help of UN Secretary General António Guterres's Honoris Causa speech, provide some key lessons.

2 The Capacity to Adapt to Change

Back then I didn't know, but my University introduce me to my (future) partners. Before exploring this subject, it is good to stress that university also taught me that being a close friend or even a girlfriend/boyfriend does not mean you can work together. Working in groups, a university design process typically teaches us that sometimes affinities in life do not succeed in work (and the other way around). Building an enterprise together demands shared values: for example, honesty, integrity, trust, and commitment—believing in achievable goals, building on other people's ideas.

Several people enjoy having a job. Several people prefer to have someone telling them what to do. Therefore, finding people that can work together and that are mentally open to risk a non-tutored path is challenging. Nevertheless, it can be done in a classroom environment, partly intuitively, partly by testing. Coming to this point, I tried to look around for other companies that started around the same time as mine, from my generation. Most of them are from a single designer, even today. The companies that were already in place when I started have the name of a founding designer. This was probably because design up until the eighties was very authorial, and when we were studying, we believed in a more professional, company-driven approach. Having my name on the door did not make sense to me, on the one hand, because it seemed like an old practice; on the other hand, having a commercial name seemed like the right thing to do, challenging the market practice. At the time, I felt like I had nothing to lose, except my time, so taking risks was not perceived as a big burden.

It is obvious from the introduction that, according to my experience, universities are a good place to meet potential partners. In trying to evaluate this assumption, I find that some of my students end up working with ex-colleagues (now called "schmoozing" in company's lingo). Strangely enough, I could not find scholarly studies in the design field about the way school teammates help each other along their professional lives. In "The Guardian", Professor Zahir Irani, Dean of the School of Professional Development at Brunel University, stated: "Students are buying the whole learning experience, and that includes the opportunity to engage with others through their university or student status" (Pozniak 2014, para. 4). This is obvious in very expensive universities and post-graduate management courses: students build a network of classmates who have parents that can afford high priced tuition. But what about simple design students? In my year, other classmates formed two other companies; however, they went bankrupt shortly after. In my experience as a teacher this kind of grouping of schoolmates does not happen very often in present days, although, it is my impression that they tend to pull friends onboard as soon as a work opportunity arises.

Companies are made of people; startups are people (although the term sounds engine-like mechanics). Companies are generally initiated by few people, so they all have to be on the same page, focused and compatible. However, being a good partner at work is not enough. A company needs to be pulled up, so it is not enough to be compatible in doing design work because there are many other types of work that need to be done in order to get to design work. To be able to collaborate in business, they have to share common beliefs and a kind of life purpose.

3 Develop New Forms of Intervention in Society

Inside a company, several practices are implemented by trial and error with real jobs (hopefully with many trials and few errors). When we began designing, we realized that talking with clients was one of the most important parts, if not the most important part, of the job. Thinking back, I have to admit that it is almost impossible to teach this (talking with clients or with potential clients), and in design it surely is one of the secrets of a successful business. However, it is something that people can develop with some major guidelines. Commercial and business dialoguing is surely something that can be taught and learned. However, there is a great part of it that cannot: namely, a person and their background. From simple things—the way a person speaks, their vocabulary, or keeping a smiley feeling–to more complex ones—body language, being very careful with what these things reveal and give to the other party, or making the other party feel like they are dealing with an expert. Many times, myself and my partners, and even competitors, spoke about how we would love to attend each other's meetings as a fly, or we say enviously that designer x or z is probably amazing in meetings. Whenever we see a senseless work, one of the first thoughts after commenting its senselessness is that we would love to have been in the meeting where that work was presented. I am inclined to think, accordingly, that a lot of value goes in this subject: the meetings. Not necessarily focusing on the client, but on the meeting itself, which is seen as a kind of chess match, a game that we want to win.

When we started a new company, we came up with a way to optimize these skills: we would always go in pairs. One of us would be the main participant in the meeting, whereas the other would be much less concerned with the issue itself and more attentive and focused on observing and taking mental notes about the meeting, with little intervention. When the meeting would end, we would discuss errors (from small to big ones) while sitting in the car or in a nearby cafe (this has to be done quickly and ideally straight after the meeting, when everything is still fresh in one's mind). A common mistake I did was talking too much, and the amount of talking is very important, as is the amount of opinions about others, or about issues surrounding the job. I could not find comparing examples of this practice, but my professional experience tells me that most of my competitors have exactly the same concern. It is very hard for me to be insightful about this issue apart from my own experience; however, I am convinced that discussing the projects roughly, raising questions, making students feel the need to focus when they talk to the teacher is important for the future. This task is increasingly more difficult. Students arrive with a huge self-esteem, because of a childhood filled with positive reinforcement. After childhood, the period when everything they do is amazing, they are given another tool—social networks—and friends keep the task of reinforcing that they are amazing, and beautiful, and loved. Consequently, when we make harsh criticism, we are probably the first ones to have ever said that their creation is not good. Finding the right balance is hard but it is surely necessary, and so is to do it with the utmost honesty and transparency. One of my students this year told me that after a Friday meeting with me she took the bus, travelled 150 km to go to her parents' house for the weekend to have some love, because she was in shock with my criticism. I just told her that her illustrations were still a bit childlike.

I am totally convinced that the observation and critique made by my partner helped me a lot to be a better designer and to perform better in meetings. Having critical inputs from clients is surely much better than the silence of "you're gone."

4 Are We Going Together?

Every company has internal discussion of goals. It is inevitable when there is more than one partner to discuss what to do next, where to go, where to cut, and where to invest. Springing from the knowledge developed inside the company, we drew some questions that we thought we should ask when someone requested us to design a project (this is more evident in branding projects, and extremely useful when in a startup). We ended up with a questionnaire for clients, to use in branding jobs or to develop new businesses. This questionnaire was firstly informal, conversational, and then became, sometimes, physical, printed. While talking to a competitor from another company, I realized they had the same idea (of the questionnaire) and did it in a much more developed and consistent way than us (probably with different questions we never shared). They really felt it was a fundamental tool. This competitor even told me that one of her clients ended their collaboration because of the questionnaire: after the partners took it home and gave it serious consideration, they realized they had very different aims and decided not to go ahead with her company. Olins (2008) talked about an "intensive interview program" in Stage One of a brand development, also stating that it should not aim to be statistically relevant but should be representative. The author referred mainly to interviews on branding programs in big companies, so the method, the aim, and the purpose are different from what I am suggesting, and he recognized that people who are interviewed struggle often with abstract issues, such as vision or core idea. In sum, designing a questionnaire that leads people to envision a future without asking them directly is a difficult task, but one that should be considered and developed with time. Questions like: if your company was a car, which car would it be (brand/model/cost)?; to whom would you feel it would be less painful to sell your company, if in 5 years it grows to be an interesting business?; who would you like to present your project to that would make you feel proud of your idea/achievements?; which Hollywood actor do you feel could represent and advertise your company now, and which would you like to be the face of the company in 5 years?; is your company male or female and why?; etc. Catharine Slade-Brooking (2016) also mentioned interviews and questionnaires in brand developing, but in a much more market-driven analysis, with university research terminology.

This led me to think that discussing goals and long-term ideas is always a good thing, not only for the company, but also to know how to ask others about it. Partners lives tend to change a lot since first starting a company, and they live together 40 hours per week, every day (sometimes more that the hours you live with your lover). One might discover after a while that the business is not his main goal, whereas the other—not married at all—wants to devote all his time and energy to the company. Alternatively, perhaps one might consider money as his main goal, while the other can be more open to losing money for a pleasurable job. One will have family pressures to perform better financially, while the other will not. One might develop personal problems, whereas the other might not. Consequently, aligning goals is always necessary and a good way to keep the decision team together. It is good not only for the entrepreneurial adventure but also to know—by experience—what to ask and how to ask others.

5 Leather Boys

In my company, clients tended to have nicknames. One could be very stiff and conventional in dressing but to us he was "the leather boy." Nicknaming came by accident but become a frequent practice. This was because we like to imagine a client's life, personal tastes, his private life after he abandoned the suit and tie. In a survey to professionals with long professional careers (Falcão 2015) only 7% answered yes and 51% answered "sometimes" when asked whether they had the same habit of nicknaming. Having this in mind, we cannot say that tagging clients is common in design studios, but we can argue that is not a strange practice since 58% of the professionals in the study recognized that, at least "sometimes," they would use it. It was much later in my life that I was aware of a research method called "personas," in which researchers use representative human profiles based on archetypal descriptions. Described by Courage and Baxter (2004), the "personas" research method has some similarities with the nicknaming process, but works in another direction. While in the "personas" method the purpose is to find fictional characters that represent typical users, the "nicknaming method"—never described, never studied—has the opposite purpose. It aims to go deep into a client's life, trying to imagine their more hidden and darker facet. This helps us to frame visual references, aware that this framing is completely based on a totally fictional "persona." It was not a formal design method—it is absolutely intuitive—but it had some impact on the work we did. The "second lives" of our clients were developed along the business meetings and were very detailed, from sexual preferences to the CD's they had in their car. Sometimes we would even get it totally right, as we would discover later. The "nickname method"™ helped us to design better while imagining the client's whole life and not only their business persona.[1]

As with business partnering, design deals a lot with people. A design team is people and a client are people. And if this is true in every commercial area is important to acknowledge that in design this is of extreme importance. Designers work with data, but they work a lot with personal tastes, expectations and culture. So these skills are important and working with students with anthropological methods, interviews, and ways to involve real people in projects is important in school.[2]

---

[1]in the same inquiry of the nickname, I've asked a second related question: "It is common to elaborate a first social portrait of the client from the way he/she dresses/speaks/car/workspace, in an attempt to find a universe of visual references for the work to be developed (which is later confirmed or changed with continuity). of the relationship)?" 58.6% of the designers answered yes.

[2]One of my colleagues in university also from a professional background used to design for IKEA. He would present is design proposals finished, with names and packaging. I asked him how did he coup so well with the IKEA naming for their products. He answered: I would use the name of the wife of the commercial director every time I can.

6 Out of the Door. Line on the Left. One Cross Each. Next. Crucifixion?

Today, design students live in a fast word where much information available from many different media. From cell phones apps to internet videos and tv series, designers have too many images in their minds. We can say it is a wonderful time to be a design student, with the huge array of information available, which is true, but it is also true that this overpopulation is sometimes damaging. I can say that with such an amount of information around (not necessarily images) a designer should never be blocked. "Design as an activity, isn't fueled by personal inspiration, but by knowledge of the world. If we run out of knowledge, we can always go out and find some more. A real designer can't be blocked, just badly informed" (Downs 2011, p. 73).

In Europe (except probably in France, Germany, Italy, and Spain), a lot of design students mix their mother tongues with a proficiency in English that makes them able to learn, explore, and use vast amounts of information in online media and services. This availability is excellent and empowers students. However, it also makes them more susceptible to copy and adapt similar solutions in similar contexts, and of seeing design as a remixing job. An excess, not only of images, but also of words. Buzzwords constantly repeated in popular culture—young, dynamic, proactive, seduction, etc.—make designers use them as if they were a kind of design culture. Thinking about a design solution from scratch requires collecting information and benchmarking, yet it also requires thinking. Designers (and consultants, and managers, and so on) think with drawings and with words. Drawing has a huge tradition in design and there is a lot of critical writing about this topic (Downs 2011). Drawing seems like the "natural" thing to do. Thinking with words is the other side of the designer's work and words do not seem to be the main issue in design project. Getting words back on board might seem like a caustic retaliation.

When I entered university, the institution had a vast tradition in fine arts and had recently started the design studies. It was in a transition period with design as a new study, about 10 years-old. Many curricula were based on fine-arts course units. Design was seen as part of the fine arts and would differ in just a few classes. This transitional situation (half fine arts, half design) put me in contact with semiology and that was something that helped me immensely in my design practice.

7 Learning to Learn

"Structuralism, unlike the other approaches discussed here, is (…) quite indifferent to the cultural value of its object: anything from War and Peace to The War Cry will do" (Storey 2018, p. 116). Structuralism does not teach me to value information, but it did catch my attention as a way to understand the interplay of words and to understand how these can be useful in designing communication. This idea is important to explain why structuralism and its expansion in post-structuralism was so important to me and, I believe, to designers. I never realized this until I started teaching and discovered that students have a lot of trouble using words to convey the right message. It is probably the most important step in a designer's education to realize that the task of organizing things visually is just too easy. If they want to earn money, they have to be able to design communication and, after that, put it into images. The "craft" is only a small part of the story.

> For the most part, it's a hard-working service field that sees itself more occupied with translating speech into visual language than speaking. (…) Many of its corporate-client practitioners are instructed to provide order and to clarity, to give their clients' companies the look, sheen, and promise of a clean new world. It's a fairly neurotic expectation, since designers can't really clean — they just cover, wrap, accent, or put into a clean envelope some messy realities. (Lavin 2001, p. 2).

Swiss linguist Ferdinand Saussure theorized about the concept of Structuralism, which was later developed further by other philosophers in different fields—Louis Althusser in Marxist theory, Roland Barthes in literary and cultural studies, Michel Foucault in philosophy and history, Jacques Lacan in psychanalysis, Claude Lévi-Strauss in anthropology, among others. Structuralists divide words in two parts, the signifier and the signified. The signifier is the word "cat", for example, and the "signified" is the cat, the four-legged feline animal. According to the structuralist model the relationship between these two units is "arbitrary"; it argues that the connection between words and reality is built by language, and that language organizes and constructs our sense of the world. Each language, therefore, makes different "mappings of the real" because it uses words in different ways (Lavin 2001). The classic example that helps explain this idea is the word "snow." In Europe, there are few root words for "snow": the phenomenon of solidification of water temperature is described with the word "snow" and a few others. In Eskimo languages (Yupik and Inuit), there are a large number of words for the same water phenomenon (more than 40).This means that the signified "snow"—natural element, iced water—is, for an Eskimo, a much more complex and detailed element than for a English or Portuguese, where most of the people have to travel north in winter if they want to see snow (https://en.wikipedia.org/wiki/Eskimo_words_for_snow).

Post-structuralism noted that the relationships between words and their signified is not a single, but a multiple one. The word "cat" for example, is used to refer to a feline; but it is also used between jazz musicians, for example, to designate a good player. As such, a jazz cat has nothing to do with the feline animal cat. In Portuguese, "gato" (cat) can signify also a pretty and desirable young man. "Post-structuralists rejected the idea of an underlying structure upon which meaning can rest safe and guaranteed. Meaning is always a process" (Storey 2018, p. 131). From 1977 on, Barthes and Foucault, and others would follow the work of Jacques Derrida (1973) in a further development of the Structuralist idea into post-structuralism, with other contributions by François Lyotard and Jean Baudrillard. Baudrillard's work is, eventually, even more important for designers because he "effectively shows through a semiological analysis that consumer objects constitute a system of signs that differentiate the population. This system of signs cannot become intelligible if each sign is related to each object, but only through the play of difference between the signs" (Poster 1988, p. 3).

Learning these concepts was extremely important to me and I believe that it can be important for future designers, too. This is why very recently I began using it in design teaching. The reason why I believe this is essential for designers is that very soon in our practice we realize the importance of this interplay of references, not only in the written world but also in the image world. As an over-simplified illustration, we can use an example that occurs every year with every design teacher. A teacher might ask the student why he/she used the colour red; and the student will frequently reply that he/she used the red because it signifies blood. And then the teacher answers back that the same color could also mean sex or intensity, Coca-Cola or Nestlé, Canada or Japan, depending on the shape of the red and the colours that interplay with it. The colour red has no fixed signification; it depends on the context in which it is used, or the system of signs surrounding its usage. So do words, as they move around in a web of signification that need to be carefully addressed in the planning phase. Studying these ideas in theory would accelerate designers' understanding of the way their elements move in a world of significance, and help them become more critical about it. In this way, disciplines such as esthetics or linguistics can have a lifetime impact on designer's education. Without university I would have never came across those issues.

8 Camarillo Brillo

Most of the times while designing visual communication professionals try to summarize it in a quintessential idea. Usually, this is a short sentence or a pyramid of ideas expressed in sentences. This is because in many cases designers need to summarize complex ideas about institutions, companies, products and services in ways people can understand and memorize and, most importantly, perceive differences between similar products, companies, institutions or services. A good example is Avis's slogan "We try harder." This sentence succeeds in condensing a company's attitude and business model. Designers often need to put into words a line of thought, which may become explicit or implicit in the final work—that helps him in the process of designing. They need to find images, words, sounds, and feelings that signify that main idea (nicknaming also helps…). If words are ubiquitous, design will be, most likely, generic. Design students live in a world filled with generic words: dynamic, young, fresh, vivid, futuristic, retro, deep, authentic, etc. Inside these words, there is a world of signifiers that, most of the times, are not fully scrutinized. We may ask students to move away from this generic world and pay a closer attention to other words.

The communication design professional and the visual communication design teacher might want to save some time to work with words. In her bestselling book about business, Barbara Minto (2009) explains the importance for the business world of having the ability to summarize ideas into words, and explains how to do it. The author stresses the role of words in getting the right information to be able to work with at a managing and consulting level. The final chapter of the book is dedicated to describing how to put it to words ("Putting it into Readable Words", p. 159). As the book is mostly destined for managers, the author reverses the problem: "Create an image" (p. 161)/"Copy the image in words" (p. 163). Although Minto never talks about post-structuralism (the book is a method for logic in thinking and writing), she uses this philosophic conceptual framework to bring design students to think about words at the initial stages of their design work might prove to be a valuable experience. Once the problem has been defined and the ideation has started, there should be a moment of getting it into words and building a quintessential idea to work with. Exercises—both individual and collective—with words, to understand the multiple meaning a word may get by putting it in different contexts, are a helpful way for design practice to apply and understand, and for further development of visual solutions. A closer understanding of post-structuralism, even if at entry level, helps students go much deeper in their work with ideas expressed with words.

9 Three Ideas from the Secretary General of the UN

This paper aims to be a reflective writing about my work experience and how my background education and training transformed my practice, and how practice can be used in a transformative manner back again in education. Having said that, and to be able to scale all the small issues discussed so far, I recall the speech of the United Nations Secretary General, António Guterres, on February 19, 2018 (Lopes 2018) at the award ceremony for the Honoris Causa degree awarded to him in the University of Lisbon. Guterres presented a set of ideas for the future of university education based on his own education and professional practice. Guterres left two messages about the role of the university. The first is that the fundamental meaning of today's universities and educational systems is not focused on the kind of things that are taught and learnt in those settings, but giving the students the possibility of learning to learn. The idea of an open education teaching the love of learning focuses on an investigative and questioning dimension, as a mission of the university. Referring to the education of his grandchildren, Guterres said that their success would depend essentially on the educational opportunities they are going to have and on the ability they will have to be able to adapt to changes, to develop new forms of intervention in society, and new professional activities. The reason for this, Guterres added, is that the concrete content that they will discuss in school will be out of date when they carry out their professional activities or other forms of intervention in society.

The Secretary-General recognized that higher education will not be able to prepare anyone for a future profession, for the simple reason that we have no idea what future professions will be. The technological evolution of the last years has reduced the life span of a profession to two or three decades. The concept of a university that can always be ahead of its time is difficult to realize, so is the expectation become a specialist in a field. The university will not be able to restructure itself completely—teachers, courses, etc.—every 10 or 20 years, and must provide the capacity for students to be able to adapt to changes. Educational contents will be completely out of date when students pursue their professional activities. At the end of his speech, Guterres left another interesting idea: the Secretary-General deals today with problems that were unimaginable at the time of his graduation, and works with tools that were impossible to conceive when he graduated with the highest grade, 30 years ago. Guterres ended his\intervention by saying that if he knew what was going to happen to him in life, he would still have re enrolled as a student at his school. His training, made without computers and with embryonic calculators, resorting to electrotechnology that is now completely obsolete, continues to serve him for the complex and enormous task at hand. This speech inspires confidence in the generalist: all of us who are active in a professional activity somehow have this feeling that we have our training, made in a world that could not even imagine the existence of the Internet, to deal with the present. Thus, as a starting point, the idea of a generalist education, more concerned with the exercise of reasoning, research, and problem solving in different contexts, still makes sense. The issue starts and ends with university and with the idea of higher education. A "superiority" that helps individuals find the right people to work with, to trace goals, to want to go their own way. A "superiority" that helps individuals become generous enough to be able to work in groups and believe in others people's ideas when they differ from their own. It helps to like people so much that we want to nickname complete strangers and imagine their lives; to like words so much that one feels they have the license to be in the driving seat of building communication. To like society so much that one can really enjoy an amateur bad designed pretensions restaurant menu. This is because one loves to learn and likes to adapt. No university can teach every student how to become a lawyer (probably the most creative profession), a writer, a choreographer, a composer, or a politician. However, sometimes, to transform normal people into designers is possible. The "capacity to adapt to change", "learning to learn", and developing "new forms of intervention in society" are the three main ideas that António Guterres pointed out in his speech, and that somehow I would like to use as a background to all the short stories of my professional life.

References

Downs S (2011) The graphic communication handbook. Routledge, Oxfordshire

Falcão G (2015) Criação Livre e Criação Dedicada, a Encomenda e o projecto em design de comunicação [portuguese]. Dissertation, University of Lisbon

Gibbs G et al ([1988] 1994) Developing students' transferable skills. Oxford Centre for Staff Development, Oxford

Lavin M (2001) Clean new world: culture, politics and graphic design. MIT Press, Cambridge, MA

Lopes M (2018) Guterres, honoris causa, avisa para a falta de regulação das novas tecnologias. Available via PÚBLICO. https://www.publico.pt/2018/02/19/politica/noticia/guterres-honoris-causa-avisa-para-a-falta-de-regulacao-das-novas-tecnologias-1803626. Accessed 9 Aug 2018

McLuhan M ([1964] 1994) Understanding media: the extensions of man. MIT Press, Cambridge, MA
Minto B (2009) The pyramid principle: logic in writing and thinking. Prentice Hall, Essex
Moon JA (2006) Learning journals: a handbook for academics, students and professional development. Routledge, London
Olins W (2008) Wally olins: the brand handbook. Thames & Hudson, London
Pipes A (2007) Drawing for designers. Laurence King, London
Poster M (1988) Introduction. In: Poster M (ed) Jean baudrillard: selected writings, 1st edn. Stanford University Press, Stanford, pp 1–12
Pozniak H (2014) How to establish your professional network while at university. Avaiable via The Guardian. https://www.theguardian.com/education/2014/jan/20/establish-your-network-at-university. Accessed 9 Aug 2018
Slade-Brooking C (2016) Creating a brand identity: a guide for designers. Laurence King Publishing, London
Storey J (2018) Cultural theory and popular culture, an introduction (8th ed). Routledge, Oxfordshire
Watton P, Collings J, Moon J (2001) Reflective writing, guidance notes for students. Exeter University

The (Trans) Disciplinary Alternative for Design

Hermes de Andrade Júnior and Tamar Prouse de Andrade

Abstract People are more demanding about the opportunities offered by the globalized market. There is a wide variety of products and services that border exclusivity and that, due to the advancement of technology, are now within the reach of many. To meet this voracious audience, a new profile of professionals with differentiated backgrounds is emerging. The purpose of this theoric chapter is to bring transdisciplinarity as a scientific and cultural approach presented as an alternative and as a methodological and epistemological suggestion for transdisciplinary practice in the daily life of the designer. We emphasize that the solution of the itinerant chair, transdisciplinary coaching or the consolidation of a transdisciplinary design curriculum for technical, ethically responsible and environmentally conscious high standards are valid alternatives to the designer's educational model. In order to incorporate the transdisciplinary nexus into the daily routine of design, one needs attitude, research and transdisciplinary actions in a continuous way.

Keywords Transdisciplinarity · Transdisciplinary coaching · Transdisciplinary design · Design thinking · Itinerant chair of transdisciplinarity · Education · Philosophy of science

1 Introduction

Education in the twenty-first century receives increasingly globalized influences, which are sometimes far removed from people's original cultural matrix. Thus, it undergoes a process of transition, in which it is necessary to search for new ways

---

H. de Andrade Júnior (✉)
The University of Vigo (CREA SI2), Vigo, Spain
e-mail: handradejunior@gmail.com

CEFH/Catholic University, Porto, Portugal

T. P. de Andrade
University of Vigo, Master's Program in Leadership and Public Direction, Pontevedra, Spain
e-mail: tprouse@gmail.com

D. Raposo et al. (eds.), *Perspective on Design*,
Springer Series in Design and Innovation 1,
https://doi.org/10.1007/978-3-030-32415-5_4

that allow diverse transformations between the relationship of these people with their environment, without one part overlapping the other, but complementing it.

In order to correct this scenario, pedagogical practices are needed that integrate the different areas of knowledge and effectively deal with cross-cutting themes for their broader exercise. Practices and knowledges for solving complex and unknown problems are required by new knowledge-depleted topics.

People are more demanding about the opportunities offered by the globalized market. There is a wide variety of products and services that border exclusivity and that, due to the advancement of technology, are now within the reach of many. To meet this voracious audience, a new profile of professionals with differentiated backgrounds is emerging.

Transdisciplinarity has been extended to design in areas such as Human Computer Interaction (HCI), service design, digital mix media, smart cities and systems, among others whose transdisciplinary essence has led to the creation of specific masters courses as well as the growing study of their cognitive aspects, which are of direct interest to several areas such as neuromarketing, healthcare systems, experience design and cyber physical systems (CPS) (Van der Vegte and Vroom 2013; Vieira 2018).

In these transdisciplinary practices there are no boundaries between disciplines, and this multiple look allows an interaction of knowledge considering the many faces of understanding the world. The look is also a movement of reflection and cooperation in order to arrive at the main objective, which is knowledge wisely constructed. Now, the transdisciplinary theory presents itself as a proposal of transformative and alternative education at this moment, capable of relating the different areas of knowledge, assuming a position of respect for differences, solidarity and integration with nature.

The purpose of this work is to present transdisciplinarity as a scientific and cultural approach, presenting it as an alternative and as a methodological and epistemological suggestion for the transdisciplinary practice in the daily life of the designer. It would be like a new way of seeing and understanding nature, life and humanity in the construction of knowledge necessary for the expansion of citizenship with solid, lasting and sustained experience. It would also challenge the construction of its creative process as it encompasses increasingly critical references to its own process, once diversity is confronted.

We present here a theoretical reflection on two main points: (i) a new pedagogical and methodological approach is needed that brings new disciplines in the curriculum that will enable the designer with arguments, knowledge, skills and experiences to design and construct objects and services for the consumer of the digital age; (ii) if this refers to the need to insert transdisciplinarity in the teaching and practice of design, there will be a transdisciplinary elaboration of contents, dynamics and interventions that integrate art, the sacred and science in different contexts, where the coaching alternative would act as facilitator of this creative process.

2 The (Trans) Disciplinary Thinking

The transdisciplinary proposal affirms that the reality of life is not fragmented, nor divided into disciplines, and that a quality education requires a teaching that involves the understanding of being, life, culture in their relationships and interrelationships (Nicolescu 2000).

Still in this rationale, Morin (2007) would affirm that reforming a thought is a paradoxical problem, since to reform the thought it is necessary, above all, to reform the institutions that allow this new thinking.

In pedagogical practice, the proposal would be an opportunity that would improve the process of teaching learning, making classes more attractive and efficient, encompassing all the disciplines of the school context, articulating them with the many faces of understanding the world and its micro-reality. But for this, it would be necessary that a renewed thought existed, and then the dilemma to be lived would be that of educators' reeducation: a milestone of a new paradigm directed to the reform of thought.

Thus, it breaks with the traditional process in which education was restricted to pre-determined and incontestable concepts, because in transdisciplinary thinking, disciplines are no longer addressed in a fragmented way and isolated from others, making it imperative to share ideas, thoughts and opinions.

The application of transdisciplinary thinking emphasizes that self-knowledge is as important as knowing and discussing ideologies in view of social transformation; emphasizes working with the body, knowing how to deal with emotions, reason and spirit; develops ecological awareness; makes respect for personal, collective, and racial differences; knows how to transit between the world of interiority and socio-political-economic exteriority; considers realities where all dimensions are interconnected (Santos Neto 2006).

2.1 Following the Design Thinking Process

Design thinking involves the process of generating ideas in a multidisciplinary group as a focus on problem solving. Thus, the use of design goes beyond the appearance of products, being applied in the design of solutions that encompass the strategic aspects of the business (Bonini and Sbragia 2011).

We want to look at the multidisciplinary group set up for the problem solving cited by Bonini and Sbragia and for the design thinking aspect to be used as a means to develop innovations in organizations (Brown 2008). Design thinking is understood by Brown as a method that permeates the activities of human-centered innovation in three phases: inspiration, ideation and implementation.

But there is the social context involved in the process, with design thinking being a co-creation process involving multidisciplinary teams. In this case, it is characterized as an open, collaborative innovation process involving a series of participants and knowledge, both internal and external.

The role of innovation in the process can be seen in Brown's (2008) discourse, when he says that design thinking should propose technologically viable solutions, consistent with the business strategy and naturally open.

In this way, the tendency of the creative process is to bring the aforementioned multidisciplinarity to transdisciplinarity, due to the growing complexity of innovation problems. To deal with process innovation, the designer can not only have a single process of acquisition (thinking).

In order to examine some aspects of the contribution of transdisciplinary thinking to the design course in the design and construction of consumer connected objects of the digital age, we use the reconstruction of the knowledge production path by a designer directed to a project to develop an application (app) for mobile communication devices that reviews part of the literature that interests our study (Ribeiro 2015).

Donald Norman, Marc Hassenzahl, Arrow Walter, Noam Tractinsky, Pieter Desmet and Paul Hekkert were authors used to understand concepts such as user experience, emotion design, aesthetics and in what ways the user relates to the product. As for usability, Jacob Nielsen, Ben Shneiderman, Patrick Jordan and Bertini et al. were the most important authors to understand this principle. Nielsen and the 10 heuristics of usability, Ben Shneiderman and its five factors that determine the usability quality and Jordan (1998) with his description of the various usability testing methods. Bertini et al. (2006) were among the few authors to attempt to propose heuristics for the evaluation of mobile device interfaces (Cf. Ribeiro 2015).

Usability are the quality attributes that define the user's ease in using the interface. The word also refers to methods for improving ease of use during the design process (Nielsen 2012). Usability should also enable the creation of transparent interfaces so as not to hamper the process, allowing the user to control the environment without becoming an obstacle during the interaction.

Krug (2006) uses a methodology to simplify the process, stating that interviewing only three or four users representing the target audience. For example, it is possible to identify in the first test the vast majority of significant usability issues, allowing, on the same day, to test and question and pass them on to the development team to correct. After the fix, it is possible to run a second and a third test with three users and find out twice as many new problems that would not be identified with the triple of respondents in the first and second tests.

Usability is normalized into three concepts: efficacy, efficiency and satisfaction, where efficacy is related to goal analysis and with what accuracy can be achieved; efficiency is the ratio of effectiveness to the amount of resources spent and satisfaction is established by the comfort and acceptability of the product by the users. It can be calculated by means of subjective and/or objective methods. (ISO 9241-11 1998).

Nielsen (1993) argues that usability is not a one-dimensional concept, but rather a set of five attributes: learning ability, memorability, usability, error prevention, and satisfaction. The system needs to be easy to learn so that the user can quickly begin to interact. According to Nielsen, this is the most important attribute of usability since it is related to the first experience any user has of a system. This factor is evaluated based on the time that the user takes to become experienced in the execution of their tasks.

2.2 Design Thinking Counterpoints

The concept of usability, so important for design conceptions, can also be considered an empowerment, but not necessarily freedom of choice. The relationship of usability and reference can define the authoritarian direction to consumer behavior, in the sense pointed out by Fuchs (2018).

It can be seen that there is an effort to compose a whole custom heuristic (Nielsen and Molich 1990) to the consumption profiles; whether they are pleasing, user satisfaction or empowerment to direct a consumer to "choose" the most conformal mobile device to be a functional companion. Excluding the real, the device is emotionally endowed with a life of its own and it qualifies as a mascot of the user who chose it.

Shneiderman (1998) considers five quantifiable factors for the usability quality of a system: (1) learning time, (2) performance speed; (3) user error rate; (4) retention over time and (5) subjective satisfaction. Learning time deals with the time the typical user takes to learn how to use the relevant functions to fulfill their task sets, while the performance speed handles the time resources required to perform the reference tasks. Users' error rate is related to the number and type of errors they commit in performing the reference tasks; the retention over time is the ability to maintain knowledge continuously and subjective satisfaction deals (using interviews or forms) aspects related to the user's enjoyment in the use of the various aspects of the interface.

As we have seen, the last quantifiable factor that is subjective satisfaction emphasizes adherence in a Human-Computer Interaction (HCI) perspective that sets the user's choice on the object (and vice versa, if artificial intelligence allows it?). It remains on a scale of satisfaction that should be the one with the highest possible efficacy possible on a tripod of concepts that support usability (efficiency, effectiveness and subjectivity). Thus, given the importance of human attributes being strongly represented there, it is difficult not to project the digital revolution back to the stage of humanity without the machine perceiving. It is the natural laws of the market that impose the rules, without much resistance.

The user experience phenomenon has attracted attention and acceptance by the HCI community (Barbosa and Silva 2010). This interest of HCI researchers seems to be linked to the fact that the concept of usability only focuses on user cognition and performance, while the concept of user experience goes beyond external factors (Roto et al. 2011).

This concept is quite comprehensive and results from the interaction of several factors, both intrinsically and extrinsically. At the intrinsic level is the result of a complex interaction between cognitive, affective, motivational and behavioral factors. Given the possibility of using the system/product, the user develops expectations about what the product can do and to what extent it allows it to reach its objectives (efficiency), resulting in how the user is likely to shape his attitude towards the system (Ribeiro 2015).

The experience of using the system allows the user to develop their perception about the product and to register the essential aspects of the interaction in memory, from the learning done at the moment of interaction. Interaction with the system also leads the user to experience emotions (positive or negative) according to the proposed goals (emotional satisfaction), providing a response to the needs of the user, according to their expectations (effectiveness). At an extrinsic level, the user experience is shaped by the social influences to which the user is subjected, as well as by the socioeconomic and technological context in which the user is at a given moment.

According to Hassenzahl's user experience model, each user attributes some characteristics to the products/services at the time of their use, the user experience being the consequence of these attributes, which includes contexts or situations that the product/service will be used (Fredheim 2011).

Such attributes can be divided into two main aspects which are pragmatic and hedonic. The pragmatic attributes are related to the practical and functional use of the product, hence the term manipulation. Normally, we relate these attributes to the term usability. The consequence of the pragmatic qualities of a product is satisfaction. Satisfaction arises if the user uses the product or service to achieve goals and if those products or services functionally meet those goals. The hedonic qualities are related to the aesthetic aspects. Hassenzahl subdivides these qualities into three categories: identification, stimulation and evocation and it is based on our day-to-day life that we are connected with varied products and services that convey to us the idea of who we are, considering some more important than others. Thus, it is necessary for these products to identify us, to stimulate us and to evoke memories and past experiences (Hassenzahl and Tractinsky 2006; Fredheim 2011).

A third important aspect of the user experience is its emotional behavior (Minge 2008; Ribeiro 2015). In Hassenzahl's model, the consequences of the pragmatic and hedonic aspects are the pleasure, satisfaction and attraction of the user by the product. But in Mahlke's model it is emphasized the emotional behavior of the user that arises from the perceptions of pragmatic qualities (instrumental qualities) and hedonic (non-instrumental qualities) that are directly influenced by interaction characteristics. These characteristics that the user attributes to the products are not only influenced by their perception, but also by other factors such as the system's own characteristics and the context of the use (Mahlke and Lindgaard 2007; Hassenzahl and Tractinsky 2006).

The user experience can change with the context in which it is inserted, even if there is no change in the product (Roto et al. 2011). Using a product at home, on a public transport, individual or in groups can generate different sensations and perceptions of the product, as well as the momentary state of the user.

Motivation influences your perception and experience. Their perception of the system's properties also influences the user's experience, because what is important for the experience are the properties projected for the system (in the sense of an aesthetic functionality) and the perception of the product's own brand (Roto et al. 2011).

However, the user experience can not be described by describing only its factors, but obviously the factors involved in the process and its main characteristics can be used to describe the situation in which the individual experiences something, contributing to clarify the reasons behind experience.

Let's put ourselves in the shoes of the voracious urban consumer by technology of communication and information! In increasing and unbounded sophistication, he would seek a maximum action-response interface with the machine in order to satisfy his emotional and spiritual needs.

Near the limit, transdisciplinarity would act as a bridge, by which the designer would reach the deepest need of the client. In pedagogical terms, it would be open to the designer student, through his teachers who raised the reform of thought and a facilitating system at least interdisciplinary, integrated access to the arts, science and the sacred. The answer would be pleasant to this faithful consumer.

Analyzing the course of the designer's review in research, it is noticed that the potential digital customer provides data of their perception and experience to the researcher, but also of their sensations; of the instrumentality of the platform; of their satisfaction with the interactivity of the technological resource; the aesthetics of the apparatus and the software; of the speed at which results are achieved; the emotion that belongs to it, and the pleasure of it all (in the most hedonic sense possible), among other constructs that can be used by the Fuzzy theory, for example.

The main inspiration behind the introduction of the Fuzzy Theory is the need to model real world phenomena that are inherently vague and ambiguous. Human knowledge about complex problems can be successfully represented using the imprecise terms of natural language. Theories of fuzzy sets and fuzzy logic provide formal tools for graphical representation and efficient processing of such information (Prokopowicz et al. 2017).

It is interesting that behind the attributes raised from the perspective of the user there is always a direction to spirituality, where life and humanization of the whole process is sought, in which our voracious client may be trying to fill an existential void not material transposed to the material plane. At this point, the transdisciplinary approach to the curriculum is applied, which allows the designer to dialogue with the existential meaning of his potential client and to contribute to making the "essence" of the client visible and perceptible to his study.

Could we think of interactive systems as a theater stage that allows the user to experience the world in a digital revolution like virtual reality? Thus, the goal of the developer as a designer would be to discover how to design interactive systems so that users can enter this world and thus become part of the action imagining themselves in the scenes portrayed. From this point of view, designing interactive systems is particularly coming from the person's own experience (Lauren 1991).

Lauren states that there is an individual experience that sustains the designer in his work, but we want to emphasize that there is a paradoxical (dis) humanization in the process of bringing exclusivity to the demanding client. The person's experience will also serve to enhance the automation of users' remote lives.

The empowerment of the proposal to encourage the study of design from palettes of preference for high-performance mobile devices is already a notable indication that the digital society may be moving towards individualized robotization and for this relativization of consumption.

If the process moves to the compulsory and not to the choice, authoritarian capitalism as pointed out by Fuchs (2018) is thus marked and authoritarianism as well. The first one establishes itself in the sense that it brings the illusion of control on the part of the user. The second is that communication systems (communication and information technology networks) are directed and oriented so that people consume and feel pleasantly possessed of the capacity to choose and to be possessors of the object chosen for their own affirmation and externalization of wealth and power.

2.3 *Transdisciplinary Coaching*

In addition to the varied theories about the origin of coaching, the literature also presents different concepts about what coaching actually represents. An aggravating factor for the development of a consistent theoretical basis has been the practice of coaching itself, which is much more expressive today than theories and research on its scientificity (Reis and Nakata 2010).

The growing injection of ideas and techniques creates confusion about the precise nature of coaching and what it intends to achieve, contributing to the difficulty of defining it (Stober and Grant 2006; Cf. Oliveira-Silva et al. 2018). One definition that most closely approximates our intention-study is that of Whitmore (1992, p. 8) where "coaching is the unlocking of an individual's potential in order to maximize performance. It's helping the other to learn rather than teach it."

As a support of psychology, coaching has adopted therapeutic and personal development elements. Schmitt Neto (D. Sc.) developed a transdisciplinary coaching initiative at a Transplant Hospital in São Paulo, Brazil (http://htejz.spdmafiliadas.org.br/). In its methodology, it built 6 modules of construction and transdisciplinary experience of 4 h a month aimed at reducing labor tensions, for self-knowledge and to emphasize team spirit.

Although its personal development aspect is evident, we see no obstacle to coaching directly in the designer's curriculum in a cross-curricular and formal academic environment, to enhance professional and personal performance.

A recently published study (Andrade and Júnior 2018) focuses on the case of the Atelier de Artes in Vila Verde, Braga District and in Portugal where citizens of the village practiced artistic works of painting and drawing and analyzes the development of people from practice of the arts.

From a course offered to the community it was possible to create "experiences" in the so-called Atelier de Artes: a didactic space provided by the local parish board that resulted in an exhibition of drawings, paintings and sculptures.

The experiences showed that cultural and age barriers influenced this practice. There was a sensitive adaptation of work dynamics towards the creative process and individual and collective development of the members, through the introduction and treatment of tutoring through transdisciplinary coaching as a transition to self-knowledge, which was directed to each individual journey, in their confrontations and aspirations.

The dynamics of the Tamar Prouse Atelier of Arts, in Vila Verde Parish Council, resulted in a project, the exhibition Art in Colors, in the Municipal Library Professor Machado Vilela.

The project involved not only presenting artistic disciplines, but also addressing the potential of the creative exercise and the need of new artists to deal with creativity in a continuous process, as a field open to the imagination, in daring to try new ways.

In the context of culture, arts and society the cultural barriers were analyzed in the direction of the blockade of the artistic production and the peculiarities of the individuals, who were worked as specific searches, instigated and understood throughout the process.

Also the motivations of the art production and the development of the members are emphasized in the technical aspect as well as in the personal aspect. For feed-back control, coaching techniques and moments of individualized conversations were used.

It was sought to understand the transitions between the habit of reproduction of image bank as a tool for the development of technical knowledge of drawing and painting in oil and individual expression, as to the image choices that happen by identity/interest and the process of self-knowledge.

The question of self-knowledge would be present in the choice of color, in the identification of individual limits and through the necessary adjustments in this search and also in the physiological factor, which marked the peculiar characteristics of each age group and concomitant conviviality.

The discursive experiences of the Tamar Prouse Atelier of Arts and the community of Vila Verde and Barbudo Town Council took place in weekly meetings of two (2) hours and were published in the newspaper Semanário V, mentioned there as members "from 8 to 80 years", characterizing a journalistic call that pointed out the different age groups represented and the diversity of the local community.

There are several significant results gathered in the atelier from this experience of transdisciplinary coaching as a transition to self-knowledge. Transdisciplinary coaching was a way of opening people's eyes to other reality plans and throwing a little light on our blind spots. Expand our mental models and dialogue with the other's eyes with an apprentice attitude. It is an opportunity to elevate our disposition to knowledge and to forget that since school it is placed as a commodity, when in fact it has always been inside us and only needs to be awakened, as by a Socratic maieutic, to gain contours of creativity and of hope to deal with the conflicts and problems that the game of life provides.

The coach had stimulated the generation of personal choices by problematizing with the students artists on what it would be like to produce art with cultural limitations in the sense of age, tradition and habits. With the gradual application of the technique, the student artists' searches became more sophisticated and personalized.

For example, the religious image became a setting where biblical personalities were portrayed as day to day people. The image of a person's face that was initially a portrait of a politician became an image that portrayed a woman crying, reflecting the emotional moment of the author of the work. The simple mangá design (a japanese design) became a thematic production to compete in a teen-age contest.

In view of these characteristics presented by the students' work, it was observed that practical and collective experience with individualized support/coaching is really effective for student development, both at the technical level and in the search for aesthetic-individual emancipation.

After six months of experience, from project to maturity, the members of the studio were able to compose their projects with more freedom of expression. Thus, the transdisciplinary coaching experience of the arts becomes valid for design when it encourages self-knowledge to improve the creative process in the design and development of projects.

3 Transdisciplinary Chair or Transdisciplinary Design?

Would it be appropriate to seek solutions that formalize transversality through an itinerant chair, in which seasonal contributions from outstanding proponents of the proposal to the students of design, or seek solutions to create a massive and direct pedagogical approach characterized by a curriculum of transdisciplinary design?

3.1 The Transdisciplinary Chairs

Let's look at an example of the intention of the chair. The Institute of Advanced Transdisciplinary Studies of the Federal University of Minas Gerais (UFMG, Brazil) recently launched a call[1] for researchers to transdisciplinary chairs.

The project aims to stimulate the presence, within the UFMG, of researchers of international reference, characterized by the quality of their training and by the advanced (cutting-edge) nature of their studies; to promote the interaction of these researchers with UFMG groups, organized around common research programs; favor

[1]The Institute of Advanced Transdisciplinary Studies (IEAT/UFMG) made public the inscriptions, from December 1, 2017 to March 15, 2018, of the nominations for the Chairs Program administered by IEAT and sponsored by the Research Development Foundation—FUNDEP. See https://www.ufmg.br/ieat/wp-content/uploads/2017/11/Chamada_Ca%CC%81tedra-FUNDEP_2018.pdf.

the advancement of transdisciplinary research in the University, allowing the articulation of different areas of knowledge; promote collaborative research at UFMG with researchers and research groups from abroad; establish actions that generate publications and international events, aiming at the internationalization of IEAT. The emphasis of the transdisciplinary chairs of the edict is in three major areas of knowledge: Humanities, Letters and Arts; Life Sciences (including Health) and Natural Sciences (including Technology). From what can be deduced from the edict, the Design course is formally included in the proposal.

The IEAT initiative is based on the Global Transdisciplinarity Chart (Morin et al. 1994), with signatories of the importance of Edgar Morin. The I World Congress of Transdisciplinarity (Arrábida, Portugal, 1994) had not yet contemplated the itinerant chair, but had affirmed that transdisciplinarity is complementary to the disciplinary approach: it brings from the confrontation of the disciplines new data that articulate them among themselves; offers us a view of nature and reality and that it does not seek to dominate the various other disciplines, but to open them to what passes through them and surpasses them (Article 3).

It also stated that the point of support for transdisciplinarity lies in the semantic and operative unification of meanings across and beyond disciplines. It presupposes a rationality opened by a new look, about the relativity of the notions of definition and objectivity. Excessive formalism, the rigidity of definitions and the absolutism of objectivity involving the exclusion of the subject lead to impoverishment (Article 4) and that transdisciplinary thinking is resolutely open and strengthened insofar as it goes beyond the realm of exact sciences through its dialogue and reconciliation not only with the human sciences but also with art, literature, poetry and spiritual experience (Article 5).

Alvarenga et al. (2005) argue that although it may seem precocious to consider the movement engendered by a distinct group of thinkers as a new chapter in the history of modern science, the present hypothesis is that this movement, as well described Nicolescu (2006), in his presentation at the II World Congress of Transdisciplinarity (Vila Velha, Brazil, 2005),[2] in the text entitled "Past, Present and Future Transdisciplinarity", allows us to state that the search for transdisciplinary thinking already has its own history. Then, The II World Congress establishes the commitment of an itinerant chair for dissemination and affirmation of the methodological and epistemological proposal (Cf. recommendation 1).

The future perspectives for the advancement of this and similar transdisciplinary thoughts that may emerge enriching this history are, apparently, broad, considering that, regardless of the difference of theoretical perspectives, the search for articulation of the scientific field with other branches of knowledge, such as philosophy, the humanities, art, literature, will gain expression in a context where the new traits

[2]The first author participated in the event and he was an institutional signatory of the document drawn up at the end of the meeting, which marked the formalization of the transdisciplinary itinerant chair. See The original full document in Portuguese in http://www.ufrrj.br/leptrans/arquivos/MENSAGEM_VILA_VELHA_VITORIA_2005.pdf.

that characterize knowledge are complexity, hybridity, non-linearity, reflexivity, heterogeneity and transdisciplinarity. However, such traits may be the challenges of designers' natural skills.

3.2 Transdisciplinary Design in Practice?

Transdisciplinarity was extended to design at the international workshop "The future of transdisciplinary design", organized by the University of Luxembourg in 2013. Today, the option for transdisciplinary design is already consolidating among the leading design schools.

It arises from the identification of the necessary integration of perspective evidenced by the need to overcome difficulties such as: ineffective collaboration; understanding of the different languages associated to the different basic formations and cultures inherent in the project practice (engineering, architecture, graphic design, product, interaction, services, experience); (in) transparency in communication and explanation of specific vocabulary and terminology; barriers to inter, multi and transdisciplinary collaboration, in the uncertainty of each one's role and contributions; to awaken the perception of diverse perspectives of people with backgrounds and functions in various design disciplines or beyond (marketing, sales, business strategies, biology, neurosciences, among others); creation of new knowledge; understanding of complexity, management and collaboration (Cf. Vieira 2018).

Transdisciplinary practice is worth it. Ramos (2014) had noted the benefits gained throughout the process of methodological renewal in a visual arts course where there was transdisciplinary inclusion in the curriculum. In their study, the importance and the fundamentals of transdisciplinarity as an instrument of evolution of pedagogical practices were analyzed through transdisciplinary projects carried out in secondary school.

The three projects developed in ESMAVC, described by Ramos (2014), illustrate progressively the application of the four pillars of theoretical foundation: project pedagogy, transdisciplinarity, inclusion and action research. In the first project, the pedagogy of the project was applied through a transdisciplinary approach between the disciplines of Portuguese and drawing (two groups of arts). Inclusion was one of the pillars of this project and had as its final product the production of a book with 224 pages of illustrated texts, the edition of which was sponsored by a large company (Portugal Telecom), which perpetuated the efforts of all stakeholders and projected the image of the school in a solemn session of presentation and launch of the book, which was attended by official entities (Ministry of Education, Autarchy and Assembly of the Republic) (see Ramos 2014).

The transdisciplinarity of this first project allowed us to capture students' interest, develop a taste for writing and interpretation of meanings, and the pleasure of illustration through a variety of languages and techniques. The information technologies played a preponderant role in the construction of the texts and in the preparation of the models.

With the second project, inclusion was once again a concern that guided the intervention of all teachers. The impact of this project on the community was so significant that it deserved a televised report by the channel SIC2.

The third project entitled "Drawings: transdisciplinary dynamics" resulted from the concern of adding to the pedagogy of design and transdisciplinarity the notion of a contemporary artistic manifestation, maintaining the action-research methodology. There were contents taught simultaneously by teachers from different disciplinary areas in the same classroom space that went beyond the classroom boundaries, a fact that represented a true laboratory of didactic-pedagogical experimentation (Ramos 2014).

The great applicabilities of the three projects for the area of design were: (a) exploration of differentiated pedagogies, close to the individual and consequently more efficient in learning; the need for content became better understood by students; increased interest and enthusiasm in the tasks to be performed; self-determination and collaborative spirit; the teacher valued himself as a knowledge manager as an alternative to that of the transmitter and the school came alive and updated (Cf. Ramos 2014).

It is essential to reflect on the validity of these types of practices and how they have or did not bring quality to learning, as well as the difficulties experienced in its implementation, so that the creation of spaces of transdisciplinarity worthy and consistent can be based.

4 Conclusions

The creation and development of methodological solutions, techniques and new technologies for improving human conditions, focused on crucial and latent issues such as reducing environmental, economic and social impacts is urgent and can only happen in transdisciplinary environments and teams (which will be the receptacles of a culture of receptivity, openness and innovation) that can overcome and solve the complex problems of the present and the future.

In addition, when applied in this way, the curricular proposal for the designer tends to be broad, flexible and integrated transversally with micro-realities, experiences and knowledge that are naturally linked to multiple dimensions, as Nicolascu (2000) imagined at the origin and will enable professionals to better serve individual and collective preferences; seeking to satisfy people with added ethical responsibility.

We present that the traveling chair, transdisciplinary coaching or the consolidation of a transdisciplinary design curriculum for technical, ethically responsible and environmentally conscious dimensions are valid alternatives to the educational model.

The understanding of the transdisciplinarity of design represents a more complex structure, but also necessary for the interested party to be able to embrace the field of design. There is a clear feeling and the manifest that it is important to illuminate the relations between the parties, for which it will be necessary to assume the study of the dynamics of the transdisciplinary structure of design, as pointed out by Vieira (2018).

Through the insertion of transdisciplinarity into the teaching and practice of design, consequently there will be a transdisciplinary elaboration of contents, dynamics and interventions that integrate art, sacred and science in different contexts, where the coaching alternative would act as facilitator of this creative process.

There is already a considerable amount of experience that values transdisciplinary thinking as applied to design, but recognition of the levels of reality in the metier is no easy task.

Knowledge niches within the design curriculum can be clusters of power and influence, not just content apposition. There are different realities that the particular spaces of each discipline of design, schools, companies and cultures, allow to delimit.

Even subject to criticism and resistance in its pedagogical project, it is undeniable that transdisciplinary space will ultimately cross all disciplines of design, in the midst of the innovation society. This is a methodological advantage as it allows the extension and application in other areas of human activity and integration by other scientific disciplines, academic and practice of other domains and other references, in a cumulative effect.

Although there are doubts about when and how this will be done, it is clear that the opportunity to innovate in this process can not be missed, as there is evidence that it is in the transdisciplinary space that the opportunities for new design applications, emerging trends, invariant and where the products of design reach the masses. In transdisciplinary space, factors such as the extension and consequent absorption of new technologies, laws and regulations that call for new approaches in classic disciplines of design, as well as the translation of specific characteristics of the design in other states for the exchange, expansion and inclusion of transdisciplinary design, are crucial aspects to develop. Changing methods, thinking patterns and the ability to communicate throughout the design professions are identified as well as represent competencies to be developed for the understanding and receptivity to other and different perspectives of the design processes (Vieira 2018).

5 Recommendations

In order to incorporate the transdisciplinary nexus into the daily routine of design, one needs attitude, research and transdisciplinary actions in a continuous way. The transdisciplinary attitude must seek to understand the complexity of our universe, the complexity of the relations between subjects, subjects with themselves and with the objects that surround them, in order to recover the senses of the enigmatic relationship of the human being with reality (like that which can be conceived by human consciousness) and the real (as absolute and always veiled reference). For this, it is proposed the articulation of the knowledge of the sciences, of the arts, of philosophy, of traditions and of experience.

Transdisciplinary research presupposes an epistemological plurality and requires the integration of dialectical and dialogic processes that emerge from research, keeping knowledge as an open system.

Transdisciplinary actions should be encouraged to articulate the formation of the human being in his relationship with the world, with others, with himself, with being, and also with formal and non-formal knowledge. Such actions should also focus on the mediation of conflicts that emerge in the local and global context, aiming for peace and collaboration between people and between cultures, but without disregarding the contradictions and appreciation of their expression.

References

Alvarenga A, Sommerman A, Alvarez A (2005) International congresses on transdisciplinarity: reflections on emergencies and convergences of ideas and ideals towards a new modern science. Health Soc 14(3):9–29. https://doi.org/10.1590/S0104-12902005000300003
Andrade T, Júnior H (2018) Coaching in the Creative process of visual arts: the case of the Atelier de Artes in Vila Verde. Retrieved from http://convergencias.ipcb.pt
Barbosa S, Silva B (2010) Human-computer interaction. Publisher Campus-Elsevier
Bertini E, Gabrielli S, Kimani S (2006) Appropriating and assessing heuristics for mobile computing. In; Proceedings of the AVI'06—working conference on advanced visual interfaces, vol 2006, pp 119–126
Bonini LA, Sbragia R (2011) The design thinking model as an inducer of innovation in companies: an empirical study. J Manage Proj 2(1)
Brown T (2008) Design thinking. Harvard Bus Rev 86(6):84–92
Fredheim H (2011) Why user experience can not be designed. Retrieved from https://www.smashingmagazine.com/2011/03/why-user-experience-cannot-be-designed/
Fuchs C (2018) Authoritarian capitalism, authoritarian movements and authoritarian communication. Media Cult Soc 40(5):779–791
Hassenzahl M, Tractinsky N (2006) User experience—a research agenda. Behav Inf Technol 25:91–97
ISO 9241-11 (1998) Ergonomic requirements for office work with visual display terminals (VDTs)—Part 11: guidance on usability
Jordan P (1998) An introduction to usability. CRC Press, London
Krug S (2006) Do not make me think! a common sense approach to web usability, 2nd edn. New Riders, Berkeley, CA
Laurel BK (1991) Computers as theaters. Addison-Wesley, Boston
Mahlke S, Lindgaard G (2007) Emotional experiences and quality perceptions of interactive products. In: Jacko J (ed) Human-computer interaction, Part I, HCII 2007, LNCS 4550. Springer, Berlin, pp 164–173
Minge M (2008) Dynamics of user experience. Technische Universität Berlin, GRK I promised Franklinstraße 28–29, Sekretariat FR 2-6 D-10587 Berlin. Retrieved from https://www.researchgate.net/publication/228982807_Dynamics_of_User_Experience
Morin E (2007). he head well set (trans: Mahler P). Buenos Aires, Argentina: Ediciones Buena Visión
Morin E, Nicolescu B, Freitas LD (1994) Letter of transdisciplinarity. Retrieved from http://inters.org/Freitas-Morin-Nicolescu-Transdisciplinarity
Nicolescu B et al (2000) Education and transdisciplinarity (trans: Vero J, Mello M, Sommerman A). UNESCO, Brasilia

Nicolescu B (2006) Transdisciplinarity - past, present and future. Moving worldviews - reshaping sciences, policies and practices for endogenous sustainable development. 142–166

Nielsen J (1993) Usability engineering. Academic Press, Boston, MA

Nielsen J (2012) Usability 101: introduction to usability. Nielsen Norman Group. Retrieved from https://www.nngroup.com/articles/usability-101-introduction-to-usability/

Nielsen J, Molich R (1990) Heuristic evaluation of user interfaces. In: Proceedings of ACM CHI'90 conference, Seattle, WA, 1–5 Apr 1990, pp 249–256

Oliveira-Silva C, Werneck-Leite N, Carvalho F, Anjos C, Brandão M (2018) Unveiling coaching: a review from the perspective of psychology. Psychol: Sci Prof 38(2):363–377. https://doi.org/10.1590/1982-3703000942017

Prokopowicz P et al (2017) (eds) Theory and applications of ordered fuzzy numbers. Stud Fuzziness Soft Comput 356. https://doi.org/10.1007/978-3-319-59614-3_2

Ramos C (2014) Visual arts and transdisciplinarity in the age of complexity: a continuous pedagogical practice. Revista Lusófona de Educação 26:103–122. Retrieved from http://www.scielo.mec.pt/scielo.php?script=sci_arttext&pid=S1645-72502014000100008&lang=en. Accessed on 28 Apr 2019

Reis G, Nakata LE (2010) Transtheoric model of change: contributions to executive coaching. Brazilian J. Prof. Guidance 11(1):61–72

Ribeiro AF (2015) The study of user experience and usability in a mobile context: development of a mobile application entitled Think an App. Dissertation of Master (Design), Faculty of Fine Arts, University of Porto

Roto V, Law E, Vermeeren A, Hoonhout J (2011) User experience white paper: bringing clarity to the concept of user experience. University of Helsinki. Retrieved from http://www.allaboutux.org/files/UX-WhitePaper.pdf

Santos Neto E (2006) For a transpersonal education: the pedagogical action and the thinking of Stanislav Grof. Lucerne, Rio de Janeiro

Shneiderman B (1998) Designing the user interface: strategies for effective human-computer interaction, 3rd edn. Addison Wesley

Stober DR, Grant AM (2006) Evidence based coaching handbook: putting best practices to work for your clients. Wiley, New Jersey, NJ

Van der Vegte WF, Vroom RW (2013) Considering cognitive aspects in designing cyber-physical systems: an emerging need for transdisciplinarity. In: Proceedings of the international workshop on the future of transdisciplinary design TFTD'13, Luxemburg. Presentation (ppsx), pp 41–52

Vieira S (2018) Transdisciplinarity of design: different levels of reality, vol 13, no 1. Project Management and Technology, São Carlos, pp 101–114. http://dx.doi.org/10.11606/gtp.v13i1.110646

Whitmore J (1992) Coaching for performance: a practical guide to growing your own skills. Fun Meister, Wallingford

Exploring Climate Changes Through LSP: A Learning Experience

José Silva and João Neves

Abstract The present article focuses a exploratory qualitative research on the development of a learning experience on the issue of climate change; developed on the United Nations Sustainable Development Goals—in specific the Sustainable development goal 13—"Take urgent action to combat climate change and its Impacts." The project developed under the course unit of Information Design 4d aimed to develop a perspective on climate change that could trigger a greater emotional affectation on climate change among potential consumers. The project, included an LSP (Lego Serious Play) methodology. Students used a strategy to make perceptions of the problem tangible in order to develop a perspective according to the importance and priorities of the so-called reptilian brain, in reference to Paul D. MacLean's triune model following the interpretation by Clotaire Rapaille. From the learning experience carried out with the LSP strategy, in the discovery session, students identified Nature as a commodity and from this perspective formed interesting possibilities in the development of Speculative Design solutions that reinforce the assumption of Timothy Morton concept of artificiality. The learning experience was carried out within the Master of Graphic Design, Association Master between the School of Applied Arts of the Polytechnic Institute of Castelo Branco and the Faculty of Architecture of the University of Lisbon, 2nd semester of the academic year of 2018/19.

Keywords LSP · Creativity · Climate change · Speculative design · Learning experience

J. Silva (✉) · J. Neves
IPCB, Av. Pedro Alvares Cabral 12, 6000-084 Castelo Branco, Portugal
e-mail: gago.silva@gmail.com

J. Neves
e-mail: joaoneves@ipcb.pt

D. Raposo et al. (eds.), *Perspective on Design*,
Springer Series in Design and Innovation 1,
https://doi.org/10.1007/978-3-030-32415-5_5

1 Introduction

According with The Sustainable Development Goals Report (United Nations, 2018) the year 2017 was one of the three hottest already recorded and was 1.1 °C above the pre-industrial period. An analysis by the World Meteorological Organization shows that the global average temperature of five years from 2013 to 2017 was also the highest recorded. The planet continues to experience sea-level rises, extreme weather conditions (the North Atlantic hurricane season was the most expensive ever recorded) and rising concentrations of greenhouse gases. These changes require urgent and accelerated action by countries in implementing their commitments to the Paris Agreement on Climate Change. According to Norgaard (2011) although climate change is present everywhere, however, it is still difficult to discern its psychological impact on the individual.

1.1 The Evolutionary Perspective

The psychological impact on the individual is pertinent in a design thinking approach. The climate change subject is closely related with the concept of sustainability. According to Luzio (2012) Clotaire Rapaille points out that the issue of sustainability must be in line with the growth characteristics. Clotaire Rapaille defends the idea that the world is changing all the time and for the author sustainability means that you are able to anticipate the changes and to be in harmony with the changes. He proposes the word evolutionary rather than sustainability. Retrieving the concept of Triune brain model, Rapaille explains that the reptilian brain prefers movement, growth, likes innovation, progress. The author defends the notion of harmony in the long-term and reciprocal growth. In the long-term, means that a person takes care of another, part of an Confucius's approach in a permanent growth perspective concerning the environment and the Planet, and even more important in an era that we call anthropocentrism, because for the first time humans are changing the Planet and we are not paying attention to this fact.

The perspective presented by Clotaire Rapaille uses a somewhat controversial logic of the central interests of the individual instinct, the reptilian "fight or flight" which at times may not be compatible with the benefits of a broader community.

In the Triune model, according to Cory and Gardner (2002) from the infant nursing, care-giving, and social bonding initiatives and responses of the mammalian modifications and elaborations arise the motivational source for nurturing, empathetic, other-interested experiences and behaviors. Here are the warm-blooded, passionate, body contacting, bonding behaviors identified with the lion, the wolf, the primates. In a world in which nearly single-minded self-preservation is simultaneously complemented and counterpoised by the conflicting demands of affection.

According to MacLean (1994) the experimental work in animals as diverse as lizards and monkeys shows the reptilian complex is involved in displays of agonistic and defensive social communication. Also it is noteworthy that partial destruction of the reptilian complex eliminates the aggressive, territorial.

The division of function between the protoreptilian complex and the limbic system is not clear cut, but rather entangled. The lower structures of the limbic node have been shown to augment the self-preservational behavior of feeding, fighting, and self- protection (Reiner, 1990).

1.2 The Acceptance of Artificiality

The Clotaire Rapaille perspective is compatible with the approach of Timothy Morton (Blasdel, 2017), the idea that "You think ecologically tuned life means being all efficient and pure" is wrong, there must be an acceptance of artificiality and human pleasure, as he himself says "Wrong. It means you can have a disco in every room of your house." According to the author the Anthropocene teaches us that we can not transcend our limitations or our reliance on other beings. We can only live with them. If we give up the delusion of controlling everything around us, we might refocus ourselves on the pleasure we take in other beings and life itself. According to Lewis and Maslin (2015) Adopting the Anthropocene may reverse this trend by asserting that humans are not passive observers of Earth's functioning.

1.3 Psychological Impact

When studying and understanding the both how and why the denial of risks, importance, and need for collective action on global warming takes place, in the transitory moments in which people talk about climate change seriously, Norgaard (2011) identifies four core reactions: fear, guilt, helplessness, and crisis of identity. These reactions lend themselves to silent endurance, cultivated distraction, or disassociation via humor or hyper-facticity, because, she argues, the practical organization of the lived social process provides them no alternate purchase.

When designing new solutions related to a new mindset it is necessary to know how to reconcile new technical or conceptual solutions with a narrative logic that is often influenced by an ecosystem of contexts that are predictable or not, but which belong to a field of speculative investigation.

1.4 Speculative Design

There are already very interesting solutions in terms of technical approaches, developed by High Schools and University Institutions, innovative solutions from an evolutionary perspective, focusing on the use of electrical energy, by photosynthesis, directly from the plants (Marques, 2008), turning the plants in a visible part of the technical solution (Chuet-Missé, 2019).

According to Auger (2013) speculative design has much in common with other areas of research associated with the Design project, such as critical design, discursive design, design probes, and design fictions. Although not referring directly, but associated with the concept of Speculative Design, Dunne and Raby (2001) refer the role of Critical Design as a practice focused not on solving problems, but asking "carefully crafted questions".

There are overlaps between these practices, the differences are subtle and mostly based on geographic or contextual use: they all remove the constraints of the business sector that define normative design processes; to use models and prototypes in the research center; and fiction to present alternative products, systems or worlds. A physical object as a "design fiction" may be identical to a "speculative drawing" or a "design probe," and so on; however, the use of the modifier gives the cultural object a substantially different value. To outline a methodological proposal, design speculation requires a bridge between the public's perception of their world and the fictional element of the concept. The inspiration and influence for this "perceptual bridge" can come from a variety of fields, such as observational comedy, psychology, ecology, horror films, and illusion, to provide insight into the complex functioning of the human mind and how it can be carefully manipulated to provoke reactions. All these perspectives meant to be ways to provide critical futures scenarios.

For Dator (2002) the critical futures study combines rational intelligence with intuitive and visionary abilities to provide a forward-looking context in which some of the "big questions" can be posed and answered: Where are we going? How do we get there? What problems need to be solved? Why take this path rather than another? Such questions are central to Future Studies and vital to the well-being of society.

Auger (2013) warns that one of the main factors responsible for the success of a speculative project is the careful management of speculation, its proposals should not go far in the future in presenting implausible concepts or alien technological habitats because the public will not relate to the proposal, due to lack of connection.

A vital factor in the success of a Speculative Design proposal is the careful management of the speculation, specifically the information about the use of technology, aesthetics, behavior, interaction and function of the designed artifact.

These perceptions can be manipulated in a precise and informed way. These exist as plausible, tangible, and accessible demonstrations or, more specifically, hypothetical translations of disruptive technological innovations for future products that could become. The main benefit of this approach is the removal of the trade restrictions that normally drive the creative process. This decoupling allows the goals to be based on issues and discourse, not on agendas; hypothetical possibilities of non-real products;

utopian concepts and dystopic counter-products. They can inspire an audience to think not just about what they want for their future selves but also what they do not want.

The Lego brick system is a very interesting solution as a way of work perceptions and logics that are not explicit. For Dyke (2018) the Lego not only supports the constructivist theory of deeper learning that occurs when constructing something external to itself, but also suggests a significant element of psychological security for the process. The author also states that cognitive development and psychological development occur together.

1.5 LSP

Lego, the company, patented the brick design in 1958 what it is today known as the bricks' clicking together feature, the clutch power, that makes the brick system such an endlessly expandable toy.

According to Rasmussen (2014) in the business world, designers use Lego bricks for rapid prototyping, simulations, and visualizations. Consultants use Lego bricks for team building, modeling of factories, and logistic processes. The company vision for the assemblage brick is that it is "more than a toy", it is a language for systematic creativity. The models that people build with the Lego bricks are meant to represent or to make visible something tangible, whether a product or a process. The model or the process of constructing should resemble or mirror the real-world topic as much as possible. In doing so, the user is using the Lego brick as a metonym. Rasmussen (2014) asserts that a metonym is "a figure of speech consisting of the use of the name of one thing for that of another of which it is an attribute or with which it is associated" (as cited in "Merriam-Webster", 2019).

According to Dykes (2018) the Lego Serious Play (LSP) sessions can be considered a training event at some level considering the kinesthetic nature of the experience and LSP process steps that require individual and collaborative learning around the organization problem being addressed.

The creative process of LSP not only supports the constructionist theory that building something external of oneself produces greater learning, but it also adds the unexpected and profound element of psychological safety to the equation. This phenomenon provided freedom to the participants to explore and engage in learning at a transcendent level. Dykes (2018) asserts that "psychological safety is defined as a team's shared belief that the current environment is safe for taking interpersonal risk" (as cited in Edmondson, 1999). The author states that the LSP allows for enough structure, responsibility, and rational thought while still stretching individuals to think, build, and create outside of their comfort zone.

According to Blair and Rillo (2016) and reporting to the possibilities developed in this article, LSP, can be defined as a method, a process, a communication tool or a language, a service, a framework, a product line, or a meeting point. It allows a systematic method that enables people to use Lego bricks to solve problems, explore ideas and achieve objectives.

It is a structured process where participants through a series of steps, think, build, tell a story, reflect and refine, to develop a shared understanding of a research topic. As a Communication tool or language, the LSP enables three modes of communication: visual, auditory and kinaesthetic. The models allow enhanced expression, deeper listening and better memory. Shared model building allows teams to understand each other's interpretation and create deeper shared meaning of key ideas. The LSP is a more tangible strategy allows for faster learning times and faster engagement with the research subject.

According to Marshall (2007) a body of research within education and psychology has emphasized the role of physical materials and manipulatives in supporting learning, this paper points to the potential of tangible systems in supporting learning. For Anderson (2015) recognizing the physicality of learning, and that multimodal learning is certainly preferable, we should strive for manipulatives and environments that encourage embodied learning.

1.6 The Homunculus

The LSP reinforces the "physicality" of manual interaction, present in the homunculus model. Dr. Wilder Penfield and his co-investigators Edwin Boldrey and Theodore Rasmussen are considered the creators of the sensory and motor homunculi. Althought they were not the first scientists to attempt to objectify the function of the human brain through the representation of a homunculus (Cazala et al. 2015). According to Marieb and Smith (2007), the motor homunculus represents a map of brain areas dedicated to motor processing for different anatomical divisions of the body. The primary motor cortex is located in the precentral gyrus and deals with signals from the pre-motor area of the frontal lobes. The sensory homunculus represents a map of areas of the brain dedicated to sensory processing for different anatomical parts of the body. The primary sensory cortex is located in the post-central gyrus and deals with signals from the thalamus. Penfield noted that his "creations" resulted in grotesque creatures because of their strange proportions. The sensory nerves arriving from the hands terminate in large areas of the brain, resulting in correspondingly large homunculus hands. In contrast, the nerves emanating from the trunk or arms cover a much smaller area, so that the trunk and arms of the homunculus appear comparatively small and weak.

This model clearly expresses a predominance of the sensorial area of the hands and possibly its implication in the construction of a concept and idea. Speech and gesture are elements of a single integrated process of utterance formation in which there is a synthesis of opposite modes of thought—global-synthetic and instantaneous imagery with linear-segmented temporally extended verbalization (McNeill, 1992).

The hypothesis of the homunculus is currently of fashion, but this is, after all, how everyone thinks of themselves. It would be surprising if this overwhelming illusion did not reflect in some way the general organization of the brain (Crick and Koch, 2003). The Homunculus representation reinforces the importance of tangible interaction in learning processes.

According to Manches and O'Malley (2011) tangible interaction does present unique learning benefits. Furthermore, by distinguishing two kinds of processes, offloading cognition, where manipulatives may help students by freeing up valuable cognitive resources during problem-solving, and conceptual metaphors, where perceptual information or actions with objects have a structural correspondence with more symbolic concepts. This type of interaction frees space for intuition.

1.7 The Role of Intuition

According to a previous article focusing the relationship between drawing and intuition (Silva, 2014) there is a body of research that presents the idea that intuition uses implicit knowledge. A growing number of results ranging from the biological (mainly, neural) to the cognitive (Lieberman, 2000; Jung-Beeman et al. 2004; Luu et al. 2010) suggest that pre-existing expertise that requires years of practice to achieve may not be an essential requirement for intuitive decision-making processes (Dreyfus, 1980; Ericsson et al. 1993). These authors suggest that intuitive decision-making processes share some of the same underlying neural structures and cognitive processes as learning known as implicit learning (Lieberman, 2000; Kaufman et al. 2010). So the hunch moment and its importance could have a relationship with the immersion level in a particular problem, the narrative could empower the immersion level in a specific problem-solving in a learning environment.

2 Development

The learning experience followed a strategy of constructionism. For Papert and Harel (1991) the simplest definition of constructionism evokes the idea of learning-by-making, The example of someone assembling Legos like a painter is a description of an activity but evokes a "way" of doing through the use of metaphors. For Maxwell (2006) community and constructionism go hand in hand. Constructionism allows two perspectives; one more related to the actual physical experience and staying close to the object can do as well as those who prefer a more analytic formal style.

The briefing that supported the development of the learning experience comprised three phases. The first phase, that took place in one session with a duration of three hours, focused on the theme of climate change. The second phase that took place in two sessions, with a length of six hours included the development of the strategy using the LSP, followed by the extraction and discovery of information. The third phase that took place in two sessions, with a duration of six hours involved the development of proposals. In the second phase, after presented the project briefing, each

Fig. 1 Students engaged in the final composition

student received a small bag containing fourteen pieces. Students carried out two exercises, repeated twice, before the collective discovery exercise. Firstly, based on an image of a pre-developed montage, the students were asked to assemble the construction corresponding to the image, the students had five minutes to assemble the construction. They were then asked to develop a composition on their understanding of climate change within ten minutes. At the discovery session, now collectively, students developed a construction on the question: "How to increase the psychological impact of climate change?" according to the assumption of Norgaard (2011) about the psychological impact of climate change (Fig. 1).

Each of the people involved in an LSP session puts in place a personal creativity strand. For Stierand and Dorfler (2014) personal creativity relates to the ability to (re) interpret the experience of the objective world into ideas that are original and useful. Intuition plays a significant role in the dynamics between creativity and judgment of a new idea. Intuition can also be understood as direct knowledge.

The composition made by the student's highlights domains of implicit knowledge made explicit by the tangible experience of externalization with the LSP. The students analized and identified the structural elements presented in the composition. In the final LSP montage assembly, at the structural level, it is evident that the elements are arranged in a row as if it were a moving shape. An item illustrative of the pollution, represented by the green wires connects the different "sections" of the montage (Fig. 3).

It is of interest to emphasize the "direction" of the composition, student described as a representation of the sea, in one end, and in the other end a representation of the more developed societies. The composition expressed implicitly in its shape, illustrates a direction towards the more advanced societies (Fig. 2).

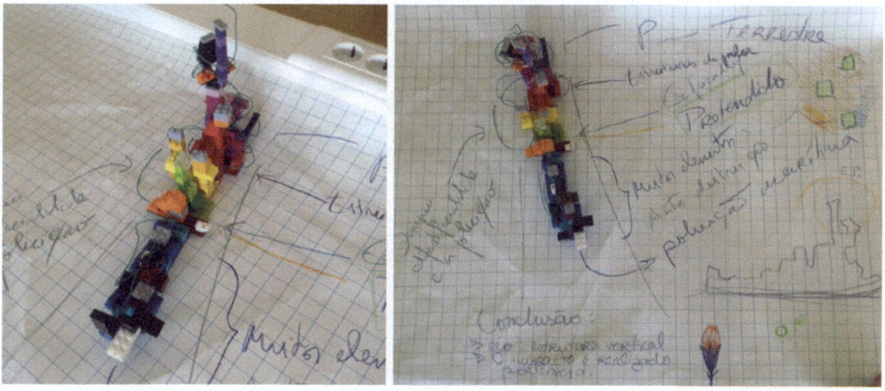

Fig. 2 LSP final composition overview on analysis notes

In the identification of structures, resulting from the discovery session, students relate the structure "3" to Nature. This identification is very interesting because the "representation" of Nature (made with green Lego bricks) doesn´t occur throughout the composition, it is particularized and designated as a structure, within the sequence, composed of more groups overall distinguished by different color groups (Fig. 3). The most interesting of this information, resulting from the discovery session is the identification of Nature as a commodity. The group discussed the idea that the group concentrated nature into a segment as if it were something apart from everything else. Something that "should" be present in all the composition, but in reality, it was not. It was as if society placed it on a depot, something that became scarce but it was needed. From this relationship between scarcity and appreciation came the idea of value and Commodity. According to Huang (2019) commodities are things of value and uniform quality, this homogenous nature of commodity quality provides possibility for exchange and distribution. Commodities are uniform to the extent that they are considered to be equivalent and interchangeable to those of other producers.

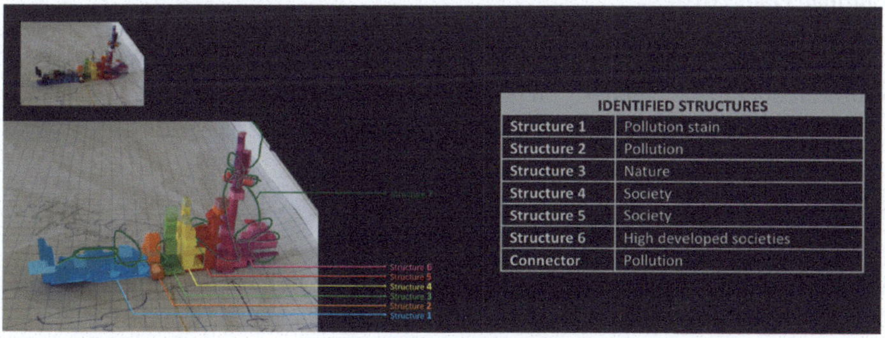

Fig. 3 Montage identification containing the different structures

Fig. 4 (Left) Gabriela Silva, hanging garden. (Right) Luisa Varela, self-powered sound flower pot

Commodities are things of value that are traded in large quantities. Commodities in turn are transformed into different consumer products with different commercial placements.

This question is of great interest in understanding a guideline for a concept according to the perspective of Clotaire Rapaille and Timothy Morton. Although controversial, the perspective of Nature as a commodity allows us to envision innovative solutions at the level of speculative design, solutions that include product development from an evolutionary instead of a sustainable perspective following the concept presented by Clotaire Rapaille and to understand how the narratives associated with these products could act at the level preparation of consumer practices.

The effectiveness of this strategy is made explicit by the students' findings through the analysis of the composition structure developed in LSP. Discovering very organized ways of understanding processes and phenomena that would otherwise be difficult to make explicit, the discovery session puts in evidence that there is a significant source of preexisting intuitive information latent in students involved in the learning experience according to decision-making processes (Lieberman, 2000; Kaufman et al. 2010).

The sketches developed by students explore the strand of consumer product, where an evolutionary perspective is present where artificiality is framed in a logic of comfort reducing the emphasis on productivity logic (Fig. 4).

3 Conclusions

Although the limits of the experience carried out, regarding the subject/question explored in learning experience "How to increase the psychological impact of climate change?" the perspective of Nature as a commodity allows to create a connection of positive "evolutionary" dependence on Nature, and in this specific case, through solutions that employ Nature "elements" as consumer products. Achieved through an appropriation that assumes the artificiality of design solutions.

When developing speculative solutions such as the use of energy from Nature, where plants can be "used" as an aesthetic element of comfort, compatible with the perspective of Timothy Morton, a perspective of consumer products also occurs. That perspective allows a better psychological impact, both positive and active, consistent with the understanding of an "evolutionary" connection with Nature according to the perspective of Clotaire Rapaille.

The formation of this concept constituted by an evolutionary relationship with the nature and comfort of a consumer product that, although artificial is in its artificiality an appropriation of Nature, allows several possibilities in terms of the development of design solutions and is useful to the level of attraction of consumption.

Resulting of this learning experience, we ask if two-dimensional visual cultural artifacts do not delimit visual reasoning, if the reasoning limitations are the result of an operational limitation felt by students in the quality of their visual representations. In those representations, students seem to be limited in understanding the real extensions and repercussions of their discoveries.

In the learning experience the LSP offered a tangible strategy, enabling a stronger engagement with the research subject and allowing richer discovery sessions. As a strategy, it offers interesting solutions in testing perceptions and logic that are not explicit and also empowers intuitive insights from implicit learning experiences. This learning experience will be reinforced with further studies of qualitative nature to "reinforce" the strategy of tangible discovery, associated to the exploration of concepts in a perspective of speculative design focused on the issue of climate change.

Acknowledgements The authors would like to thank the collaboration of the students from the course of Information Design 4d, Master of Graphic Design, course of the Association Master between the School of Applied Arts of the Polytechnic Institute of Castelo Branco and the Faculty of Architecture of the University of Lisbon, 2nd semester of the academic year of 2018/19.

References

Anderson S (2015Learning and thinking with things. In: Designing for the internet of things

Auger J (2013) Speculative design: crafting the speculation. Digital Creativity 24:11–35. https://doi.org/10.1080/14626268.2013.767276

Blair S, Rillo M (2016) Serious work how to facilitate meetings & workshops using The Lego® Serious Play® method, 1st edn. Promeet

Blasdel A (2017) A reckoning for our species: the philosopher prophet of the anthropocene. In: The Guardian. https://www.theguardian.com/world/2017/jun/15/timothy-morton-anthropocene-philosopher. Accessed 25 April 2019

Cazala F, Vienney N, Stoléru S (2015) The cortical sensory representation of genitalia in women and men: a systematic review. Socioaffective Neurosci Psychol 5:26428. https://doi.org/10.3402/snp.v5.26428

Chuet-Missé J (2019) Las plantas también pueden ser fuente de electricidad. In: La Vanguardia. https://www.lavanguardia.com/natural/20170118/413415278331/planta-electricidad.html. Accessed 22 April 2019
Cory G, Gardner R (2002) Algorithms of neural architecture, Hamilton's Rule, and the invisible hand of economics. In: Cory G (ed) Algorithms of neural architecture, hamilton's rule, and the invisible hand of economics, 1st edn. Praeger Publishers, Westport, pp 345–381
Crick F, Koch C (2003) A framework for consciousness. Nat Neurosci 6:119–126. https://doi.org/10.1038/nn0203-119
Dator J (2002) Advancing futures: futures studies in higher education, 1st edn., p 101
Dreyfus S, Dreyfus H (1980) A FIve-stage model of the mental activities involved in directed skill acquisition. Operations Research Center–University of California Berkeley, California
Dunne A, Raby F (2001) Design noir: the secret life of electronic objects, 1st edn. Birkhauser, Basel
Dykes W (2018) Play well: constructing creative confidence with LEGO serious play. Doctor of Philosophy in Human and Organizational Systems and Organization Development and Change, Fielding Graduate University
Ericsson K, Krampe R, Tesch-Römer C (1993) The role of deliberate practice in the acquisition of expert performance. Psychol Rev 100:363–406. https://doi.org/10.1037//0033-295x.100.3.363
Huang W (2019) Commodity finance: principles and practice, 2nd edn. Harriman House Limited, Hampshire
Jung-Beeman M, Bowden E, Haberman J et al (2004) Neural activity when people solve verbal problems with insight. PLoS Biol 2:e97. https://doi.org/10.1371/journal.pbio.0020097
Kaufman A, Kornilov S, Bristol A et al. (2010) The neurobiological foundation of creative cognition. In: Kaufman J, Sternberg R (eds) The neurobiological foundation of creative cognition, 1st edn. The Cambridge Handbook of Creativity, New York, pp 216–232
Lewis S, Maslin M (2015) Defining the Anthropocene. Nature 519:171–180. https://doi.org/10.1038/nature14258
Lieberman M (2000) Intuition: a social cognitive neuroscience approach. Psychol Bull 126:109–137. https://doi.org/10.1037//0033-2909.126.1.109
Luu P, Geyer A, Wheeler T et al (2010) The neural dynamics and temporal course of intuitive decisions. Public Library of Science
Luzio F (2012) Founder and CEO of archetype discoveries worldwide interviewed by Fernando Luzio. Novos Intérpretes 8
MacLean P (1994) Human nature: duality or triality? Politics Life Sci 13:107–112. https://doi.org/10.1017/s0730938400022358
Manches A, O'Malley C (2011) Tangibles for learning: a representational analysis of physical manipulation. Pers Ubiquit Comput 16:405–419. https://doi.org/10.1007/s00779-011-0406-0
Marieb E, Smith L (2007) Human anatomy & physiology laboratory manual, 7th edn. Pearson Benjamin Cummings, San Francisco
Marques N (2008) A festa da Ciência aberta a todos. Expresso
Marshall P (2007) Do tangible interfaces enhance learning? In: Proceedings of the 1st international conference on tangible and embedded interaction (TEI '07). https://doi.org/10.1145/1226969.1227004
Maxwell J (2006) Re-situating Constructionism. In: The International Handbook of virtual learning environments, pp 279–298. https://doi.org/10.1007/978-1-4020-3803-7_11
McNeill D (1992) Hand and mind: what gestures reveal about thought, 1st edn. University of Chicago Press, Chicago
Norgaard K (2011) Living in Denial, 1st edn. MIT Press, Cambridge
Papert S, Harel I (1991) Situating constructionism. Constructionism
Rasmussen K (2014) Building a better business using the Lego Serious play method, 1st edn. Wiley, New Jersey
Reiner A (1990) The triune brain in evolution. In: MacLean PD (ed) Role in paleocerebral functions. PPlenum, New York, xxiv, 672p. Science 250:303–305 (1990). https://doi.org/10.1126/science.250.4978.303-a

Silva J (2014) The drawing and the intuitive action on knowledge acquisition—empowering design though reasoning with drawing. In: Proceedings of the 9th Conference of the International Committee for Design History and Design Studies. https://doi.org/10.5151/despro-icdhs2014-0099
Stierand M, Dorfler V (2014) Researching intuition in personal creativity. In: Handbook of research methods on intuition
United Nations (2018) The Sustainable development goals report. In: Unstats.un.org. https://unstats.un.org/sdgs/report/2018. Accessed 8 May 2019

A Systematic Review on International Design Research, in Order to Ensure that the Findings Can Compare Design Capabilities in the Portuguese Reality

Maria João Félix

Abstract This chapter will present a first phase of a contextual study, which will initially focus on a systematic study of internationally produced thinking about the most pertinent contours, challenges, methods and results of design research. This study will make it possible to establish a reference framework, both for the collection and classification of data related to the Portuguese reality and for its subsequent analysis. Design research has a recognized trajectory, but the discourse built so far is not visible to the academic community. Researchers, creators, and theorists began documenting design when it was recognized as something that could be taught. Despite the isolated attempts of theoretical depth by some researchers, the degree of relevance between research and the different areas of design has been quite diverse, such as engineering, architecture and product design, still touching on territories such as arts and crafts. To summarize the available information on the evolution of research in the area of Design and the problematization that the international scientific community has generated around the establishment of its outlines and methods, namely on practice-based research, it will serve as a contribution to the clarification of aspects that are very relevant to a rapidly expanding area.

Keywords Design research · Practice-based research · Relationship between science and design · Design research methodology

1 Introduction

Design is still quite recent as a subject, which hasn't allowed the conceptual and tendencies updating, creating an impressive lack of understanding and unification, that other subjects have already acquired. As stated by Charles Owen, Design isn't a science, nor an art or any other subject and it has its own goals, values, measures and procedures. Through used analogies and comparisons in science and arts, we can find

M. J. Félix (✉)
School of Design, Polytechnic Institute of Cávado and Ave, Barcelos, Portugal
e-mail: mfelix@ipca.pt

CIAUD, University of Lisbon, Lisbon, Portugal

D. Raposo et al. (eds.), *Perspective on Design*,
Springer Series in Design and Innovation 1,
https://doi.org/10.1007/978-3-030-32415-5_6

reference points, but Design hasn't been widely investigated, formalized, codified or even included in the literature created for this area. Summing up, the theoretical knowledge base for Design is scarce for the ones looking to work in a more rigorous way, the academic and scientific models in a search for orientation and references for the research in Design.

The history or past of this activity, has come since the beginning of the twentieth century and it is present in the subject development, in the position and academic recognition and in the search of a "science" and of a methodology that goes on till our days. In this work, examples of some worldwide points of view will be given, related to Design research. The goal is to show that there are, on one hand, many reliable and different positions in the design research and on the other hand, that they can well coexist. That can make the design research look like a confusing area, but it is also a sign that there are ongoing processes in its formation as an academic domain. The research in Design is halfway through an historic definition process as an established area of knowledge production. This is a process, through which, many sciences have gone through before.

2 Problem

Bayazit (2004), publishes in the credited magazine "*Design Issues*" (vol. 20.1) an article entitled "*Investigating Design: A Review of Forty Years of Design Research*" focused mainly on the methodological development efforts and excluding right away other topics and investigation lines under the typologies that Frayling (1993/1994) identifies as: Research *in* Design, *about* Design and *through* Design.

These efforts seem motivated by the desire to attribute a scientific status to design, desire that had a prominent echo in the school of Ulm and opened doors, in a certain point of view, to the technoscience initiatives in a way to rehearse "intelligent" computational systems to turn Design deterministic or at least, algorithmic. In those days, research allowed Design the possibility to work based on safe data, and, by doing that, it allowed the possibility to avoid subjective choices as well. It also allowed Design to go further the manufacturing level of knowledge, the knowledge acquired by mere experience and established the base for an additional disciplinary development of a professional activity. Frayling (1993/1994) observed that differences between the world inside the academy and the professional world have always existed in the Design and Art area, but that it is also necessary to preserve these areas' autonomy.

The academic world needs to form people and generate knowledge, but it is also true that connections between the two worlds are necessary to increase the tasks performance and at the same time, find narrower collaboration methods. But the investigation efforts result equally of the desire to systematize the Design process in a way to turn it capable of being taught and managed. Other perspectives allow to question if each Design studio won't be an investigation lab (Laurel and Lunenfeld 2003), following Universities in the production of the new knowledge, promoting its reutilization.

If experimentation and research in Design processes itself through an artifact project (in the investigation of new means and languages as a poetry channel, many times interpreting new processes, materials or programs), through theoretical studies or in the convergence of the two previous approaches, so the academic investigation can guarantee that its coverage to the different listed typologies, is evaluated by national and international peers and without commercial pressures.

It is important to mention that the contour definition of what Design research is and, consequently, of the configurations that a Ph.D. in this area can assume, continue to raise discussion, in the international community, though the pressure to qualify people with this degree, namely in European countries, has made overcome, in a pragmatic way, these doubts using the outlines established by other disciplines.

The concern with the qualification research in the design area has had, in the last years of this decade, an extraordinary development by the recognition of its importance in the innovation matrix and companies' competitivity, but also for its importance as a strategic partner, for example, in the search for solutions for the planet's sustainability. The research in Design should demonstrate, within the academic world, that it is a subject which aims to produce the body of knowledge of a specific subject and thus, be in conformity with the university standards. On the other hand, through this body of knowledge, it should conquer its own space and become an independent subject that is recognized as such.

3 Methodology

For the realization of this work it was necessary to proceed with the theoretical ground, which sustained itself on the collection and bibliographical analysis, through which we studied and systematized, in a deeper way, the internationally produced thinking, about the contours, the challenges, the methods and the most pertinent results of the research, in the Design area.

The contents analysis was done through instruments of primary information collection which consisted of the analysis of official decrees, photographs, letters, articles, etc. And as secondary sources, books, theses and monographs were consulted. The analysis of documents was done in libraries, official organisms, institutions, as well as some reference works acquired for the purpose and relevant for the study at hand.

The whole instruments elaboration and conception process, previously mentioned, resulted of the research, analysis and information processing collected during the bibliographical review. Later, it was defined the problem that ended in the deduction of our hypothesis that could contribute for the scientific knowledge of the studied phenomenon. Being a continuous and interactive process, we present the principles of our scientific procedures and stages, describing the performed operations in a way we can progress step by step, in this investigation process, until we reach our objectives. From this context, and to translate the general hypothesis into operational hypothesis, we use the inductive process (formation of a theory through already known data) and the deductive process (prediction of new data through already existing theories).

We continue with the non-interventionist methodology to support our evidences and translate them in a language and a way that allows us to do a systematic investigation work for the collection of observation data. In this way, we used written and unwritten documentation sources, planning the data collection methods through the documentary research (books reading, articles, investigation reports, through the opinion of experts with developed work on investigation, and testimonies), defining our analysis model. In the analysis of the written documentation, coming from the documentary research, we present a brief explanation of the main approached authors and that concentrate the aspects and proposed reflections in the wording of our study object.

To perform the analysis of the theories directly related with the approached phenomena and in a way to reach our proposed goals, we stand out, in our research, the works and theories developed by Bayazit (2004), Margolin (2002), Archer (1981), Frayling (1993), Simon (1969), Cross (1993), Schön (1983), Dorst (1997), Friedman (2000), Findeli (1999), Downton (2003), Laurel (2003), Jonas (2004) and many others. At last, some conclusions were drawn, through the systematization and analysis of the collected information, which had as objective, the production of a strategic reflection that could be useful in the development of the investigation, in this area.

4 Results

For the theoretical construction of the research in Design and the consequent systematization of the internationally produced thinking, it will be important to acknowledge that there's a moment in the evolution of each area or subject, on which the intellectual questions focus, as those in which the passage of an ambiguous and irregular territory is established for one of the substantiated analysis.

Many of the articles about research in Design are related to the history or past of the activity (Bayazit 2004, p. 17), since each investigation report should prove its roots. The research field in Design is much wider than most researchers acknowledge and covers many stakeholders. This area is constituted by multiple discussion groups or networks, each one with their interests, based on their own criteria, and whose goals result on the best practices and on significant results, usually, published in magazines, on internet sites and conferences promoted by the different groups. These groups have many different purposes and are of different typologies. For some, the goal of the investigation is to create new products, while others intend to acquire a bigger understanding of design as a cultural phenomenon.

These ideas overlap frequently, thus, usually, the researchers of these groups belong to several networks. However, according to Margolin (2010, p. 80), is through the cross pollination that an investigation field, with several networks, can expand itself and produce results that transcend the interests of any group. It will also be necessary to recognize that the concept of "research in Design" has different meanings, depending on who uses it. According to the definition of L. Bruce Archer (1981), stated in the Portsmouth DRS, research is the systematic inquiry whose goal is knowledge. "*Design research is systematic inquiry whose goal is knowledge of, or in, the embodiment of configuration, composition, structure, purpose, value, and meaning in man-made things and systems […]*" (Archer 1981, p. 30) This definition is very simple, but it indicates what research means, being the word "systematic" perhaps the most important one, because it raises the question of the relation between science and design, as we will see later on.

Regarding the concept of research, Downton (2003) refers that some disciplines have felt obliged to migrate to the science paradigm, in a kind of validation for the respectability and trust science gives.

Frayling (1993/1994) bases himself on the idea that there is too much confusion in the discussion on the theme Design Research, concerning words' meanings that become concepts and especially, at the level of understanding that most people does of words (Costa 2005), that constitute the base vocabulary of the discussion. It starts by the very own word *research,* which deconstructs as *re-search.* This separation takes it in two ways, the first indicates the main meaning of the word, the search, the second one suggests that that search implies going back to the past, a "return" (re) looking at what exists. This second way, problematic, reveals itself analysis worthy: in what way can we—namely, in Design, which has as plan the search for something that doesn't exist—"research" the new?

We consulted the *Oxford English Dictionary* (OED) and the history of the same word usage, to prove that the usage of *research,* goes back to the sixteenth century, and it was done in daily contexts, by individuals in search for some information (with small r) and not exclusively in specialized on professional contexts (with capital R), as it is mainly understood nowadays. It also proves that the word was used in artistic contexts, namely in music, which strengthens the theory that the word and its meaning aren't property of the scientific activity.

Research, concluded Frayling (1993/1994) from this collection, implies a search whose object is exogenous to the researcher. Moreover, than the person or group that carries on the investigation is capable to communicate with others what it's being looked for. Proved the pertinence of the use of the word in artistic activities, the author goes to a more complex field: what, in art, can constitute research? Referring to a Picasso interview (1923), where the artist claimed not to be a researcher and not even wanting to be associated to any kind of search; despite Picasso using some sources to paint and sculpt, he refers that that doesn't constitute any kind of research, once his last goal is not to collect information to reach a conclusion; his goal is to paint and that, according to Picasso, can't be searched, it is found. He even says that the artist is a maker, not a researcher. "*In my opinion*" *said Picasso,* "*to search means nothing in painting. To find is the thing. Nobody is interested in following a man who, with his eyes fixed on the ground, spends his life looking for the pocket-book that fortune should put in his path* […]" (Picasso 1985, p. 2). One of the goals of the research in Design is to study deeply the artificial world, built by human hand, and the way activities implied in it have been directed to academic studies or to the industry. According to the concept of Herbert Simon, the global activities of design can be called of "The sciences of the artificial" (1969), in opposition to the natural sciences.

The term "artificial" is used by Simon to indicate the systems that have a determined shape or behavior, just because they adapt themselves (or are adapted) to the environment, bearing in mind certain goals or purposes to achieve. Some art areas, from handicrafts and design call investigation the activity performed by an artist, when creating a piece of art or handicraft. This kind of investigation is not the theme of this work, thus we consider that the performed activities by an artist while creating a piece of art or handicraft can't be considered research. However, it is possible for an external observer, to do research through observing the way the artist works. This way, he's giving his contribute for common knowledge. These can be observable phenomena.

As Christopher Frayling states *"Research through art and design is less straightforward, but still identifiable and visible*" (Frayling 1993/1994, p. 5). Still in "*The sciences of the artificial"*, Herbet Simon went even further when stating that *"the proper study of manking is the science of design"* (Simon 1996, p. 138). Simon suggested that the Design study be a fundamental and interdisciplinary study, accessible to all involved in the creative activity of making the world artificial, which includes the entire humanity. Besides, he suggested that, to understand the creative activity, we should start by sharing the Design creative and professional process experiences.

This seems to be another challenge in the Design research, helping to build a "conversation about Design" mode (Cross 1999, p. 8), which is, at the same time, interdisciplinary and disciplined. The paradoxical task of an interdisciplinary discipline creation proposes a conversation between common disciplines to reach an understanding that allows to create new knowledges and Design perceptions. According to Cross (1999) about some of the main examples of that developing "conversation", we can refer to the series of articles, that in past years, have won the annual prize "*Design Studies Award*" regarding the best published article in that newspaper. This prize was set for the first time in 1987, and in an analysis of winners, we can see, Schön (1988) (Urban Planning, MIT, USA) *Designing: Rules, types and worlds,* an analysis of the Design protocols to identify reasoning standards based on rules derived from concepts-type; Burr and Andreason (1989) (Engineering, Lyngby University, Denmark) *Design models in mechatronic product development,* an analysis of the design models properties, leading to architectonic Design adequate models proposals in the mechatronic area; Downing (1992) (Architecture, Texas A&M University, USA) *Conversations in imagery*, study of the memory role (mental image of memorable places) in the architectonic design process; Roy (1993) (Design and Innovation, The Open University, UK) *Case studies of creativity in innovative product development,* study of creative designers to obtain information about the creative process and of innovative products development;

Goldschmidt (1995) (Architecture, Technion, Haifa, Israel) *The designer as a team of one*, a designer's work comparative analysis and of a small team facing the same design problem; Purcell and Gero (1996) (Design Science, Sydney University,

Austrália) *Design and other types of fixation experimental,* studies of design problems resolution, aiming the understanding of the fixation causes and effects; Janlert and Stolterman (1997) (Computing and Informatics, Umea University, Sweden) *The character of things*, a consideration of how things (hardware and software), as well as people, can have a "character". These examples come from the investigation in design performed in different domains and with distinctive methodologies, but each individual contribution had something to say to the members of the biggest research community in Design.

These contributions of "good practices" of the research in design have the following characteristics in common: "***Purposive,*** *based on identification of an issue or problem worthy and capable of investigation.* ***Inquisitive,*** *seeking to acquire new knowledge.* ***Informed,*** *conducted from an awareness of previous, related research.* ***Methodical****, planned and carried out in a disciplined manner.* ***Communicable,*** *generating and reporting results which are testable and accessible by others*" (Cross 1999, p. 9).

These characteristics are, naturally, the normal characteristics of good investigation in any subject. Such criteria, normal and academic, don't inhibit or prevent the investigation from being "designerly" (Cross 1999), always related with Design, in its origin and intention. Still according to Cross (1999), we should establish a distinction between the practical works and the investigation ones, not considering that the practical works are investigation works. The most important, while researching, is to extract reliable knowledge, whether from the natural world, whether from the artificial, and provide that knowledge to others in a useful way. That doesn't mean that the Design practical works should be completely excluded from the Design research, but, to qualify as research, there must have, first, a reflection by the implied professionals and, then, a communication of the obtained results from that reflection.

According to Silva (2010, p. 84) the purpose of the research in Design is to orientate the practice and contribute to the improvement of the area and for the development of theories and methods.

The practice of design may involve research, but it's not the research in itself. Without investigation, we are left to our intuition, based on the existing knowledge. The investigation involves the creation of new knowledge, which is different from accumulating information, or the organization of the already existing knowledge.

However, the collection of information can be the beginning, or first step, of an investigation process. One of the dangers, in this new research area, is that other subjects' researchers, apart from Design, import inadequate methods and approaches to develop design understanding. Another danger is that researchers can join underlying paradigms, of which there's only one slight perception.

It is necessary to develop this intellectual perception in the Design community. A good example is Kees Dorst's work, who analyzed, with explicit comparisons, the paradigms that justify Herbert Simon approach, on one side, and Donald Schon, on the other side, which represent, respectively, the positivist philosophy and constructive philosophy. Simon's positivism transmits a vision of Design as being the rational problem solving and leads to the idea of design as a reflective practice. These two visions seem to get into conflict, but the use of both paradigms to analyze Design activity leads Dorst to consider that different paradigms complement each other, in order to obtain a general vision of all design activities.

Until very recently, the Design area was a complement of art and handicraft. The Design transformation in one industrial subject brought responsibility that Design studies only recently started to deal with. Design is, now, becoming a generalizable subject which can be applied to several areas. Understand design as a subject that can function this way means to develop a design general theory. This general theory should support the application theories and operational programs. Going from one Design general theory to the problems' resolution task involves a conceptualization mode significantly different and an explicit knowledge management (explicit knowledge) more than adapting tacit knowledge of the individual design experience. According to Friedman "*The challenge of any evolving field is to bring tacit knowledge into articulate focus. This creates the ground of shared understanding that builds the field.*" (Friedman, Creating design knowledge: form research into practice 2000, p. 13).

Until now, the majority of Design theories involve clinical situations or supported theories which are necessary, but not enough. In social sciences, the supported theory has become a strong and sophisticated theory formation system in different levels. Basically, these theories take to higher levels of understanding, whose informed literature is rich in discussions about the theory construction and theoretical awareness.

One of the big problems of research in Design is the inability to develop a supported theory in practice (Friedman 2003, p. 520). Instead of that, Designers, many times, confuse practice with investigation, instead of developing theory in practice, through inductive articulation and research, some designers argue that practice is research and that research based on practice is in itself a way of building theory. Design theory isn't identical to design practice tacit knowledge. Although tacit knowledge is important to every practice field, confusing tacit knowledge with design general knowledge involves a categories' confusion. According to Polanyi, who makes the distinction between tacit knowledge and theory construction, we can consider the theoretical knowledge more objective than the immediate experience. "*It seems to me,' he writes, 'that we have sound reason for … considering theoretical knowledge more objective than immediate experience…A theory is something other than myself. It may be set out on paper as a system, of rules, and it is the more truly a theory the more completely it can be put down in such terms.*" (Polanyi 1974, p. 4) Only the explicit articulation allows us to test, consider or reflect about the theories we develop. For that reason, the failed attempts, to connect Design reflective practice to design knowledge (Friedman 2003, p. 520) and to propose tacit knowledge or direct creation, as theory building methods, must be inevitably considered as a "one-way path". All knowledge, all science and all practice are based on a wide cycle of knowledge management that moves from tacit knowledge to the explicit one and vice versa.

Until now, with its handmade tradition, Design has counted mainly with tacit knowledge, but it is time to consider the explicit ways in which design theory can be built and recognize that, without a body of knowledge based on the theory, the job, in the area of Design, won't be ready to face the challenges Designers find in a complex world like today's.

According to Findeli et al. (2008) the construction of a consistent and coherent methodology, of investigation in design, has been a constant concern. This phenomenon isn't unrelated to the tendencies and oscillations that methodologies have been going through in the past years and that have affected the scientific community. This excessive focus on the search of the correct or ideal method tends to be unsuccessful, if it's not supported by a necessary epistemological, preliminary and insurmountable investigation. Unless we know exactly what the goal of our investigation is, it's not worth arguing, debating and discussing about the correct or ideal method. Nowadays, the epistemological question of the investigation in design is far from being solved. We think that it is this way because the doubt between the investigation objectives and the design projects still persists. As strange as it may sound, the main question of what the investigation goal in design could or should be is still under consideration and can be divided in the following set of questions: "*What exactly are the objects of design considered as a scientific, academic discipline? What are the phenomena of the world we are interested in observing and understanding, that are not already the "property" of other disciplines? What do we intend to say about these phenomena that is not known yet and that other disciplines cannot know or at least that design claims to know better?*" (Picasso 1985, p. 2).

We believe that these enlightenments are useful to identify what we consider to be the main conditions for an investigation that matches the design subject or area specifications. The vulgarized reflection of the Royal College of Art director, Christopher Frayling from 1993/94, about the research in Design and which is considered the founding formula of the discussion between practice and theory, is still the possible structure of this scientific activity.

Frayling invokes Herbert Read, who had devised a distinction between the three types of investigation in the teaching of art: research for art, research about art and research through art. It is from this distinction that Frayling proposes his own formula, the one that, according to the author, derives from his practice, with the established, but still controversial terms, and this time applied to design (Findeli 1999; Downton 2003, p. 2).

The concept of Education through Art, which is a literal translation of the expression that gave the title to the work of Herbert Read- Education through Art, is marked by the idealism of someone, who had lived the war and wanted a society where it had no place. Hence the importance of art in education. In the same year that the third edition of Read's work was published in 1956, the Portuguese Association of Education through Art appeared, of which were part, among others, Calvet de Magalhães, João dos Santos, Almada Negreiros, João de Freitas Branco, António Pedro, Adriano Gusmão, Breda Simões and Arquimedes Santos. A few years later, also under the influence of Herbert Read's work, in 1971, it was created in the National Conservatory, the Educational Teaching through Art Course, which had as teacher, among others, Arquimedes Santos who stated: education through art serves mainly the personality formation.

According to Christopher Frayling typology, the research is divided into three models (Frayling 1993/1994, p. 5):

(a) research into art and design, in this typology the traditional studies about history, aesthetics or perception about a variety of theoretical perspectives related to design, social, economic, political, ethical, cultural, iconographic, technical, material and structural are included; (b) Research through art and design, the second typology focuses on the project and connects with what is known in England as practice-led research. As examples, are quoted the study of materials' behavior, the development of technology to fulfill a new task or the methodological reflection about the project development; (c) Research for art and design, the third typology is the hardest to characterize in the way that its results are made concrete in the object, or in a objects' body in which, in a certain way, the thought is materialized in the artefact, in which the goal isn't, in the first place, communicable in the sense of verbal communication, but in the sense of visual, iconic or imagery communication.

According to Costa (2005), it is still weird, in a research context, to join the terms "art" and "design". It is necessary to understand that in the scope of the teaching in England, design and art had, since the middle of the 19th century, an identical path, thus both were taught in the perspective that knowledge was only knowledge when connected to the know-how. The Portuguese reality, where design only comes in higher education, not yet academic, in the last quarter of the century, turns it hard to join the terms, because, since then, in fact art and design still with common roots, differed significantly in several points of their economic, social and cultural realities.

This way, what matters to us from Frayling formulation (Frayling 1993/1994) and that many authors (Laurel and Lunenfeld 2003, p. 11; Downton 2003, p. 2; Jonas 2004; Schneider 2007) have already stated, is the following: Research into Design; Research through Design; Research for Design.

There are, however, other perspectives based on this typology, Findeli et al. (2008), for example, proposes the three areas again, where the research projects' conclusions, in Design, should be evaluated, but yet, and as he considers that these research practice typologies, in Design, have received many interpretations, sometimes, contradictory and not always useful, wanted to clarify his vision, reminding that his method definition of "Research through Design" comes from a methodological critic to both most used current methods in the area, the research for and into Design. The argument divides itself in the following way:

(1) The first, research for design is common to any research project in any subject. It is an original and significant contribution for knowledge, and in our case, for design knowledge. This kind of research is relevant for the practice of design, given that its purpose is to certify that several parameters, on which the design process accomplishment depends (technological, ergonomic, aesthetic, psychological, etc.) are treated adequately, which is, that the design project be responsible and properly informed. However, they aren't scientifically acceptable for many reasons, such as (Findeli et al. 2008, p. 70):

    – It is generally based on an already available knowledge.

- When new knowledge is produced (for example, after interviews, field observation, comparative analysis, etc.), generally, is not accomplished with the expected rigor by the scientific standards, or because the "researcher" isn't qualified enough or (more and more) because the timing limitations don't allow it.
- It is mainly tacit and there's no intention to publish it or discuss it in the design research community. Actually, in many cases, it is really confidential.

This means that the research and the design practice differ in their respective goals, in the validation and evaluation criteria, in the public and contexts, hence one shouldn't be evaluated by the criteria of the other.

(2) The second, research about design, corresponds to an expected improvement in the design practice and, consequently, in the level of the user's satisfaction. This kind of investigation, according to the scientific standards, is normally performed by several subjects, besides Design. Actually, the fact of being published is, again, the proof of its rigor and acceptance by the scientific community. The problem found in this kind of research is its relative lack of relevance for design. "Design", here, means the design practice and the teaching or research in design. Research about design (which is, about its objects, its processes, its actors and participants, its meaning and its importance for society, industry, culture, etc.). performed by researchers of other areas (like anthropologists, archeologists, historians, psychologists and many others), has, as main goal, to contribute for the advance of its own subject and not of design in particular. More precisely, these researchers aren't orientated to find out how they can certify that the produced knowledge, in their research, is relevant for Design, even if it is how it is expected sometimes. In fact, designers are the ones in the best position to decide if such knowledge is relevant and, if it is the case, how it could be implemented in the respective practices.

(3) The third, research through design, brings some useful consequences for the teaching of design. The previous observations give us two important criteria that the research must follow: it must be rigorous, this is, it should regulate itself by the usual scientific standards and it must be relevant, which is, contribute for the improvement of the design practice. The research through design must be understood as having properties, not only from the research for design, as from the research about design.

On one side, the idea of research, through research, earned enough credit for the creation of research projects, also called by practice-based research, practice research, action research in design, clinical research or project-grounded research. On the other side, there is still a struggle for the methodological solidity and scientific recognition.

The relative dispersion in the semantic space indicates that the interpretations of the idea and research principles, through design, can differ and that a real consensus has not yet been reached. The main obstacle, here again, is of epistemological nature. The majority, if not all promoters, of the research through design, agree that the design project must have its place in the scope of the research project, but that this last one shouldn't be confused with the first one. Where some researchers differ, or stay in silence, is about the function (epistemic) to be attributed to the design project within the research, that is, in the restrictive scope of a rigorous and relevant knowledge production process in design. This is certainly one of the crucial questions in the design subjects and, in fact, in all professional subjects.

Behind this question lays the eternal philosophical enigma of the relation between theory and practice, thus one thing is claiming that the practice is important and necessary for the construction of the theory, and other, more challenging, is explaining how this contribution of practice for theory must be planned and operationalized.

In his work, *Creating design knowledge: from research into practice,* Ken Friedman asks "*How does new knowledge move from research into practice*", to which he answers: "*Concrete research results become visible to practitioners in a myriad of ways. Journal results, conferences, corridor talk among colleagues, knowledge transfer in shared projects, Internet discussion groups. The important issue is that a field must grow large enough and rich enough to shape results and circulate them. As this happens, the disciplinary basis of the larger field also grows richer. This leads to a virtuous cycle of basic results that flow up toward applied research and to clinical applications. At every stage, knowledge, experience and questions move in both directions…Practice tends to embody knowledge. Research tends to articulate knowledge*" (Friedman 2000, p. 23).

Based on the Friedman thinking, (Frankel and Racine 2010) represent the positions regarding the categories and subcategories discussed so far. They refer the flow between the research for design, the research through design and research about design as a circular process, in which each action informs the other. Although these research categories in design are inter-related, they also represent different levels of knowledge in design. Downton writes about three types of knowledge: "how-to-knowledge" (Downton 2003, p. 62) demonstrating that it is known how to draw, "knowing-that" (Id., ibid.), learning how someone draws and "knowledge-of" (Id., ibid.), having peripheral knowledge about what people can draw. Dowton states that "[…]" the theories of design are concerned with what design is, what should be and could be […]" (Downton 2003, p. 79). Based on this perspective we can wait that the research in design fits in any of these levels by a variety of combinations, depending on the initial cause or hypothesis. As mentioned above, design knowledge is connected to epistemology. "*What is design Knowledge about and after?*" (Findeli et al. 2008) which sums up the question. Taking up the question of the design research purpose, the criteria that any design research project should follow aims at three of the research final users in design, who are interested in its production: the design research community, the design practice community and the design teaching community. "Three "end-users" of design research are interested in its output: The design research community, the design practice community, and the design education community" (Findeli et al. 2008, p. 74).

The kind of knowledge and the part of information considered relevant and valued by each of these communities differ. For this reason, there's the need for design researchers to consider these aspects, by building their protocols and by writing their conclusions and final reports. Expressed in the conventional terminology, the researchers' community is interested in the fundamental knowledge or theoretical, while the practitioners' community focus on the applied and useful knowledge and the educators' community, in the teachable and applicable knowledge. This means that it's not worth performing design research if it's not to improve the act of design and consequently, the life of the people involved, that is, most likely the entire humanity. Consequently, the purpose of the design research is directly connected to design's, which means to improve or keep the world inhabitable in all its dimensions, physical, psychological, spiritual), while the conception focuses on the act of design itself, that is, on the description and understanding of the act to keep or improve these relations.

Bearing in mind this model structure, we see that, behind each project and design situation, there is potentially; an anthropological question, which is, a question related to the way an individual or a community inhabits or intends to inhabit the world. Summing up, nothing that concerns the relations between the human being and his environment must be strange to design; One opportunity to contribute for knowledge already available about the act of design. If performed in a reflective way, each design situation is an opportunity to reveal something new about what's at stake in the design practice, or, at least, to confirm an aspect of the "*designerly way of knowing*" (Cross 1982), which was already observed, described and theorized. Cross used this term in a 1982 article, and has been applied in several contexts ever since, in the area of research design, Cross states that there are ways of knowledge peculiar to a designer's conscience and capacity and believes that we get that knowledge from three sources, people, processes and products.

In other words, in the research based on the project, we should look beyond the immediate product of the design project. Without, however, neglecting the last one. The challenge for the researchers and research teams is to understand both potentialities referred above and the adequate and accessible research problematics. In this context, there are naturally countless potential and available research objects and projects. The contextual circumstances will help determine which are the most appropriate for the programed research.

According to Findeli et al. (2008), this is, actually, a vast research program, but it will be acceptable such ambitious statement? No, if the design scholars pretend that this is their private territory. But yes, if they believe that they have a word to say about this serious subject about the description and understanding of the way we inhabit the world and how some intend to improve it.

A possible entry in the puzzle is to understand that the two potential questions, referred above, aren't actually, design private property. It is certain that almost none scientific subject is completely unfamiliar to our human condition and the way the human being, individual and collectively, connects with the world. If we adopt a wide definition of the design objects, then we definitely share these questions with many other subjects. Trying to isolate a specific part of the world phenomenon and adequate it on behalf of the design scientific statements, maybe it isn't the only possible way to guarantee a specific knowledge domain. We believe that, instead of that, it is the kind of question Design asks about these phenomena that constitute this subject's originality.

For Costa (2007), in the academic context as well as in the studio practice, the research in design tends to be developed to and with the users. The limitation of the object study will be, therefore, an obvious consequence of that tension, decided from universities and labs, according to their own criteria and properties, and whose strategies go, naturally, by their geographical, cultural, social and economic contexts. There will be, however, at the university level, the need to think that path as something coherent, this is, as a part of a whole where each project or research can add something to the previous works and anticipate the future ones (Costa 2007, p. 168), creating a body of knowledge over which other paths will be drawn. Being, apparently, a detail, this is, however, one of the factors that have prevented a bigger cooperation among universities and the social and economic base, namely in Portugal.

The search for new and different ideas by the industry will be the evidence of a split, hardly surpassed, between the academy and the market, about what constitutes the subject. This split shouldn't be solved by unilateral yielding, by the academy, otherwise, in a near future, design won't be able to assert itself as a research object and field. This question is especially relevant in the current context of the higher education reorganization and the consequent need to raise funds by the labs and investigation units, as the only way to guarantee the survival. According to Cross (1999, p. 10) we are still building the adequate paradigm for the research in design. The construction of such paradigm can be useful, in long term, for the practice and education in design. We still know relatively little about the skills in design and that restricts our adequate study to mankind, this is the goal of the research in design.

Despite the several approaches related to this theme, we conclude that the various adaptations of the Frayling typologies, have contributed to the refinement of the questions that circle around the types of research in design, however, we consider that not to misrepresent the initial reflection, we adopt in this work as basis the Frayling thinking, where briefly a phasing of three research modes go from the traditional theoretical observation about the design subject and its material productions (referring to other subject areas like History, Psychology or Sociology…) to the ethnographic observation of the designers production processes (seeking clarification about their production and creation methodologies), to the most common of the projects practices in design (watching their own product projected as manifestation of knowledge).

The first model deals with the Design Theoretical Studies, though directed to the practice, the products, the speeches or the Design (meta speeches).

The Design Studies (Margolin 1998) are divided in four topics:

(1) The study about the Design practice (which includes the activities related to conception, planning and products manufacturing); (2) The study about the design products (in which we should highlight the identity and interpretation of the products and their cultural role); (3) The study of the Design speech (about the Design ontology, the different arguments of what is or may be this subject); (4) The study of the studies in Design or meta-speech (the epistemological reflection about this field of knowledge about itself).

In the second model we can find the research guided by practice, the focused reflection on the project and directed to the methodological understanding of its development. And in the third, the results of the research will be applied in the new drawn product, resulting the research whether on the work, whether on the convergence of both.

5 Conclusions

In this work, it has been done a revision of the internationally produced thinking and the most pertinent results of the research in the design area. This contextual study points its importance for the design methodology, as well as the scientific research. The majority of the studies about design research, have been made based on the post-world war II society demands. The scientific developments, during the war and the lack of resources in the society of the time, demanded and gave an impulse to the creation of new ways that allowed the resolution of the existing problems. These new developments would come to contribute to deep alterations in the design teaching and research.

Design would then be understood as a process and a systematic vision originated from these alterations. Research in design, included the work of Horst Rittel, at Ulm University, together with Bruce Archer. This was a second generation of design methods that acknowledged the participants and the argument as parts of the process. Rittel understood that the planning problems as "wicked problems", given that while science deals with tame problems, most life problems are inexorable. Bruce Archer acknowledged the importance of education in design, in the general education, at schools, for children, for all. "Design has its own things to know, ways of knowing and ways to find them out" and conveyed the idea that science shouldn't be imitated. Progressively, it has been recognized that design should be understood in its own terms. The typical research included a theoretical analysis about what design is, the descriptive studies and participative methods.

Schon would take the way designers work as a starting point and recognized the contribution of the professionals of this area, having researched the design cognition, especially in architecture. However, a split started to occur between the research in architecture and in engineering. The typical research included a protocol analysis (protocol studies Delft. 1976), parallel methods to design, product engineering design and computer assisted methods. Comparing the numbers of official studies, the remarkable events and activities in specific subjects, over time, there has been a visible change in the design research activity focus. During the 60s and 70s, many activities in the area of architecture and engineering were performed. Design studies in mechanical engineering were more useful in the 80 s and more recently, software and electronic design studies were developed.

The idea of research, directed to the practice, reflects the orientation of these subjects to develop distinctive practices that explore their capacities and ways of knowledge. The quick expansion of Ph.D. programs in design created the need to form research units or groups that can publish and spread a growing volume of work in the area. Until this moment, some of those works have contributed for the development of new design practices, such as interaction design, sustainable design, medial design, services design, organization design, universal design and development design. Bruce Archer recognized, right away, the relation between two types of research, one directly related with making things and the other, more concerned with understanding, not only the things in itself, but also the environment in which they are conceived, made and used. Archer was worried with the practice (praxeology, modelling, technology and metrics), the understanding (taxonomy, history, axiology, philosophy and epistemology) and teaching.

References

Archer B (1981) A view of the nature of the design research. In: Jacques R, Powell JA (eds) Design: science: method. IPC Bussiness Press Ltd, Guilford, pp 30–47
Bayazit N (2004) Investigating design: a review of forty years of design research. Design Issues 20:16–29
Burr J, Andreason M (1989) Design models in mechatronic product development. Des Stud 10(3):155–162
Costa R (2005) Investigação em/por design, Tese de Mestrado em Arte e Comunicação. Faculdade de Belas Artes da Universidade do Porto, Porto
Costa R (2007) Estruturas Triangulares na Investigação em Design, Convergência na Dispersão. Obtido de http://ria.ua.pt/bitstream/10773/6066/1/MONO_RuiCosta.pdf
Cross N (1982) Designerly ways of knowing. Design Stud 3(4):221–227
Cross N (1993) A history of design methodology. In: de Vries MJ, Cross N, Grant DP (eds) Design methodology and relationships with science. Kluwer Academic Publishers, Netherlands, pp 15–27
Cross N (1999) Design research: a disciplined conversation. Des Issues 15(2):5–10
Dorst K (1997) Describing design: a comparison of paradigms. Ph.D. thesis, Faculty of Industrial Design Enginnering, Delft University of Technology, The Netherlands
Downing F (1992) Conversations in imagery. Des Stud 13(3):291–31
Downton P (2003) Design research. RMIT University Press, Melbourne, Australia. ISBN: 0864592671
Findeli A (1999) Introduction. Des Issues 15(2)
Findeli A, Brouillet D, Martin S, Moineau C, Tarrago R (2008) Research through design and transdiciplinarity: a tentative contribution to the methodology of design research. In: Swiss design network symposium. Berne, Switzerland, pp 67–91
Frankel L, Racine M (2010) The complex field of research: for design, through design, and about design. Obtained from www.designresearchsociety.org; http://www.designresearchsociety.org/docs-procs/DRS2010/PDF/043.pdf
Frayling C (1993/1994). Research in art and design. Royal Coll Art Res Pap 1(1). ISBN 1874175551
Friedman K (2000) Creating design knowledge: form research into practice. In: IDATER (International conference on design and technology educational research and curriculum development), Loughborough University, UK
Friedman K (2003) Theory construction in design research: criteria: approaches, and methods. Des Stud 24(6):507–522
Goldschmidt G (1995) The designer as a team of one. Des Stud 16(2):189–209
Janlert J, Stolterman E (1997) The character of things. Des Stud 18(3):297–317

Jonas W (2004) Obtido de Forschung durch Design, Vortrag gehalten beider Tagung des swiss design network. Scweiz, Basel: http://8149.website.snafu.de/wordpress/wp-content/uploads/2011/08/2004_Basel.pdf
Laurel B, Lunenfeld P (2003) design research: methods and perspectives, Mediaeworks Pamphets Series. MIT Press, Cambridge
Margolin V (1998) Design studies: proposal for a new doctorate. In: Heller S (ed) The education of a graphic designer. Allworth Press, New York
Margolin V (2002) The politics of the artificial. University of Chicago Press, Chicago
Margolin V (2010) Design research: towards a history. Design Research Society, International Conference, Design & Complexity to be held, School of Industrial Design, Université de Montréal, Montréal (Quebec), Canada, p 80
Picasso P (1985) Reprinted from The Arts, New York, May 1923. In: Goldwater R, Treves M (eds) Artists on art. John Murray, London, pp 416–417
Polanyi M (1974) Personal knowledge. University of Chicago Press, Chicago
Purcell T, Gero J (1996) Design and other types of fixation experimental. Des Stud 17(4):363–383
Roy R (1993) Case studies of creativity in innovative product development. Des Stud 14(4):423–443
Schneider B (2007) Design as practice, science and research. In: Michel R (ed) Design research now, essays and selected projects. Birkhäuser Verlag AG, Basel, pp 207–217
Schön D (1983) The reflective practitioner. Basic Books, New York
Schön D (1988) Designing rules, types and worlds. Des Stud 9(3):181–190
Silva FM (2010) Investigar em design versus investigar pela prática do design– um novo desafio científico. Revista Ingepro, Inovação Gestão e Produção, pp 82–91
Simon H (1969) The sciences of the artificial, 1st edn. MIT Press, Cambridge. ISBN 0262190516
Simon H (1996) The sciences of the artificial, 3rd edn. MIT Press, Cambridge

Communication Design

Graphic Design as Visual Arguments: Does This Make a Reliable Appraisal Possible?

Karel van der Waarde

Abstract The article shows how assessments of graphic design can be made more reliable when graphic design is approached as a visual argument. Each designed object makes a claim that it is improving a situation thereby implicitly stating that a current situation was not satisfactory. This article uses two warning pictograms as examples to show how visual information can be assessed. The pictograms warn against the risks of taking medicines while pregnant, and the potential affect of medicines on driving cars. These pictograms claim that they warn effectively about possible unwanted effects of a medicine. Toulmin's diagram describes the relation between a claim and its supporting evidence and reasons. An application of this diagram to both pictograms shows that just assessing the claim is not sufficient, and that it is necessary to question the available evidence and reasons too. At least six different fields provides rules and principles that can be used to assess the evidence and reasons. Designers, clients, legislation and standards, professional peers, people/patients, and society can all legitimately assess the design of pictograms. One of the major challenges for graphic design is to find relevant evidence and reasons, and to consider these in such a way that a balance between the different fields is achieved.

Keywords Graphic design · Visual arguments · Warning pictograms

1 Assessing Graphic Design?

Due to the practical nature of the profession, it is hard to define exactly what graphic design is. Practicing designers will give very different descriptions of their activities, aims, and results (van der Waarde 2012). The shelves in a bookshop under the heading of 'graphic design' show a similar variety. It is likely that there are books on corporate identity, posters, pictograms, signage, typography, historical reviews, biographies of designers, webdesign, and books on tattoos, streetsigns, and calligraphy. It is hard to

K. van der Waarde (✉)
Swinburne University of Technology, Melbourne, Australia
e-mail: waarde@glo.be

Graphic Design—Research, Solariumlaan 15, 1982 Elewijt, Belgium

D. Raposo et al. (eds.), *Perspective on Design*,
Springer Series in Design and Innovation 1,
https://doi.org/10.1007/978-3-030-32415-5_7

see what these topics have in common and why these are all classified under 'graphic design'. And it is not only the bookshops and practicing designers that define and describe graphic design. Researchers, museum-staff, teachers, publishers, librarians, job centers, and governments all have their own definitions and descriptions, and there can be substantial differences within these groups.

It is surprising that most of these groups look mainly at the 'designing' of visual information. The emphasis is on the 'making', which are the activities to design and develop visual artefacts. The 'interpretation' of visual information receives much less attention. A description of graphic design that relates both these areas could be: '*graphic designers make information visual in such a way that the connections of a client are enabled to notice, read, learn, consider, and act*'.

Any new design claims that it will change an existing situation into a preferred one. Inherently, any new design therefore criticizes the existing situation and makes a claim that the suggested change is an improvement. Making such a claim requires a form of an '*argument*' in which the claim is motivated by comparing the existing situation with a changed situation. Both the 'designing' as well as the 'interpreting' can be seen as an argument because both change towards a new situation. In other words '*graphic designers produce visual arguments for their clients*'. Visual arguments aim to verifiable change situations for both clients and people by enabling, empowering, and equipping beholders to act (van der Waarde 2018).

Both sides, the 'development of an argument' and the 'interpretation of an argument', are essential to assess graphic design. The lack of clear boundaries and a clear core of the graphic design profession makes it hard to discuss these assessments in general, and it is therefore necessary to reduce the scope to a manageable topic.

2 Example: Two Medical Pictograms

Pictograms to inform patients about medicines provides a suitable example. A first example warns against the risk of taking medicines while pregnant, and a second example warns against the risk of a negative influence of medicines on driving behaviour. Both pictograms appear on the outside of packaging for medicines that are prescribed by a doctor and dispensed by a pharmacist. These pictograms are mainly used in Spain and France.

There is ample evidence of the negative consequences of the use of medicine during pregnancy. And it is very clear that some medicines do influence driving behaviour because they make patients drowsy and sleepy. However, it is not exactly known how many people are affected in reality. Road accident statistics rarely record 'medicine-use' as a possible cause, and medicine use during pregnancy is usually avoided as much as possible.

Example 1: Pregnancy warning In France, since October 17, 2017, two pictograms draw attention to the risks related to the use of medicines during pregnancy. Figure 1 shows a 'warning' in a red triangle, and a 'prohibition' in a red circle with a diagonal

Fig. 1 Pictograms that appear on medicine packaging in France. Warning: 'may influence', and prohibition: 'don't take'. Based on: https://servier.com/en/new-pictograms-for-drugs-that-may-present-a-risk-during-pregnancy/

slash. Figure 2 shows four alternative pictograms that are used in different types of patient information, but these do not appear on the outside of medicine boxes.

The pictograms in Figs. 1 and 2 show a range of approaches to indicate 'a pregnant woman' and 'do not'. A pregnant woman is illustrated in profile because this perspective is most characteristic. This can be visualized as an outline (Fig. 2a, c, d) or as a solid form (Figs. 1 and 2b). The 'do not' instruction shows even more variety: a diagonal red line (Figs. 1 and Fig. 2b), a diagonal grey line (Fig. 2d), a red cross in a circle (Fig. 2a), and a black cross through a face taking a tablet (Fig. 2c). The outer shapes also vary substantially: a red triangle, a red circle, an orange circle, and a black square in a variation of line-thicknesses. Only one pictogram (Fig. 2c) is supplemented by a text 'Do not take if pregnant'.

Example 2: Driving warning A second example is a pictogram that indicates that 'this medicine might impair your driving'. The French health authorities have made the visual information in Fig. 3 obligatory in 1999 (AFSSAPS 2005).

The colour coding of the pictograms indicates the level of severity. There is a visual inconsistency here too. The triangular shape of the pictogram in Fig. 3 on

Fig. 2 Alternative symbols or pictograms for 'Do not take during pregnancy'. These are not used on medicine packaging. Based on: **a** https://www.apotheek.nl/kunt-u-dat-even-uitleggen/beelden, **b** https://www.iso.org/obp/ui#iso:grs:7010:P042, **c** https://www.usp.org/health-quality-safety/usp-pictograms, **d** http://www.bijsluiterinbeeld.nl/

Fig. 3 A three-step approach that is obligatory in France

Fig. 4 A visual warning on medicine indicating that 'this medicine might impair your driving'

niveau 3 implies that these are warnings, indicating that the use of a medicine 'may influence'. This seems to be in conflict with the accompanying text which states it is a prohibition: 'do not drive'.

The pictogram in Fig. 4 is legally obligatory since 2011 in Spain (AEMPS 2012). An experimental study (N = 1385) showed that "85.7%, correctly related the symbol with the possible effects of the medicine on driving" (Fierro et al. 2013, p. 1059).

The Druid project (EMCDDA 2012) investigated the use of pictograms in warnings for driving. These pictograms have been evaluated in a study with 270 drivers and concluded that '*pictograms can be seen as a valuable means to reinforce both written and spoken information given to patients by health care providers at the time of consultation*' (Monteiro et al. 2012). The visual inconsistency of the French pictograms in Fig. 3 is avoided here by colour coding the 'level of risk', and not by differentiating between 'a warning' and 'prohibition' (Fig. 5).

The literature on medical pictograms provides a range of reasons to use pictograms (van Beusekom et al. 2018). The main reasons are that pictograms are faster to interpret than words, that they are language independent, and that they are more visually salient and therefore attract attention. The consistent use of these pictograms since 1999 (Driving) and 2017 (Pregnancy) also seems to suggest that pictograms are an effective way to inform and warn patients.

These pictograms were designed to change a situation in which patients were insufficiently warned against the risks into a situation in which patients are sufficiently warned. The existence of these pictograms on medicine boxes makes a claim that the addition of pictograms at least amplify the spoken warnings of doctors and pharmacists, and could in optimal circumstances reduce or prevent the risks.

The selection of these two pictograms as examples makes it possible to formulate a question about the assessment of graphic design on a more feasible scale: '*Which evidence is needed to show that these two pictograms effectively warn about possible*

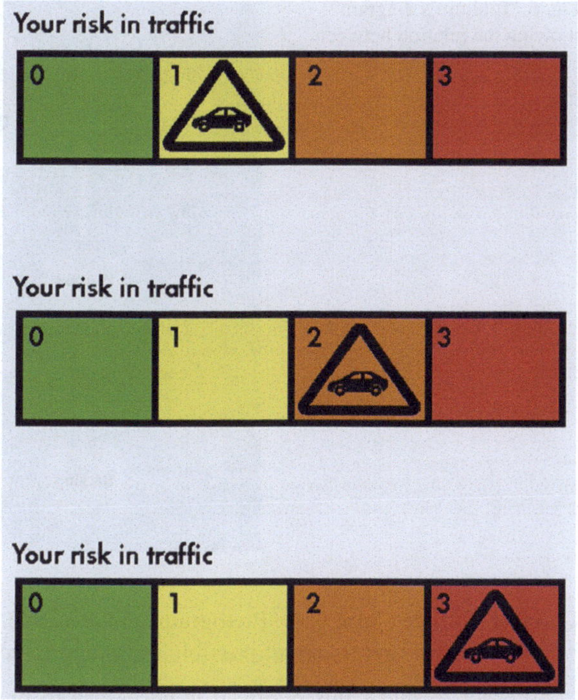

Fig. 5 A rating that shows 'your risk in traffic'. This visual was most effective in an experiment (N = 1006) because it provided good insight into the different levels of driving risks (EMCDDA 2012, p. 30). This is not (yet) used on medicine packaging

unwanted effects of a medicine?' Or, in other words, '*could the claim that these pictograms make be correct?*'

And this brings us back to the inconclusive definition of graphic design activities and results. If the focus of the assessment is solely on the development of the visual results, the discussion leads to topics like the illustrations, the colours, the shapes, and the line-thicknesses of the pictograms. If the focus expands to include the interpretation and actual effects of these pictograms, a different approach is required.

3 Do Graphic Designers Really Develop Visual Arguments?

In order to discuss this claim about the example pictograms, it is necessary to have a look at the kind of evidence that is required. Stephen Toulmin (1922–2009) published in 1958 five essays in '*The Uses of Argument*'. The third essay is on 'the layout of arguments' and discusses the relation between 'claims' and 'evidence' (Toulmin 1958). In this essay, Toulmin related the components of an argument in a diagram (Fig. 6) where the Claim (C) directly follows from the evidence or Data (D).

Applying the structure of this diagram to both medical pictograms, the argument seems to develop like this:

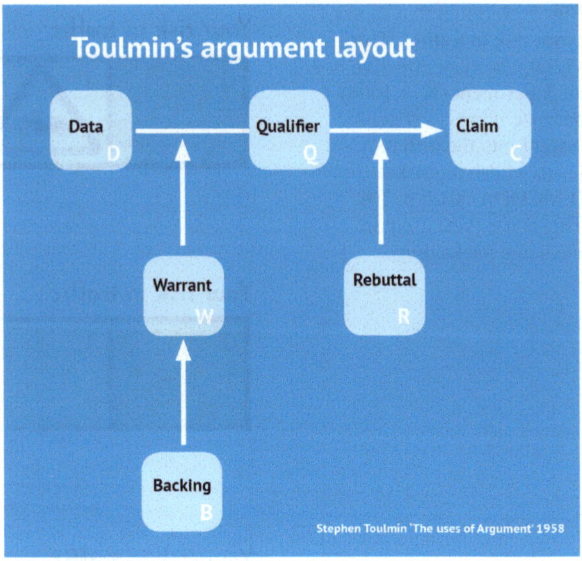

Fig. 6 Toulmin's diagram showing the relation between components of an argument

Claim (C): The claim C is 'Pictograms are effective to warn pregnant women and car drivers about possible unwanted effects of a medicine'.

In order to make this claim, it is necessary to provide some form of data that supports this claim.

Data (D): Evidence suggests that current warnings are not sufficiently adhered to. This leads to high risk events in traffic and pregnancy, and it is therefore necessary to give additional visual warnings to patients to augment this message.

Qualifier (Q): The claim needs to be qualified to limit the scope because it is unlikely that this claim is applicable in all circumstances and in all contexts. The pictograms only warn for two events, and appear only on medicine packaging in France and Spain.

Warrant (W): A warrant provides a reason why evidence (data) is relevant to the claim by providing 'general rules and principles'. In this example, it is in general important to warn patients about the risks of taking medicines to avoid further harm. And in general, people understand pictograms easier and faster than words. Pictograms are furthermore language independent which makes them easier to interpret for non-native speakers and for people who have difficulties reading texts.

These general principles might be questioned because they are based on a combination of assumptions. Asking why the warrants are selected leads to a—usually implied—authority that can provide a backing.

Backing (B): The rules and principles are backed by an authority. For pictograms, there are several types of backing, as we will see later.

Rebuttal (R): The rebuttal considers the conditions of exceptions in which the claim cannot be correct. Example of rebuttals are: 'Why are these pictograms only used in France and Spain?', 'Are there more effective alternatives to warn patients?' and 'Are we sure that additional warnings are more effective than for example selecting another medicine that does not have these effects?'.

The question '*Which evidence is needed to show that these two pictograms effectively warn about possible unwanted effects of a medicine?*' can now be formulated even more specifically by asking: '*on which rules and principles (W) is this evidence based, and which authorities (B) back this?*'

The application of Toulmin's diagram might seem as an unnecessary complication. The question could simply be answered by interviewing designers and their clients, and by observing and interviewing patients who might be at risk, or patients who took the risk. But this straightforward approach would not provide a complete picture of the context in which graphic design needs to be assessed. It would not do justice to the complexity of the situations in which both pictograms are designed and interpreted.

4 Who Can Assess the Quality of Visual Communication Design?

Until here, the text has mentioned 'designers of pictograms', 'clients' (pharmaceutical industry), and 'patients' (pregnant women and car drivers). It has also mentioned 'French, Spanish, and European legislation'. In addition, there are the views of a professional disciplines, and of a society. Each of these six fields has a specific perspective on the effectivity of both pictograms. Each considers specific evidence to evaluate the claim and assess the graphic design of the pictograms. And, according to Toulmin, each field refers to specific backing authorities (B) that provide a basis for specific rules and principles (W). As Toulmin stated: "*The kind of backing we must point to if we are to establish its authority will change greatly as we move from one field of argument to another.*" (Toulmin 1958, p. 96). This idea of 'field-dependence' in arguments is recently developed further into a more practical approach (Jackson and Schneider 2018). The six fields below are based on this approach, and start from the idea that the assessment of claims in arguments is based on different fields that are related, but are independent from each other.

Field 1. Design: The Visual Qualities

The visual qualities of two example pictograms vary substantially. The motivations and reasons for design decisions related to the pictures of pregnant women and a car, the colours, the line-thicknesses, and the ways to visualize 'don't' remain unknown. However, there must have been a designer who has carefully considered these visual variables, and considered design criteria like aesthetics and originality. Furthermore, the practical use of these pictograms shows that the evidence that the designers provided must have convincing enough for clients to accept and use both pictograms.

Field 2. Client: Commercial and Strategic Qualities

The pictograms were commissioned by a pharmaceutical industry, governmental institute, and a research organization. They aim to make sure pregnant women will not take particular medicines because it is safer for mother and baby. The driving pictograms might increase road safety. These considerations are based on the idea that both governments and pharmaceutical industry must cooperate in their longer term strategy related to the safety of patients. A more cynical view is that the use of these pictograms mainly aims to reduce the risk of liability claims. The commercial benefits and risks of the use of these pictograms on medicine packaging have been carefully considered.

Field 3: Standards and Legislation: The Legal Qualities

The European legislation specifically mentions 'pictograms' in relation to health education. The legislation states: '*The outer packaging and the package leaflet may include symbols or pictograms to clarify certain information …*' (Directive 2001/83/EC, 2001; Article 62). France and Spain have applied this legislation and have added pictograms. Apart from the legislation, there are also standards for the design and evaluation of pictograms by the International Standards Organization (ISO). For example, there is a standard for designing symbols (ISO 22727), and there are three standards for testing symbols (ISO 9186). If the same pictograms are used on a screen, which is highly likely, the accessibility standards of the World-WideWeb Consortium (W3C) will apply too. Adhering to legislation, and following international standards increases the likelihood that the warnings are effective.

Field 4: Professional Discipline: The Professional Qualities.

The disciplinary considerations look at the acceptability of approaches that are underlying professional practice and its education. Medical disciplines aim to 'do no harm', and use 'evidence based practice' that is disseminated through academic publications, research reports, and protocols. Design disciplines aim to promote good design through 'best practice', design awards, and endorsing the 'best examples' in trade-magazines, annuals, and on websites.

Field 5: Users: Usability, Applicability, Understandability

The fifth field that could examine the validity of the claim are the users (people, beholders, patients) of these pictograms. Especially pregnant women and car drivers need to be able to recognize the illustration ('what is depicted?'), and correctly interpret its intention ('what does it mean?'). These users decide if these pictograms are noticeable, understandable, and relevant in their personal context. Whether people actually decide to follow or ignore these warnings remains to be investigated.

Field 6: Society: Long Term Benefits and Sustainability

The sixth field relates to the expectations and requirements of a society. Do pregnant women, which is a relatively small group, and car drivers, which is a substantial group, warrant this additional warning on the tight space of a medicine packaging? How do male patients who do not drive a car react? What is further unknown at the moment is the effect of these pictograms on the costs of healthcare. For example, doctors cannot know what a specific medicine package looks like. And pregnant women can only see the 'pregnancy pictogram' after they have collected their medicines from a pharmacy. This collection can only occur after a consultation with a doctor who prescribed the medicine. If pregnancy was discussed during a consultation, than the pictogram will cause a conflict between the advice of the doctor ('take it'), and the warning on the packaging ('don't' take it'). This could lead to additional consultations, and an increased insecurity about the risks of a medicine.

Each of these six fields considers different kinds of evidence to assess if 'pictograms are effective to warn pregnant women and car drivers about possible unwanted effects of a medicine'.

5 Discussion: Applying Toulmin's Diagram to Visual Design Assessments

This discussion brings the three previous sections together. In order to assess graphic design, it was necessary to take three steps:

- choose examples to reduce the scope of the discussion to manageable dimensions (=focus on two pictograms);
- select an approach to assess these two pictograms (=Toulmin's diagram of argument components);
- determine the different fields that are able to provide assessments (=field dependent arguments)

Based on these three steps, it is now possible to determine the 'general rules and principles' in each of these fields, and describe on which authority these rules and principles are based. Toulmin called these 'warrant' and 'backing'.

This suggests that each of the six fields discussed in Sect. 4 needs different types of evidence, and that they use different warrants and backing. Table 1 shows the kinds of evidence that could be provided to support a claim in six different fields.

According to Toulmin these backing (B) and warrants (W) are essential to validate the relation between data and claim in each field. This means that it is now possible to describe which evidence is required to support the claim 'pictograms are helpful to warn against two specific risks' in each of the six fields.

Assessing design evidence. A pictogram can be seen as successful if the design is based on the expertise of a designer or a design-studio, it seems functional, it is aesthetically correct and original, and fulfills the brief of the client. The general principle (W) is that designers have a professional training and experience to consider a combination of requirements within financial and time-limitations of a client. This is backed (B) by the premise that designers have professional expertise, and apply

Table 1 The relations between field, backing, warrant, and evidence

Field	Backing (B): authority	Warrant (W) general rules and principles	Evidence, data (D)
1. Design	Individual expertise, design methods and process	Designers have professional experience and ability to design specific visual information	Experience, opinion
2. Client	Economic structure	Specific commercial interests	Financial monitoring
3. Law/standards	Legal structure, internationalization	Specific legislation and standards	Compliance, approval
4. Discipline	Disciplinary boundaries and activities	Promote best practice	Empirical evidence
5. Users	Human centered	User centered design	Benchmarking, monitoring, observations, user interviews, empirical evidence
6. Society	Long term survival	Sustainability, ecological, social inclusion. Cultural	Empirical evidence

design methods and design processes to develop an acceptable—but hopefully an extraordinary—visual proposal. The claim gets stronger if the designer increases experience, expert skills, and specialist knowledge.

Assessing financial evidence. A pictogram can be seen as successful if it was delivered on time and within a budget, and it fits into a longer strategy of the client. For both the pregnancy as well as the driving pictogram, the aim – apart from the legal obligation—is to show that the pharmaceutical industry cares and warns patients. The longer term aim is to reduce the number of patients who take risks during pregnancy or while driving. The general principle (W) is commercial because 'not warning' might lead to a reduction in sales if patients claim that a medicine has caused harm. The underlying principle (B) is the economic structure in which pharmaceutical industries produce and sell their products. The claim gets stronger if financial data can be provided that indicates potential costs and profits.

Assessing legal evidence and standards adherence. A pictogram can be seen as successful if it is developed according to the method described in ISO-standards and fulfills the legal requirements. The general principle (W) is that it is simply an obligation to adhere to relevant legislation and that following legislation and standards makes pictograms internationally acceptable. The legal system and international standards provide the backing (B) of these general principles. The claim gets stronger if a pictogram conforms to legislation and standards.

Assessing disciplinary evidence. There are several different disciplines involved in the assessment of pictograms. From the design discipline, a pictogram can be seen as successful if it wins awards, is shown at conferences, is published in trade journals, and if it generates more work. In a medical discipline, a pictogram is successful if it proves to be effective in a treatment which might lead to inclusion in protocols and best practice guidelines. The general principle (W) depends on disciplinary characteristics, but the backing (B) is likely a need to develop a discipline through best practice and education. The disciplinary claim therefore gets stronger if it helps to develop a professional discipline.

Assessing evidence of successful use. A pictogram can be seen as successful if it enables individual patients to make appropriate decisions and are able to act in the most beneficial manner. There is experimental evidence from two studies that people are able to interpret the 'driving pictogram' correctly, but it is by no means certain that all patients will actually adhere to the warnings. The general principle (W) is that the design must be 'user centered'. The backing (B) is provided by a human-centered approach that focus on the activities and behaviour of people. A claim gets stronger if empirical data about the reactions of different people are available and there is some proof that it is 'not excluding' particular groups.

Assessing societal evidence. A pictogram can be seen as successful if it benefits a society in any way. This can be financial by reducing healthcare costs through the prevention of extra costs, ecological by reducing waste and energy consumption, inclusivity by not excluding particular groups, and so on. A claim gets stronger if it really reduces the number of people who drive a car after taking medicines that affect driving behaviour, and if it really reduces the number of women who take medicines during pregnancy. The general principle (W) are related to the rules that keep a society going, which are based (B) on the need for a long term survival. These criteria can only be monitored over a longer period of time.

The description of the general principles and rules (W) and their backing (B) on which authority these rules are based shows that each of these six fields requires different kinds of evidence to assess a claim. The evidence can come from personal experience, benchmark studies, observations, empirical studies, legal approval, adherence to standards, and user interviews. It is unlikely that different fields accept evidence from other fields without a thorough discussion.

The application of Toulmin's model, and especially a consideration of the warrants and backing, shows that it is necessary to consider evidence in at least six fields to assess if a claim is correct. Focusing on a single field, and accepting the evidence in a single field only, could lead to an incorrect acceptance or rejection of a claim. The evidence that supports the use of the two example pictograms remains scattered and is inconclusive in several fields.

6 Conclusions

This article provides an overview of the criteria that can be used to assess examples of visual information. The design of visual information always makes a claim that the newly designed artefact is part of a beneficial change when compared to the existing situation. However, in order to make this claim, it is essential to support this claim by evidence and reasons. This approach is based on Toulmin's diagram of the components of an argument that shows the relation between 'a claim' and 'its supporting evidence'. Two medical pictograms were used as examples to discuss how 'a claim' and 'its supporting evidence' are related. The claim is that '*Pictograms are effective to warn pregnant women and car drivers about possible unwanted effects of a medicine.*' It turns out that there are at least six independent fields that can assess a claim. Each of these fields require specific forms of evidence and each field is unlikely to accept the evidence of other fields. In order to design visual information, and claim that the result is effective, it is therefore essential to consider and provide at least six different kinds of evidence to suit the expectations and criteria of each of these six fields.

This article is based on the following assumptions, and it is clear that each of these could be challenged:

- Graphic design as an activity can be seen as 'developing a visual argument'. This might not be true for all activities of all graphic designers.
- Both pictograms are examples of graphic design. It is not sure if both example pictograms are designed by professionals, and it could therefore be possible that the wrong examples are selected.
- Evidence based decisions are preferable above decisions based on irrational unsubstantiated fear-rooted gut-feelings.

However, it is likely that these three assumptions are correct for at least a substantial part of graphic design practice. The main benefit of this approach is that it shows the necessity to provide reasons, evidence, and data to support claims in at least six fields. The application of Toulmin's model reveals different backing authorities on which assessments are based. These are related to fields of design, finances, law, disciplinary practice, user-centered, and society.

This approach has two consequences for graphic designers. It becomes necessary to provide evidence in different fields to support a claim. The difficulty is in finding the balance between the expectations of the different fields, and providing reliable evidence for each of them.

The analysis of visual arguments also shows the complexity of the assessment of two fairly simple pictograms that have clear aims. The actual influence on behaviour of car drivers and pregnant women is unknown, and it is unlikely that any change in behaviour could be directly linked to pictograms on medicine packaging. However, the benefits of these pictograms might be in any of the other fields. Each field could reliably assess the benefits and risks of developing and interpreting pictograms.

References

Publications

AEMPS (Agencia Espãnola de Medicamentos y Productos Sanitarios) (2012) Medicamentos y Conducción. https://www.aemps.gob.es/industria/etiquetado/conduccion/pictograma.htm

AFSSAPS (Agence française de sécurité sanitaire des produits de santé) (2005) Informations sur les médicaments et la conduite automobile. https://www.ansm.sante.fr/var/ansm_site/storage/original/application/36fd6bd36152361b7a426456edabffce.pdf

Directive 2001/83/EC (2001) Directive 2001/83/EC of the European Parliament and of the Council of 6 November 2001 on the Community code relating to medicinal products for human use. Off J Eur Commun L 311/67

EMCDDA (European Monitoring Centre for Drugs and Drug Addiction) (2012) Driving under the influence of drugs, alcohol and medicines in Europe—findings from the DRUID project. Publications Office of the European Union, Luxembourg. https://doi.org/10.2810/74023

Fierro I, Gómez-Talegón T, Alvarez FJ (2013) The Spanish pictogram on medicines and driving: the population's comprehension of and attitudes towards its use on medication packaging. Accid Anal Prev 50:1056–1061. https://doi.org/10.1016/j.aap.2012.08.009

ISO 9186 (International Standard Organisation) (2014) Graphical symbols. ISO, Geneva

ISO 22727 (International Standard Organisation) (2007) Graphical symbols—creation and design of public information symbols—requirements. ISO, Geneva

Jackson S, Schneider J (2018) Cochrane review as a 'Warranting Device' for reasoning about health. Argumentation 32:241–272. https://doi.org/10.1007/s10503-017-9440-z

Monteiro SP, van Dijk L, Verstraete AG, Alvarez FJ, Heissing M, de Gier JJ (2012) Predictors for patient knowledge and reported behaviour regarding driving under the influence of medicines: a multi-country survey. BMC Public Health 12:59. https://doi.org/10.1186/1471-2458-12-59

Toulmin SE (1958[2003]) The uses of argument. Cambridge University Press, Cambridge

van Beusekom MM, Kerkhoven AH, Bos MJW, Guchelaar Henk-Jan, van den Broek JM (2018) The extent and effects of patient involvement in pictogram design for written drug information: a short systematic review. Drug Discov Today 23(6):1312–1318. https://doi.org/10.1016/j.drudis.2018.05.013

van der Waarde K (2012) Communication design education: could nine reflections be sufficient? In: Poggenpohl S (ed) Special issue of Visible language on 'Envisioning a future of Design education', vol 46(1/2), pp 8–35

van der Waarde K (2018) Research and design watch: a personal view. Inform Des J 24(1):80–84

W3C (2008) Web content accessibility guidelines (WCAG) 2.0. https://www.w3.org/

Communication Design and Space Narratives

Maria Luísa Costa and Inês Amaral

Abstract The experience with space is achieved through projects developed according to a design perspective. In addition to collaborating on the creation of spatial narratives by promoting the experience, the design also acts in the sense of enhancing accessibility in both the physical and cognitive domain. Wayfinding systems as well as informational technological systems are informative elements that communicate with citizens, fostering experiences of greater accessibility and the creation of social space. This paper discusses the concept of space from its social dimension, questioning on how spatial narratives, places, and paths can create individual or collective experiences within urban space through Communication Design by the use of technology.

Keywords Communication design · Wayfinding systems · Spatial narratives · Georeferencing systems · Augmented reality

1 Introduction

The growth of cities manifests itself as a historical trend and particularly intense since the nineteenth century, due to the progressive conversion of the rural population into an urban population. Data published by the United Nations (UN) in 2008 prove the growth of cities and reflect the choices of individuals. A study by the Population Division of the UN Department of Social and Economic Affairs indicates that the

M. L. Costa (✉)
Centro de Investigação em Arquitetura, Urbanismo e Design, Universidade de Lisboa, Pólo Universitário, Rua Sá Nogueira, Alto da Ajuda, 1349-063 Lisbon, Portugal
e-mail: design.luisacosta@gmail.com

Instituto Superior Miguel Torga, Largo da Cruz de Celas nº 1, 3000-132 Coimbra, Portugal

I. Amaral
Faculdade de Letras da Universidade de Coimbra, Largo da Porta Férrea, 3004-530 Coimbra, Portugal

Centro de Estudos de Comunicação e Sociedade, Universidade do Minho, Campus de Gualtar, 4710-057 Braga, Portugal

D. Raposo et al. (eds.), *Perspective on Design*, Springer Series in Design and Innovation 1,
https://doi.org/10.1007/978-3-030-32415-5_8

world's population living in cities will double from 3300 million in 2007 to 6400 million in 2050. In 2008, for the first time in the history of Humanity, more than half the inhabitants of the planet lived in cities.

Combining the tendency of growth to the problematic of the hegemony of the city, it is confirmed the true protagonism that assumes the contemporary urban space and the necessity of an adequate communication that allows all the citizens to live and to experience the city in its fullness. The cities grew and evolved according to the citizens who inhabit it, as well as social and cultural needs of each era. These multi-layered structures prevail visible in each city, and it is possible to glimpse in these layers the stories and memories of the citizens who were successively inhabiting them, revealing and expressing the cities through this plurality and collective experiences. The public space of cities reflects their reality, their originality, the life and the character of the citizens, that is, what can differentiate it from the others, the reason why it is essential that the cities potentiate their public space, through a close connection, communicating and promoting access.

City and Design are keywords of contemporaneity with exponential growth from the Industrial Revolution. Cities are growing and increasingly influencing people's culture. Design becomes omnipresent in everyday life, and at each moment new needs and areas of intervention are identified for this disciplinary area. The importance of Communication Design in the city context manifests itself in the readability of the city itself, as well as in its communication and projection. The designer should act as a mediator facilitating the interaction and communication of the city and in the city. Contemporary discourses emphasize the role of design and its transformative action in the production of social consciousness.

Communication Design can intervene in space to provide experiences of fruition of this through wayfinding systems, which potentiate new experiences, mediation of social practices and the creation of places. If the function of wayfinding is orientation, its use leads to paths and the production of anthropological places (Augé 1994). However, the construction of spatial narratives and places through Communication Design can be increased with the use of technology. This chapter explores the concept of space from its social dimension and reflects on how spatial narratives, places, and paths create individual or collective experiences through Communication Design using technology.

2 Social Dimension of Space

The end of the project of colonialism and the beginning of the urban revolution has made it evident that it is not only a question of the built environment but of having a critical concept of space in order to understand its social dimension. Lefebvre (1995) claimed that there is only social space because there is no emptiness waiting to be filled. The author argued that it does not make sense to consider a preexisting space for social practices. Therefore, "the dichotomy space produced versus created space can be analyzed from power relations and their spatial echo" (Costa et al. 2019).

Lefebvre's unit theory of space (Augé 1994) argued that the natural space is disappearing and societies produce their specific social space. Hence, reproductive, production and symbolic relations are evidenced in a triad of space: (i). Spatial practices: "The spatial practice of a society secretes that society's space; it propounds and presupposes it, in a dialectical interaction" (Lefebvre 1991); (ii). Representations of space: "Conceptualized space, the space of scientists, planners, urbanists, technocratic subdividers and social engineers, as of a certain type of artist with scientific bent – all of whom identify what is lived and what is perceived with what is conceived" (Lefebvre 1991); (iii). Representational spaces: "Space as directly lived through its associated images and symbols, and hence the space of 'inhabitants' and 'users'" (Lefebvre 1991). The triad proposed by Lefebvre (1991) allows distinguishing produced space (reproducible space and that is a product) of created space (collective and continuous work).

The social space derives from social practices, symbolism, and reproductive and production relations (Lefebvre 1995). The daily social practices challenge the representational spaces, allowing sociological evolutions of spaces through constructed environments that derive from emergent forms of urbanity (de Certeau 1984). Therefore, social practices structure the determining conditions of social life. Consequently, the interpretation of space production stems from the circuit of discourses, practices, and social relations.

The relations of micro-power (Foucault 1984) that establish themselves in the social space construct spatial narratives that individualize places. Places and non-places are a dominant note of the contemporary social space (Augé 1994). Contemporary social and cultural practices unfold from the perspective of trends: a plethora of events and spaces and the individualization of references. The organization and constitution of places are modalities of collective and individual practices. Non-places are not symbolized and create a unique and non-relational identity, being empty of affectivity and, therefore, identified as spaces of passage (Augé 1994).

Augé approaches a reformulation of the notion of space and proposes an anthropology of 'over modernity', opening horizons beyond what the author considers to be a kind of 'ethnology of solitude' (Augé 1994). 'Over modernity' results from a world of excesses in terms of time, space and individualization. In the 'over modernity', circulation, communication and consumption prevail. This argumentation focuses on the premise of content and identity as elements both social and spatial. Therefore, contemporary social and cultural practices unfold in a perspective of trends: the superabundance of events and spaces and the individualization of references. In his theory, Augé (1994) differentiates space of place. The first has a more abstract and, therefore, more comprehensive definition: space is the notion of extension/distance between two points. The anthropological place generates identity, is associated with culture located in space and time, and reveals a construction with objectives and the relationships that individuals maintain with them. Augé's (1994) approach presupposes a multiplicity of spaces where places and non-places are distinguished, which take on the whole as contemporary spaces of circulation, communication and consumption. According to Augé (1994), the organization and constitution of places are modalities of collective and individual practices, for which the author states "the anthropological space is both a principle of meaning for those who inhabit it and a principle of intelligibility for those who observe it" (Augé 1994). Non-places are not symbolized and create a unique and non-relational identity, being empty of affective relationship and, therefore, identified as spaces of passage. In the traditional sense, a place is a space that can be occupied. Non-places correspond to spaces whose function is to enunciate potential routes to a given destination.

Lemos (2010) argues that "places are a result of negotiations among territories. Today, new senses of places emerge from these new layers of territories" (Lemos 2010). The author considers that "space is constituted by places that are created by inside territories in an endless process of mutual influences, horizontal and vertical (inside each category and among them). Today we have to take into account a new form of territory in contemporary societies: the digital, informational one. Every territory is a place of social control of borders, of informational exercise of surveillance and violence. The territoriality is a 'cultural artefact' that shapes social relations and our relationship with the material and symbolic world. We are always immersed in territorial layers (subjectivity, physical, cultural, political, economic), and these layers constitute places" (Lemos 2010).

Lemos's perspective does not focus on the idea of the end of places or non-places (Augé 1994), but rather on the creation of "new senses of places, territories, mobility, and community" (Lemos 2010). It follows that the spaces are created by territorial dynamics and, accordingly, are "production of social meaning (emotional, historical, cultural, political, imaginary)" (Lemos 2010). It is on the basis of this assumption that the author argued: "space is a vacuum, place is produced" (Lemos 2010).

The emergent forms of urbanity that are produced and reproduced by design projects in the built environment are elements of communication that foster networks of significance and meaning sketched by subjects who draw their own subjectivity. The way each citizen relates to the city is different. The dynamics of contemporaneity, mobility and mutation are processes that will build citizenship, in micro and macro scales, in a discourse in which "I" and "Other" are a continuum in alternation. The city can be interpreted as a portrait of mental and symbolic maps that reflect a different way of appropriating the same space depending on the presence of individuals and the relationships between them (Costa and Amaral 2018). Public space, rather than the emptiness between buildings, is the place where the citizen feels and lives the city in what is most genuine (Costa et al. 2019). The public space is the place of construction of multiple memories and spatial narratives grounded in experiences both individual and collective (Costa and Amaral 2018).

3 Connecting People to Space

Wayfinding systems are informative elements that communicate with citizens, fostering experiences of greater accessibility in the public space. Mollerup (2005) considers that wayfinding systems intended for 'wayfinders', which it defines as being people walking or cycling, and whose objective is to find a specific place. The function of wayfinding systems is to enable people to complete their journeys successfully. According to the author (Mollerup 2005), finding and showing the way are activities as old as Humanity itself, asserting wayfinding systems as a means to solve a problem.

This ancestral need is inscribed in our symbolic and oneiric universe. Examples are manifested from Greek mythology where Ariadne, through the wire delivering Theseus, finds the possibility of this finding the way out of the maze; or in children's stories such as of the brothers Grimm "Hansel and Gretel", and in which Hansel marks the way to be able to return home.

The study of wayfinding systems, and the areas directly related to them such as typographic readability, color perception, systems for the blind, among others promoted the development of cognitive sciences (Berger 2009). It is common to confuse signage with wayfinding. Signage is effectively part of a wayfinding system, but this covers many more areas, such as architecture, landscaping, lighting, signage and information systems being only part of the wayfinding. Over time, wayfinding systems have undergone significant changes due to social and cultural needs, which are interconnected with the means available and technological evolution. The need to show and make accessible paths has increased, with wayfinding systems essential in the face of population growth, urban growth and the need to access increasingly complex spaces.

Lynch (1960) uses for the first time the term 'wayfinding' in his book the "Image of the City" and relates to the process of forming a mental picture of one's surroundings based on sensation and memory. It is also in this perspective that Passini (1996) states that the wayfinding concept integrates the notion of spatial orientation, which means the psycho-ability of a person to mentally represent the spatial characteristics of a given environment and to stand within that representation.

Thus, the wayfinding consists of a comprehensive concept that incorporates the notion of spatial representations, being "composed of three interrelated processes: (1) decision making and the development of decision plans also called action plans: (2) decision execution, transforming decision plans into behavior at the right time and place along a route: and (3) information processing, comprising environmental perception and cognition which provide the person with the information necessary for the two decision-related processes" (Passini 1996).

The need for in-depth study of wayfinding systems led in the 1970s to the creation of the Society of Environmental Graphic Design (Gibson 2009). The name of this association has been changing over the years reflecting the changing paradigms of society. Today, in a society where experience is prioritized, the association is called the Society for Experiential Graphic Design, self-describing itself as a community that connects people and places by way of planning, designing and building experiences.

The wayfinding is a set of processes and methods that enable information to physical movement of a point to another and wherein like elements architecture, the surrounding space, the lighting art and technology assume leading roles. According to Berger (2009), it consists of a set of activities that allow us to find the way to a certain destination. Passini (1996) states "wayfinding design concerns all features of the built environment which are related to the purposeful circulation of people and their ability to mentally situate themselves in a setting. These design features include spatial layouts, architectural features related to circulation and graphic displays including audible and tactile supports". Rosa (2012) considers that wayfinding systems allow the user to easily find their destination through the creation or use of multiple visual, physical and sensorial resources. This connection to physical and sensorial resources refers to the idea of experience and to the importance that it assumes in contemporaneity.

These definitions echo the aforementioned Society for Experiential Graphic Design, whose objectives are to create experiences that connect people with the place, creating environments that improve the experience and make them more dynamic, through technologies. Berger (2009) points out that future difficulties in the field of space graphic design are difficult to predict, and it is almost guaranteed that designers will need to embrace in their projects an even broader and improved set of transversal capabilities where new spaces will be feasible through the blending of virtual reality, cyberspace, and multiple intelligent materials.

The main means used in wayfinding systems usually consist of maps and signal systems, symbols and colors that indicate the direction of individuals. Progressively these means integrate mobile applications and automatic identification systems, paving the way for new approaches in wayfinding systems, namely the use of systems of georeferencing and Augmented Reality (AR). Amaral (2012) states that digital is not disconnected from the so-called offline world. It will be rather an intermediate space with presence simulations, through the realization of belonging in digital environments and the development of new sociabilities without territory.

4 Technology and (New) Space Narratives

The rehabilitation of urban spaces has been one of the focus of design, making it a responsible part of the creation of built environments that enhance political, environmental, social, cultural and economic awareness. Communicating the city, with its idiosyncrasies, is an integral part of the professional practice of design. Therefore, the answer has been the development of information systems that allow the spatial organizational fluidity and experiences of space enjoyment. Through the systems of georeferencing and augmented reality, technology enables a reformulation of social spaces and the creation of places of individual or collective experiences.

The influence of new technologies in public and private spheres of society has given rise to a new social field and directly interferes with how citizens perceive and relate to the city. In Bourdieu's (2001) theory, field arises as a configuration of socially distributed relations. According to the author, the field represents a symbolic space in which mediation between social agents and society takes place. In this perspective, the social field is a system of positions, which can be challenged and altered. It follows that the fields are microcosms with their laws. Therefore, agent strategies are determined by capital (cultural, social, economic, political) and the habitus (schemes of perception, thought and action) of individuals. In this chapter, we adopt the concept of the social field proposed by Bourdieu (2001), which refers us to symbolic social spaces that follow from paths proposed by technological systems. These itineraries also anchor in perceptions, identities and memories that compete for representations of the social world through interaction with digital systems of spatial representation.

In digital spaces, the social dimension of shared construction of meanings and representations replaces the physical place. The idea of territory is diluted in simulacra of presence, feelings of belonging, permanence and own codes. Since the Internet and digital spaces are eminently symbolic spaces, information and the context of interaction are assumed as preponderant elements in the process of reformulation of spatiotemporal notions (Gomes 2000) and spatial representations. Therefore, the transformation of spaces into anthropological places results from processes of individual and social cognition within the dynamics of the new forms of sociability that result from the introduction of interactive technology in individual and collective spatial narratives. It is therefore essential to understand that the scenario of the new generation of digital devices in which the user is effectively proactive, using tools that allow content sharing, worldwide dissemination and the association of users in the context of networks of interests changes the perception of the user and encourages the merger of this with the consumer (Amaral 2012).

The concepts of 'user-generated content' and 'user-generated media' make it possible to maximize the notion of participation on the Internet through the formation of social networks of contents that are in permanent mutation, where social representations are decontextualized, disaggregated and consumed collectively (Amaral 2012, 2016). In the context of this type of structures, semantics is essential for the understanding of the development of new practices and, consequently, relationships based on streaming of contents, drawn by appropriations of the technique. As Jones postulated, "Computer-mediated communication, it seems, will do by way of electronic pathways what cement roads were unable to, namely, connect us rather than atomize us, put us at the controls of a 'vehicle' and yet not detach us from the rest of the world" (Jones 1997).

The technique, through the tools of communication, creates mechanisms of interaction that, by themselves, are assumed as spaces of collective narratives and shared social representations through georeferencing systems and content indexation through geotagging (Amaral 2016). The purposes of using semantic indexing are varied. These practices refer to the formation of networks where information circulates, and social connections are established. In our perspective, and adopting the argument of Barata (1989), sociability is the result of all the interactions that occur in the social field. In this sense, it concerns the form of social relations and summarizes phenomena of interaction or the ability to found groups and build networks supported in the ties between individuals (Ferréol 2007). Therefore, semantic indexing in systems of georeferentiation can be concretized in spontaneous and organized sociability (Gurvitch 1986).

Georeferencing systems use geolocation applications that allow personalized local search experience (Amaral 2014). Through a global positioning system (GPS), it is possible to access interactive wayfinding systems or location-based social media services applications. Foursquare and Swarm are two examples of location-based social media service application. Facebook or Twitter, as social networking tools, also allow georeferencing. These applications use "global positioning-system (GPS) in mobile devices to generated locations and provides the list of places. 'Check-in' is the process of identifying the location where the user is" (Amaral 2014).

Geotagging can be defined as "the process of adding metadata with geographical identification to content. It is also defined as a form of geospatial metadata" (Amaral 2015). Geotagging allows spatial indexing of content as it is a process of recognized geographic locations which enhances the development of geographic databases, Web resources, as well as multimedia content. This practice changes how the user interacts with content and other users in the digital space. Applications and systems with geotagging capabilities often add latitude and longitude coordinates, distance, and place names to media content. The added data consist of textual and visual resources (Amaral 2015).

As a communicational practice, wayfinding systems depend on sociocultural needs and technologies. The use of maps becomes obsolete, and we hardly imagine individuals moving by car, guided by a map in paper form, as well as with whom one moves on foot. GPS systems have revolutionized the way we project into space and move into space. Thus, maps have fallen into disuse making their reading and interpretation increasingly difficult. This situation happens both with young people who favor the use of technologies and by other age groups where lack of acuity makes it impossible to read these maps.

The increasing ease with which individuals interact with mobile devices, the enhancement of their capabilities, as well as all the resources available online, open possibilities in the context of the creation of differentiated products for use in wayfinding systems, emphasizing the Augmented Reality (AR).

AR provides interactive experiences that are so valued today in our society. AR creates layers that overlap the 'real' world, allowing access to differentiated information, more in-depth or more playful, depending on the goal to be achieved. The addition of a layer of virtual information in the physical context can significantly alter the way we understand the wayfinding systems and access to different information. Many companies, such as Google, are already exploring this possibility. In addition to finding a way, individuals can experience and experience spaces through expositive content, immersive experiences using analogical media and multimedia installations.

Although entirely different, concepts such as Augmented Reality (AR) and Virtual Reality (VR), are somewhat connected, and there is some confusion among its functionalities, which it is essential to clarify. VR creates a virtual and interactive world, through which it can include the senses of sight, hearing, and touch. AR also provides interactive experiences; however aims to be a supplement to the real world rather than creating an artificial environment (Höllerer and Feiner 2004).

The use of AR in wayfinding systems consists of the provision of additional and punctual information and can be used in reading paper maps to be used both in outer and inner space, but without the need for a full immersion as will be necessary in games or other immersive activities, the technology of which may prove to be tiresome and damage to the health of users.

Azuma (1995, 1997) considers that the designation Virtual Environments is more adequate than Virtual Reality, because it allows a more precise description of the effects achieved. However, whether one term or another term is used, the difference between VR and AR is in their treatment of the real world. Virtual Environments completely replace the real world with a computer-generated synthetic environment that completely immerse the user inside this environment Azuma (1997). In turn, Augmented Reality adds virtual information to the 'real' world through user insight and never replaces it. It follows allowing an intersection of the 'real' and virtual worlds (Azuma 2017).

Azuma (2017) states that AR will succeed in the consumer market sooner if it can establish new forms of media that users find compelling. Considering that AR will make it possible to establish meaningful connections between the surrounding real environment and virtual contents, generating experiences whose value results precisely from this connection and not just from one of the worlds.

5 Conclusions

Georeferencing systems and augmented reality are intelligent and interactive systems that allow users to interact with space through technological interfaces. These systems individualize spatial experiences, foster new modalities of practices in space, and recreate symbolic representations. The construction of individual or collective paths within space allows the creation of new spatial narratives amplified by technology and communication systems.

Identity, memory, and individual and collective experiences contribute to the construction of spatial narratives through digital devices. These narratives can be individual or collective and often become public and therefore shared through social media. The same itinerary, which previously focused on individualistic logic, can now be experienced through a collective narrative to which different people contribute through the recording of their perception of space, materialized in data and metadata. Hence, different forms of sociability arise that allow the construction of collective narratives in which social bonds are established.

Individual narratives that enhance immersion and unique experiences enable the construction of spatial representations that result from a mix between the perception of the moment and the memories that give meaning to the symbols.

The issue of references and formatting of thought have been questionable elements in the production of space in contemporary times. Urban design has even come to be seen as a break with the past, enhancing emerging landscapes. Are we faced with the emergence of abstract spaces, which enable representations and imaginations of space in a reconfiguration of the relations of power and their materialization in space practices?

We conclude that emerging forms of urbanity created by urban design for sustainability, based on the question of "glocalization", point to a perspective of new structures of sociability that result from a spatialization endowed with multiple identities and memories. We assume as a premise Benjamin's (2001) argument about urban itineraries and collective memory to compete for new forms of sociability in the contemporary urban world.

References

Amaral I (2012) Participação em rede: do utilizador ao "consumidor 2.0" e ao "prosumer" [Network participation: from the user to the "consumer 2.0" and the "prosumer"]. Revista Comunicação e Sociedade 22, 131–147

Amaral I (2014) Geotaging. In: Harvey K (ed) Encyclopedia of social media and politics, vol. 2. Sage Publications, Thousands Oaks, pp 574–575 (2014)

Amaral I (2015) Foursquare. In: Wherry FF, Golson JG (eds) The Sage encyclopedia of economics and society, vol 2. Sage Publications, Thousands Oaks, pp 727–728

Amaral I (2016) Redes Sociais: sociabilidades emergentes [Social Networks: emerging sociabilities]. Editora LabCom.IFP, Covilhã

Augé M (1994) Não-Lugares: Introdução a uma Antropologia da Sobremodernidade [Non-Locations: Introduction to an Anthropology of Overmodernity]. Bertrand Editora, Lisboa

Azuma R (1995) Predictive tracking for augmented reality. Ph.D. dissertation, University of North Carolina at Chapel Hill, Department of Computer Science

Azuma R (1997) A survey of augmented reality. Presence Teleoperators Virtual Environ 6(4):355–385

Azuma R (2017) Making augmented reality a reality. In: Proceedings of OSA Imaging and Applied Optics Congress (2017)

Barata O (1989) Introdução às Ciências Sociais [Introduction to Social Sciences]. Bertrand Editora, Venda Nova

Benjamin W (2001) Sociologia [Sociology]. São Paulo, Ática

Berger C (2009) Wayfinding: designing and implementing graphic navigational systems. Rockport Publishers

Bourdieu P (2001) O poder simbólico [The symbolic power]. Difer, Lisboa

Costa ML, Amaral I (2018) Cidades e Sistemas Wayfinding. Convergências - Revista de Investigação e Ensino das Artes XI(21) (2018)

Costa ML, Amaral I, Daniel F (2019) The contribution of information design to age-friendly cities: s case study on Coimbra and public transport. In: Rebelo F, Soares MM (eds) AHFE 2018, AISC 777. Springer International Publishing AG, Berlin, pp. 485–496

de Certeau M (1984) The practice of everyday life. University of California Press, Berkeley

Ferréol G (2007) Sociologia: Léxico das Ciências Sociais [Sociology: Social Sciences Lexicon]. Porto Editora, Porto

Foucault M (1984) Space, knowledge and power. Foucault Reader 239(256)

Gibson D (2009) The wayfinding handbook: information design for public places. Princeton Architectural Press, Princeton

Gomes C (2000) Novos mapas culturais e o ciberespaço [New cultural maps and cyberspace]. Passados Recentes, Futuros Próximos, Universidade de Coimbra, Actas do IV Congresso Português de Sociologia - Sociedade Portuguesa(2000)

Gurvitch G (1986) A vocação actual da Sociologia [The current vocation of Sociology]. Cosmos, Lisboa

Höllerer T, Feiner S (2004) Mobile augmented reality. In: Karimi H, Hammad A (eds) Telegeoinformatics: location-based computing and services. Taylor and Francis Books, Boca Raton, pp 221–260(2004)
Jones S (1997) Virtual culture. Identity & communication in cybersociety. Sage Publications, Thousand Oaks
Lefebvre H (1991) The production of space. Blackwell Publishing, Oxford
Lefebvre H (1995) The right to the city. In: Lefebvre H, Kofman E, Lebas E (eds) Writings on cities. Blackwell, Cambridge, pp 63–184
Lemos A (2010) Post – mass media functions, locative media, and informational territories: new ways of thinking about territory, place, and mobility in contemporary society. Space Culture 13(4):403–420
Lynch K (1960) A Imagem da Cidade [The Image of the City]. Edições 70, Lisboa
Mollerup P (2005) Wayshowing: a guide to environmental signage. Lars Muller Publishers
Passini R (1996) Wayfinding design: logic, application and some thoughts on universality. Des Stud 17(3):319–331
Rosa C (2012) Informação Pictográfica – O universo dos pictogramas: Métodos e procedimentos de design para obtenção de coerência formal [Pictographic Information - The universe of pictograms: Design methods and procedures for obtaining formal coherence]. Ph.D. thesis, Faculdade de Arquitectura da Universidade Técnica de Lisboa (2012)

A City's Cultural Heritage Communication Through Design

Lourdes Pilay and Marco Neves

Abstract All objects, forms and patrimonial elements of cities are repositories of graphic memory and history, which account for values and meanings of cultural identity of a territory to its inhabitants and to its visitors. By review of the main literature and a case studies method, we characterize significant aspects of heritage buildings which present by visual means, unnoticed information by most observers. This is a descriptive and explanatory report, to verify common aspects. Design can take advantage of this and use such visual elements to add value to contemplated objects and to cultural interaction that emerges from them. Graphic design is presented as a strategic tool which contributes to communication of these visual manifestations which, in turn, constitutes material cultural heritage of a place. Therefore, this paper approaches cultural heritage of a territory, as treated from the perspective of strategic design and graphic design. Design acts both as an intermediate and interpreter that gives meaning to information, so it can be transmitted to citizens.

Keywords Design · Cultural heritage · Strategic design · City

1 Introduction

Territories have groups of people who share customs, traditions, languages, believes, systems and values, characteristics which, as a set encompass a common sense of belonging to a place. At the same time, cities create and recreate types of visual

L. Pilay (✉)
ESPOL, Polythecnic University, Escuela Superior Politécnica del Litoral, ESPOL, Facultad de Arte, Diseño y Comunicación Audiovisual, Campus Gustavo Galindo Km 30.5 Vía Perimetral, P.O. Box 09-01-5863, Guayaquil, Ecuador
e-mail: mdpilay@espol.edu.ec

M. Neves
CIAUD, Lisbon, School of Architecture, Universidade de Lisboa, Rua Sá Nogueira, 1349-063 Lisbon, Portugal
e-mail: mneves@fa.ulisboa.pt

D. Raposo et al. (eds.), *Perspective on Design*,
Springer Series in Design and Innovation 1,
https://doi.org/10.1007/978-3-030-32415-5_9

and communicational languages that bring people close to their own characterization of the environment. Buildings, murals, historic houses, streets, parks and crafts amongst others are all part of a unique communication system, which is made to be a conveyance method of cultural transference for the identity of those inhabiting and those who observe the place.

Cultural heritage being both a tangible and intangible cultural expression must be first construed by people to be acknowledged and then owned.

Having within its visual elements different manifestations of identity discourse, how is this cultural heritage transmitted so that inhabitants can make it their own? How does design influence cultural heritage in its broadest sense? These questions arise from the perspective of design, communication and cultural heritage education. Romero (1998) states heritage can be turned into an educational resource, valuable for the population. To understand cultural heritage, it must be taken into consideration that it works as a virtual unity for development and sustainability.

In that sense cities are presented as favorable scenarios for design to work through strategies that add value to the territory. These spaces tell tales and stories through their visual and informative data, linked to each shape, object or imagined element, so what happens to the expressiveness of such a place for both those who inhabit it and foreign visitors? What does this space pose graphically? Is it possible to make room for its objects within the cultural heritage Cosmo vision? Can these objects grant an identity? Is this a window to the past to outreach visual cultural heritage within city areas? This is what we aim to disclose in this research paper: whether communication is immersed in this debate.

People and cultural expressions are part of a city's wealth and distinction. In that sense, graphic design may serve as both a facilitator and a link to reduce the gap between cultural heritage information given to people and their access to it, allotting value both to the given data as well as to visually allowing population to embrace their cultural legacy.

2 Cities Culture and Cultural Heritage

In the Middle Age citizenship was the characteristic which singled out permanent residents of a specific area (Alderoqui 2003). The idea of a city encloses this definition of citizenship and it extends it, because of urban, architectural, political, technological, economic, cultural and social development which follows. In this case, it is the person's own conditions (such as skills or essence) and social interaction ability, which consequently, make up the collective personality of a group of people in a determined area.

Culture is an acquired practice that evolves with the passing of time and that involves many people in the process. Supported in anthropological perspective, culture is mainly defined as that which is socially acquired, and learnt through education, in the broad sense, as we grow up and become a member of a society (Goodenough 1975). Anthropology have studied culture to get to know different lifestyles, customs, and knowledge of social groups, which exposes a more daring conceptualization of the term culture. Hymes (1964) states culture is "how things are shaped in the mind of a population and the models through which they are perceived, related and construed" (Hymes 1964: 36). Hence, if this perception is socially accepted, what are those common things that affect simultaneously a person to give way to a sociocultural construction?

On his part, Leach (1979) states that, despite being binding, concepts of culture and society are in fact different and therefore worth establishing such differences.

About society it is stated that being a set of social interactions, culture is the content of said interactions. However, about culture Leach (1979) emphasizes on the component of accumulated resources, both material and immaterial, which people inherit, use, transform, increase, and communicate. This last definition of culture aims to understand how it relates to a city. We know they are strongly related to cultural studies and that both weave local identity of a society.

From a graphic design view point, the aim is to promote the existent cultural heritage information as a service available to inhabitants or visitors. Coincidentally, for Margolin (1991), 'culture of design' reinforces the idea by which design is an activity defined to a point by the social environment in which it works. This means design intervention can improve things such as: processes, techniques or services which will result in a change in society as we know it.

The cultural value of a city establishes its identity since that specific characteristic is the one perceived by both locals and tourists. According to Rivera and Ledezma (2014) local and cultural values of identity influence in the creation of a city. Said anthropological influence of design in community behavior and in social habits may contribute in teaching of cultural aspects that add value to these cultural heritage resources local as well as global.

3 Strategic Design and Cultural Heritage

Vitta (2003), an image systems researcher, states it is necessary to culturally interpret a landscape of images surrounding us, since these make up a system through which we are projected every day. Day to day life experiences reach visual culture, but what can be salvaged and valued to be turned into a cultural or transcultural experience?

If the way to decode a message transmitted through an object presented to an individual is unknown to them, the person can hardly relate to it. People must be provided with tools and instruments to encourage a pragmatic dialogue with culture, this is key to a visual understanding and training of a person.

Strategic design arises as a starting point, aiming to establish communicational strategies which enhance the territory from a design perspective. This more recent perspective of design is based on theories in which design is viewed as science, as Friedman (1997), who mentions "Design sciences emerge when skills-based professions move from traditional rules of thumb or trial-and-error methods to the use of theory and scientific method". Also for Cross (2007), the interest of research in design must focus on development, articulation and transference of design knowledge from these three sources: people, processes and products. When Cross (2007) mentions processes, he suggests focusing on strategic procedures, techniques and design methodology. When it comes to products, he alludes to shapes, materials and outcomes which bring characteristics into design (quoted in Acevedo 2011). This means we should take a multidisciplinary side of design as an action, which builds practical and experimental ways to make sense to a whole. In this case watching the territory and its content in a global way through the eye of design.

For this matter, observing Fig. 1 we will notice that there is an element located in the upper edge of the gate, which is a metallic ornament. If this element belongs to a sign, what is its meaning in a specific surrounding?

From the perspective of design how can it be addressed? Maybe from a semiotic point of view adapted to design, keeping in mind that: "Semiotics studies all cultural processes (in which human agents play parts getting in touch using social convention) as communication processes" (Eco 1989). Or from visual communication itself as the territory's graphic element, an identification symbol of the observed community and the craft of an area.

Frutiger (2007) mentions that "everyday objects, rarely have a symbolic meaning in themselves" (Frutiger 2007: 201), but when we combined them with other objects or contexts those characteristics determine a new symbolic expression. If that symbol holds symbolic adherence to a person, its perception may vary.

On the other hand, it should be considered in this same example, other immaterial cultural heritage principles immerse in the object's display: the craftsman who built the shape, the ancient technique applied in it, how the skill was developed, which

Fig. 1 Upper iconography on Patrimonial House door in the Historical Park, Guayaquil (2017). *Source* Author

tools he worked with, and what is his influence when designing the object, e.g., the use of cultural inherited resources to portray it in an ornate object.

Design can communicate these contents to citizens through some means or strategy that can spread knowledge from iconography to interaction with the presented information. Design makes use of means, resources and media which allow it to carry messages to people in a specific context. The scope of action of this discipline in urban visual language allows observers to decode vernacular characteristics in objects as well as both material and immaterial representations, which establishes recognition with space so that people can own such information and commit it to memory. Through design a systemic interactive bond can be established between image, information and culture.

For a person to be able to interpret visual information which a city provides, would improve the relationship between them (city and people), as well as reinforcing citizen's identity and sense of belonging. From a researcher point of view, strategic design is presented as a useful tool employed in the development of models or systems which would link development areas with a common goal.

Becerra (2008) point out that design is capable of managing a meta project, "strategic design considers this new scenario where the product is performing. It adopts tools belonging to semiology to set out on this expedition leading to the product and its myths….design itself is an example of all of this. Spawned as a project trade or profession, whether for products or information, it has become a myth devouring any field in this way" (Becerra et al. 2005: 10–11).

Strategy is not only set to commercial ends but focused on how each party, factors and sectors which sometimes work isolated are related and through their systemic merger they are used to make decisions that bring you closer to problem solving and innovation. This merger can be observed in Fig. 2, where we have included information design and visual communication in the term 'graphic design'. It can help us deploy a strategy to focus on adding value to contents between cultural heritage and territory, linking this system of relationships with education for the citizens.

Communication through a design perspective places citizens as main performers of this strategic system and the promotion of cultural education. The cross-cutting sectors involved are education, design and territory, which as a set build the local discourse for cultural heritage and culture empowerment.

Finally, for this system to work it must make use of technological, social and economic factors as well as others which will bring along effective communication that will add value to the city story and will allow inhabitants to relate to their cultural heritage.

4 Communication of Cultural Heritage Through Design

According to Frayling (1993), research is the most important nourishment for the practice and teaching of art, craft and design. Research through design requires some

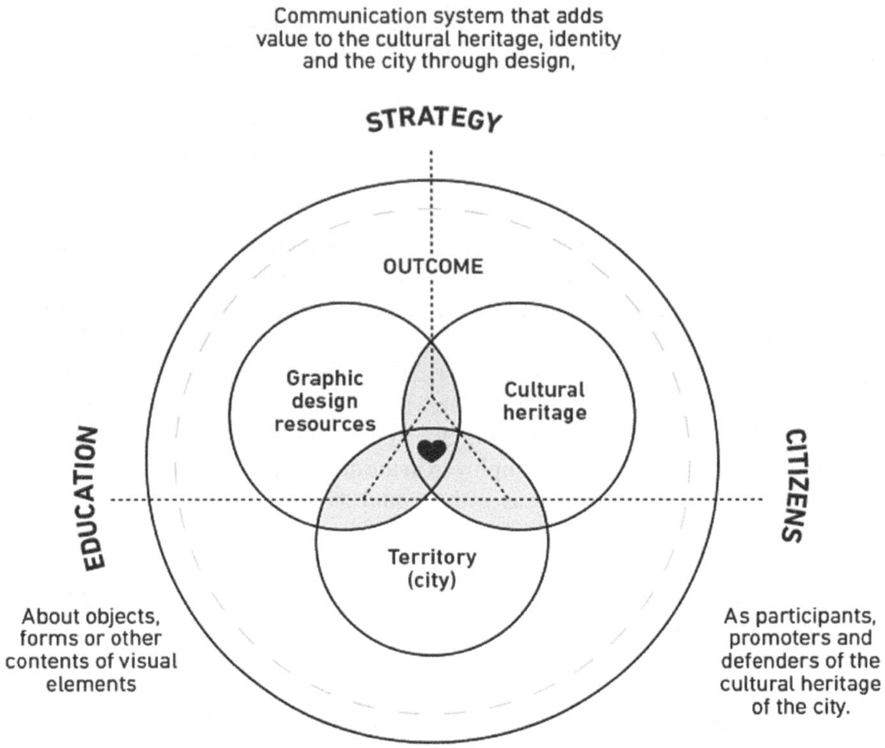

Fig. 2 Strategic system: cultural heritage, territory and design (2019). *Source* Author

characteristics of mixed methodology to be developed. Its main approach is the completion of a practical research, making it a more complex methodology than previous ones. It is mainly made up of these three stages: the search of research material (raw material and background), tasks development (field work), and applied research (theory verification). Additionally, Pontis (2009) explains that research through design, combines theoretical research with practical actions, creating a trial and error cycle, that goes from the most general to the most specific.

Design being a project-based discipline works with theory, practical components, applied research and field work to test and verify its theories to develop projects. Designers and communicators encode the world, understanding the code as a system of symbols which enables communication amongst people. Flusser (2007), also points out that the base of our culture is to inform, shape and prevent restrictions in the new communication ways between verbal and visual. To spread cultural expressions prevents the subject from disappearing, that is how design cooperates with this manifestation.

Cháves and Sánchez (2001), mentions that in the case of graphics there isn't a specific stance regarding cultural heritage, "The general notion we have of cultural heritage – unlike the undertaken when it comes to fine arts or architecture –, is always leaving out visual content, such as typography, illustrations or morphology of an area." (Cháves and Sánchez 2001). When it comes to urban graphics it refers to other elements unrelated to art and architecture. He otherwise suggests taking the focus back to objects, such as inscriptions, serigraphy or materials from a more folkloric construction, which also have a unique aesthetic identity citizens acknowledge and recognize as their own. Besides, on many occasions these materials are removed without anyone realizing they are part of a social construction of the city's visual cultural heritage (Cháves and Sánchez 2001).

Cháves and Sánchez (2001) review 'Barcelona Gráfica'[1] by América Sánchez, noticing that for the broader public, graphic design is part of consumer items, which he phrases like: "they are utilitarian symbols which perish through their uses unnoticed" (Cháves and Sánchez 2001). It's at these moments when graphics and city items conservation makes sense, since they are part of communication in a common space along with citizens, it is those works of art from graphic design (Massad and Guerrero 2009); typographies, ornaments, iconographies, editorials, and illustrations that may be admired by passersby and take them back in time, where the use of such shapes, images and styles were part of the city's everyday life before the careful and thorough look of a few.

Inhabitants recognize their essence in the environment, they watch those details, and they embrace them, making them part of their own identity. From that point of view, we take cultural heritage as a commodity available to everybody and willing to be admired and preserved.

Within the dialogue established between culture and cultural heritage, we can find graphic design working as a link for the communicational narrative in which people take part. Cultures change constantly and that changes the way people define themselves.

5 Case Studies

We undertook two case studies in which design has communicated cultural heritage. The method was used to understand and interpret graphic design projects through action research. These projects have improved the image of a territory and its heritage, to benefit local identity from those cultural resources. In both cases value is added using creative alternatives through design for safeguarding the presented objects' content, so to carry out cultural heritage transmission and to promote the territory they come from.

[1]'Barcelona Gráfica' (2005) is a photograph catalog which gathers about one thousand and nine hundred images with graphic details of Barcelona.

5.1 *Typographic Landscapes*

Since 2003 the University Center Senac leads a research in which teachers and students residing in São Paulo are involved. This project includes São Paulo's Architectonic Typography and Typographic Landscapes; both consider the typographic presence in São Paulo's historic center building facades. The project intends to identify, understand, describe and store information on the epigraphs present in these buildings. They define typographic landscapes as "a subgroup of graphic elements present in the urban environment: symbols forming words, dates and other messages made up by letters and numbers" (Gouveira et al. 2008).

The properties located in their research area contribute to the construction of such landscape with their objects, added items (mailboxes and signage among others) and epigraphs.

Once the research area has been defined, a log with the present elements is created and the epigraphs are shaped to get molds. The aim behind this is to be able to contribute to the city's memory conservation using this log and experimentation.

5.2 *Chilean Stirrups*

Undurraga and Bravo (2016), graphic designer, researched decorative items that laid on her house walls which belonged to her grandfather and held a great emotional value to her father. The stirrups are characteristic elements on the Chilean *huaso*[2] riding harness, originally made of metal and wood—for support and protection of the rider's feet—it is known that they were introduced into the country with the arrival of the Spaniards. Stemming from that interest she decided to begin an exploratory research, that led her to gather iconographic information from at least 100 stirrups from Chilean philatelists and museums.

The project aims to add value to the ichnographic richness of the Chilean stirrup. It starts with a bibliographic record which includes the stirrup's history and its characteristics, morphology and psychology behind its origin. An image index of the pieces was put together including illustrations to show the iconography of the carving. With a huge baroque influence, the evolution of this object comes from the metal stirrups introduced by the Spanish conquerors, adapting to the materials and available technologies and incorporating a strong decorative load.

Its iconography is composed of geometric and schematic elements with predominance of spikes, suns, rosettes, stars, crosses and animal elements such as ducks and deers, come up as part of the folk culture interlaced with religious influence which stemmed from the jesuit baroque.

---

[2]The term 'huaso' is used in Chile to refer to the individual who lives in the central area or part of the south of the country and is dedicated to the tasks of the old haciendas of the central valley and the coast of that area.

Because of this work, '*The Chilean stirrup. Salvaging visual cultural heritage from design*' was presented (Universidad del Desarrollo 2016). It was edited by the Design School from Universidad del Desarrollo and the Chilean Breed Horse Breeders Federation, Santiago de Chile. This project suggests the formalization of research through design as a contribution in this area, fulfilling challenges that generate knowledge and transferring them to society.

6 Conclusions

Sometimes cities, their objects and visual elements with cultural information go unnoticed by people. As we have observed, design has much to offer for visual and cultural heritage in a territory. It can contribute to the conservation of cultural assets through its recording, analysis and sharing; assets which otherwise would be doomed to be damaged, destroyed altered or forgotten.

The presented cases show that with the appropriate action research procedures from design we can get effective results for the transmission of cultural heritage. The practice of design is strategic to solve problems in this field helping retrieve the city's identity.

Several of the researched authors state the importance of educating people in regards of their cultural environment. Research through design is strengthened in this regard, since it presents a local community with opportunities to add value through chosen processes. Graphic design has been fundamental in that sense, as it contributes with its expertise to prolong the life of that existing heritage.

References

Acevedo O (2011) Módulo de Metodología de la Investigación [Research Methodology Module]. In: Strategic design master, Universidad de Valparaíso, Chile

Alderoqui S (2003) La ciudad: un territorio que educa. [City: a territory that educates]. CADERNO CRH, Salvador, N. 38:153–176

Becerra P (2008) Diseño estratégico en los sistemas de comercialización y consumo [Strategic design in marketing and consumption systems]. Fundación IDC, Mendoza, Argentina

Becerra P, Cervini A, Zapico S (2005) Through and about the product. Strategic design and SME innovation in the city of Buenos Aires. Instituto Metropolitano de Diseño e Innovación, Buenos Aires, Argentina, pp 10–11

Cháves N, Sánchez A (2001) El patrimonio gráfico y su recuperación [The graphic heritage and its recovery]. Revista Tipográfica N° 47

Cross N (2007) From a design science to a design discipline: understanding designerly ways of knowing and thinking. In: Design research now. Birkhäuser Basel, pp 41–54

Eco U (1989) La estructura ausente. Introducción a la semiótica. [The absent structure. Introduction to semiotics]. Lumen SA Editorial, Barcelona, España

Flusser V (2007) O mundo codificado. [The coded world], San Pablo, Cosac Naify. Cited in Cambariere L (2017) El alma de los objetos. [Soul of the objects]. Paidós, Buenos Aires, Argentina

Frayling C (1993) Research in art and design. Royal College Art Res Pap Ser 1(1)

Friedman K (1997) Design science and design education (No. 1997). Norwegian School of Management
Frutiger A (2007) Signos, Símbolos, Marcas, Señales. Elementos, morfología, representación, significación [Signs and symbols. Their design and meaning], Spanish Edition 10° edn. Gustavo Gili Editorial, Barcelona, Spain
Goodenough W (1975) "Cultura, lenguaje y sociedad", en El concepto de cultura: textos fundamentales ["Culture, language and society", in The concept of culture: fundamental texts]. Cited in Kahn JS (1975) Anagrama, Barcelona, España, pp 157–244
Gouveira A, Farias L, Cescato F, Indicatti F (2008) Urban Typography – modeling of epigraphs. 8° Congresso Brasileiro de Pesquisa e Desenvolvimento em Design
Hymes D (1964) Language in culture and society. Harper Row, New York. Cited in Kahn JS (1975) El concepto de cultura. Textos fundamentales [The culture concept]. Anagrama, Barcelona, España
Leach E (1979) Political system of Highland Burma. Beacon Press, Boston in Firth R (1968) Man and Culture
Massad F, Guerrero A (2009) Inventario urbano. Resenhas Online [Urban inventory. Online Reviews], São Paulo, año 08, n. 087.02, Vitruvius. http://www.vitruvius.com.br/revistas/read/resenhasonline/08.087/3047. [13.01.2018]
Margolin V (1991) Design studies and the education of designers. Temes de disseny 6:49–54
Pontis S (2009) Diseño gráfico: Un novel objeto de investigación. Caso de estudio, el proceso de diseño [Graphic design: a novel object of investigation. Case study, the design process]. Iconofacto 5(6):9–18
Rivera N, Ledezma M (2014) La ciudad como valor e identidad en ampliando ámbitos y prácticas [Identity and public space. Expanding scopes and practices]. Universidad de Extremadura, Fundación La Caixa, Cáceres, España, pp 77–95
Romero C (1998) Identidad y Espacio Público. Ampliando ámbitos y prácticas [Identity and public space. Expanding scopes and practices. Universidad de Extremadura, Fundación La Caixa, Cáceres, España, pp 13–30
Undurraga C, Bravo U (2016) El Estribo Chileno. Rescatando el Patrimonio Visual desde el Diseño [The Chilean Stirrup. Rescuing the visual heritage from design]. Universidad del Desarrollo, Santiago de Chile, Chile
Vitta M (2003) Los sistemas de la imagen [System of Images]. Paidós, Barcelona, España

Contributions to Brand Systems in Lisbon Tailoring Brands

William Cantú and Fernando Oliveira

Abstract The present article intends to articulate the subject of Branding and brand language synthesis contributing to a cross-cultural analysis in the areas of Branding, Graphic Design and Fashion. The perspectives approached here attribute a multidisciplinary method for the advantageous outcome of *insights* benefiting a more effective brand management. The articulation of the matters sought the comprehension of the current panorama of fashion brands in the tailoring section, with metaphorical names, in Lisbon, as well as its visual patterns. On the context of the analysis undertaken on this research, the model for the representation of visual identity systems (Oliveira 2015) occupied a centre spot on the concept of diagrams expressing the visual languages of brands here studied, contributing to a bigger understanding of the subjects here referred. For the development of this research, a qualitative methodology was utilized, which based itself on case studies complemented by a deep literary revision intertwined between the subjects. It allowed the conclusion of the importance of the brand language synthesis on the process of brand language analysis and cultural expression of their DNAs. On visual trends, it is concluded that there are similarities in structure and ways of communication between the brands.

Keywords Brand language · Visual identity · Model · Diagrams · Tailoring Lisbon Brands · Fashion

1 Introduction

This article refers to the process of visual language analysis on brands of a very specific market segment. Contemporary tailoring in Lisbon, on the taxonomy of

W. Cantú
Culture and Communication Program/Faculty of Arts and Humanities, University of Lisbon, Lisbon, Portugal
e-mail: williamafonsoc@gmail.com

F. Oliveira (✉)
UNIDCOM-IADE/IADE Universidade Europeia, Lisbon, Portugal
e-mail: fernando.oliveira@universidadeeuropeia.pt

D. Raposo et al. (eds.), *Perspective on Design*,
Springer Series in Design and Innovation 1,
https://doi.org/10.1007/978-3-030-32415-5_10

metaphorical brand names (Mollerup 1999, pp. 111, 114), becomes in this research the study group given the relevance and the interest of the population in the making of made-to-measure clothing, which can be verified through the existence of several brands within the same segment,[1] which will not be elucidated in this article. The objective of this text is to contextualize and map the language of the brands in this sector, allowing to understand the elements that constitute them, as well as their importance in the system. This work seeks to understand the elements of visual language as a representation of the brands' DNAs and the importance of visual design as a contribution to the control of brand language. It is intended to understand how these business units communicate by testing the applicability of the model to the representation of visual identity systems developed by Oliveira (2015, p. 367). This same model uses the languages of visual synthesis to perceive visual languages of the brands, generating comparative aspects between its elements and codes in relation to the DNA. The synthesis of visual languages is presented here as a methodology that allows the synthetic representation of communication elements of the Tailoring brands.

2 Objectives

The objectives of this article are related to the perception of brand languages through their arrangement. We try to understand if there is a readiness in the interpretation and analysis of the brands' visual systems through the aid of visual synthesis languages/diagrams. This article intends to focus on the elements of visual creation, giving light and projections about them on the level of visual behaviour and strategic management components. Thus, it aims to clarify the market niche and visual behaviour, allowing to compare the visual behaviours and to perceive the differences between the visual languages of the cases studied, as well as their respective visual elements, and to understand the value of the application of the Model in the aforementioned analysis.

3 Methodology and Research Question

Through a qualitative and analytical approach (Bauer and Gaskell 2008), the observation of the case studies (Yin 2001) presented in this article originated the following research question: *Are there patterns in the visual languages of Lisbon fashion brands with metaphorical names in the segment of tailoring?*

[1]Other tailoring brands, which have patronymic names and which for this reason will not be elucidated in this article are for example, Paulo Batista, Rosa and Teixeira, Nunes Corrêa, Brito Alfaiate, João Bento Vicente & C. Lda., Venâncio Alfaiate, among others.

The nature of this article is an applied investigation, generating knowledge with possible practical applications and referring to the solution of certain problems (Pradanov and Freitas 2013, p. 51). It analyses a segment to withdraw strategic insights about it. The methodological processes used throughout this research have emphasized bibliographical research, seeking to approach different placements in a cross-cut way and appending theoretical concepts for the development of the research (Yin 2001; Pradanov and Freitas 2013). In bibliographical research, a methodological procedure allowed the use many sources with raw data to understand relevant concepts and finally in the case study, the focus was into perceiving a business unit through a pragmatic analysis of a representative part (Gerhardt and Silveira 2009, p. 39). To these methods, visual synthesis languages/diagrams were also added (Tufte 2007, 2008, 2009, 2010) as well as DNA analysis of the brands (Mateus and Gomez 2009; Menegazzi and Gomez 2013). Regarding diagrams as an auxiliary research method, Lupton and Phillips (2008) affirm that these are systems of graphical and synthetic representation of processes or events. According to Oliveira (2015), diagrams may also be useful in the perception of brand language since they help in the comparison and synthesis of data, as Tufte (2010) refers, and that they are essential for the mapping of brand languages.

The approach given to this research reveals itself as mixed, being the first part (non-interventionist) constituted by a literature review where the subjects serve as a basis to questioning and informing about study areas (Teixeira and Neto 2017). At a later stage, through the observational procedure, an interventionist approach was used to interpret the visual language syntheses of the segment studied, using the method developed by Oliveira (2015, p. 367). The analysis, carried out in a qualitative way through a comparative procedure, sought to understand gaps and relations between the case studies. At last, in a final phase through a monographic procedure, there was an attempt of understanding if the visual languages syntheses contribute in fact to the analysis of fashion brands' visual languages in the tailoring segment.

4 Theoretical Framework

Authors such as Marshall and Erlhoff (2008), Olins (2008), Wheeler (2009) and Raposo (2012), define 'brand' as a representation of a product, service or organization and as an immaterial representation which the human-being intercepts and relates to. For Erlhoff and Marshall (2008, pp. 49–50), the brand is a name, a design or a symbol that differs products or services and adds value the consolidation of a structure (brand architecture), as the authors Olins (2008, pp. 44–53) and Wheeler (2009, pp. 22–23) state, is relevant because it helps in understanding the brand and its strategy. Wheeler (2009) and Olins (2008) emphasize the importance of the strategic and value component that are allied to brands through their behaviours, which reflects the realities where they are implemented.

The search for 'exclusive' has become a common behaviour pattern in contemporary societies, as well as customization because both have an extension in consumer-brand relations. The search for the 'singular' emerges as a trend and seems to be adequate to the reality of Lisbon through the revival of old crafts (Pimentel 2015). In relation to the graphical representations, Tufte (2009) emphasizes its value by confirming that it is possible to draw relations through trademarks, but for this, it is necessary to follow some principles of visual representation. When we mention the concepts of Tufte (2009, p. 51), we speak for example of "graphic integrity", the notions of "data-ink"; representation through information groups; proportionality of the data represented; application of subtitles; in the excellence/graphic elegance (Tufte 2009, p. 177) or in the format that can be adapted to the context (Tufte 2009). No less relevant are the cause and effect relationships provided by the information comparisons, which are useful in the perception of patterns and which should demonstrate a macro and micro reading hierarchized by priorities of information (Tufte 2010).

About visual languages, they allow better readings of complex information. Tufte (2010, p. 51) references that visual representations aid the communication and understanding of messages and also that visual systems are communicators that help transfer signification. Oliveira (2013) points out that Tufte's notions seem to be fundamental to create visual orientations that facilitate the comprehension of the language as a whole and that those graphic notions can expand the visuality with clarity and efficiency. Therefore, schemes become a set of systems since they begin to contemplate a set of meaning and syntax. As so, they are facilitators of messages and can include a large amount of information that is read with greater clarity and less complexity (Oliveira 2015).

In terms of brand languages, the elements of a brand serve to identify and differentiate it, its product and or services (Olins 2008; Wheeler 2009). According to Lupton (2011, p. 132), brand languages components (colour, typography, form, imagery) works as a system and helps to communicate the values associated with it. The same author still infers that the language of a brand should create a cultural dialogue with society by establishing communication with its public. For Wheeler (2009), the brands must transmit not only structural but visual coherence through a representative language in which its elements have the capacity to transmit immediate recognition. The definition of Oliveira (2015) tries to merge aspects of theory and practice, supporting his work not only in the ideas from authors like Mollerup (1999), Wheeler (2009), Raposo (2012) or Olins (2008) but also in the trough interviewing branding and design professionals. The result is a model for the analysis of brands' visual languages which represents a system that synthesizes and organizes its elements revealing the main component groups and the relations between them. This components that form brands visual language are defined by: Personality; Basic Elements (Name, Symbol, Color, Typography); Complementary Elements (Imagery, Sound, shape, Movement); 5th. Element; Graphic Mark and Brand Applications/communication. Other models such as Mono (Raposo 2008, p. 134), or Van Nes (2012, p. 7) are relevant for this kind of study, however, Oliveira's (2015) proposal is concerns the process and is exhaustive in the characterization of the elements.

For this reason, it has proved to be the guiding thread of this research work while dealing with the mapping of visual brand languages in the given business unit.

5 Case Study

The case study presented in this paper is exemplified by a group of Lisbon tailoring brands. In this context, the brands analysed possess metaphorical names (Rodrigues 2014). Based on the aforementioned principles (Tufte 2009) the schemes of visual language systems were elaborated not only to represent the language of the brands but also through a visuality capable of comparing the collected information. This conditioning factor formed the brand representation, so it has a neutral foundation in the diagram and in the displayed information to reveal the behaviour of the language analysed (Figs. 1, 2 and 3). The analysis of the brands has the main objective of identifying the trends in the visual language representation as well as investigating if there is a density of patterns related to some elements of these brands. For this, we have structured the analysed information by grouping the referred content in a manner that is comparable.

The brand seeks to represent a highly personalized project, comfortable suits, modern and daring, where the consumer can be identified as "unique". The audience fits in the gentleman's lifestyle, who are metrosexual men and joyful entrepreneurs in living in the scope of the city. The brand physical store is located in Principe Real, a neighbourhood in Lisbon characterized by its refinement and bohemian nightlife. This neighbourhood is in a privileged location, which welcomes designers, conceptual brands and irreverent gastronomy. UOY seeks to be elegant and modern through an appearance which converges with a revived classic and creative environment. The name of the brand originates from the word "YOU" and plays with the concept of being original (uncover the original you). The logo has three varieties, all with the same elements and within the same visual community: the logo (UOY) with the slogan of the brand (uncover the original you); the design of the letter "Y" resembling

Fig. 1 UOY—Uncover the Original You brand language synthesis. Developed by the authors © William Cantú/Fernando Oliveira

Fig. 2 - Cutter brand language synthesis. Developed by the authors © William Cantú/Fernando Oliveira

Fig. 3 Alphaiate brand language synthesis. Developed by the authors © William Cantú/Fernando Oliveira

the design of the blazers. The official typeface of the brand is Quattrocento Sans and its design is classic, elegant and humanistic. The characters are wide, and the height of x allow great legibility. The characters also have a great personality to be used in larger sizes and displays. The main colours of the brand are brown, grey and blue, worthy and prestigious colours (Heller 2014, p. 48). There is a predominance of stains of flat colour and brown. Terms such as "flat" and "clean" are emphasized in communication both printed and digital given its visual clarity. The fifth element of the brand is not yet a total reality but seems to be designed with this intent as it seeks the distinction with the "Y" of the logo. This element is also responsible for the transposing transparency stains and image cuts with a 45° angle. In the imagery, there is a relevant presence of textures and photographic dynamics. A model stars in the photographic sessions and represents the persona of the brand: someone with an aesthetic sense that is preoccupied with the formality of dressing, but which aims sophistication and irreverence. The shop is similar to a gentleman's club and the communication occurs through different platforms as Facebook, Instagram, physical store, printed materials (flyers, diptychs, photographs, vouchers, stamps, posters and bags.). To sum up, the UOY language displays avant-garde elements and praises the innovation which goes along with an exquisite visual representation. The brand tends to create a connection with the city and the search for the representation of tailoring is given by elegant fabrics with textures of character.

The personality of the brand represents the casual urban man who is in contact with the communities of the metropolis. There is a relationship between the brand, the Architect and the Planning. The representation of its approaches and concepts is related to the perfectionism of the confection. The brand seeks to transpose the message of exclusivity and experience in the suit making while creating credibility for the service provided. The locations of Cutter's physical stores are in Lisbon, Príncipe Real and Oporto, in Largo dos Lóios. These spaces are characterised by a strategic location of passage and also by the centrality, in the respective cities. They are places with personality, flooded with irreverence and sophistication. The name of the brand can be related to the very concept of Architecture and cutting. The logotype does not have an associated symbol and it is used in its fullness in the graphics applications. No specific typeface was identified, but there were very reliable approximations to the one used. It is an old-style[2] typeface with serifs and where a slight difference in the thickness of the characters is visible. It is intended to transmit stability and discipline within a controlled letterform design. The secondary font is a neutral sans-serif typeface that does not influence the legibility on documents. The brand colours live through the predominance of blue, which is related to the royalty and aristocracy as well as premium and high-quality fabrics. The brands' colours intend to represent tranquillity, the perfect seam, time-consuming and meticulous work. The dark blue can be seen as a global representation and is a source of transmission of brand trust, customer loyalty and product integrity (Heller 2014). The brand universe of images (Imagery) is related to the tailoring, store space and the confection. The whole ambience of the stores is warm and contemporary, representing the irreverence. The Form is reflected in the pop-up shop, which reflects formality and seriousness. In this brand, modern tailoring is revived with first-rate fabrics and a careful and organised image. The brand communicates through Facebook, Instagram, physical store and website. In short, Cutter's language idealizes itself through the construction of suits. It is a young brand that speaks to different age groups and with different lifestyles. It has a diverse communication, which uses different elements to represent, with great emphasis, the blue colour and the flat design approach.

The Aphaiate brand represents itself as an icon of visual elegance that conveys the quality and high class of its products in a clean way. The personality of the

---

[2]By Old Style we consider, in the light of Heitlinger's ideas (2014, p. 934) the fonts with the following characteristics: "moderate contrast, little difference between thick strokes and fine strokes. Capitals are almost the same height as lower-case ascents. The height of the x is moderate. (…) Serifs somewhat irregular, but robust and with round support. Diagonal bar in the letter 'e'. Quite long capitals (M, W, K)".

brand assumes the revival of the tailoring tradition and conveys confidence to the consumer through organization and advantages over ready-to-wear. The location of the brand's physical store is in Santos, a traditional and very characteristic neighbourhood of Lisbon, which has the presence of the creative industries and as well as a lifestyle that oscillates between bohemian, creative and business. The logo is a stylized representation of the Greek letter "alpha". It originated in the mutation of the Latin "ph" > "f", generating homophony between "alpha" and "alphaiate", the brand name. The font associated with the brand is AVENIR ® NEXT from Monotype. This is a non-serif and geometric typeface that has a good influence when concerning the readability (Heitlinger 2014, p. 220). It also allows us to clearly perceive that there are transparency and minimalism between the typeface and the communication. Alphaiate's colours are also clear associations to royalty. The beige humanizes the whole identity and conveys confidence. The use of this chromatic creates an elitist profile for the brand and to overflow the conviction and intimacy between tailor and client (Heller 2014). Charm and elegance are characteristics present in the brand language. The typology of images used belongs to the universe of tailoring and the gentleman's lifestyle. However, there is an impression of absence in representing the casual way of life, which is not that visible in the imagery. There is also a substantial amount of details represented. In this way, the image of the brand represents refinement and modernity linked with consumer satisfaction. The suits are the representation of a daily dress that should look for constant comfort. The architecture related to the brand refers to luxurious and refined surroundings and the brand communication media are Facebook, Instagram, physical store and website. Synthesizing, the language of the Alphaiate endeavours to represent a soberer universe through the clear colours that make up the brand's palette. The communication is incorporated with relevance into the photographs that portray the different fabrics, details of the products, the brand projects and its persona through a photographic universe representing the corporate man.

5.1 Global Synthesis and Comparative Analysis

Generally, we can see that there is a search for elements that transpire the concept of an old craft in both three brands, either by the presence of imagery based on photographs or by spaces decoration. The brand imagery is similar between the brands and the product is aimed at the same audience. The details of the images such as fabrics, textures, urban notes, architecture or masculine objects, breathe this urban lifestyle and creates communication within graphic refinement.

The three brands communicate predominantly with neutral colors conveying a classic and exquisite environment—UOY, brown, blue and grey; Cutter, royal blue and Alphaiate, violet, royal blue and beige. The use of metaphorical names reveals the purpose of the business through an indirect association. The associations create feeling and bring with time heritage and culture to the brands. The names tend to represent concepts surrounded to the brands. UOY—Uncover the Original You creates an association with its tailor-made concept of co-creation where the customer gives personality to his pieces discovering the client "original side. The brand Cutter - suit architect relates the suit-cutting (modelling) to the architectural designer, who performs precise measurements. Alphaiate relates his name directly to the tailor's profession. The typography of the brand is quite distinct and does not seems to reveal a pattern in this sector. The only brand that lives through the symbol is Alphaiate. The Greek letter alpha replacing the "a" creates a symbolic and verbal connection with "alpha" (Portuguese phonetics) of tailor. The alpha letter is also used as a trademark symbol, appearing alone or together with the name. The UOY is a typographic brand and having no icon uses the lettering as a representative. The same happens with Cutter. The UOY and Cutter marks are related to the concept of the service they present. The Alphaiate brand does not have an associated signature, differentiating itself in the sector through other approaches like the light imagery. The 5th. element, as we have already mentioned, identifies the marks without the need to use the other associated elements (Mollerup 1999; Oliveira 2015). It is only present in UOY, which highlighted the "y" relating it to the collars of the blazers, giving expressivity to all the communication of the mark (see Fig. 2). The communication supports of the brands are also the same, there is a site for e-commerce, a Facebook and an Instagram page. Through these supports analyzed it was possible to understand all graphic consistency presented as well as to perceive the level of investment of these brands in their communication. The trends of brands related to communication points to a similar path. In other words, is intrinsic and transversal in the analysis carried out that the brands are not revealing an innovative or irreverent personality to the public.

6 Conclusions

Apparently, the three brands communicate in the same way using the same channels and the same media. Having a similar audience, the differentiating concept does not prove to be sufficient to differentiate the way they communicate, which is very similar. Not only the communication is similar to the brand structure, it does not allow differentiation. The same material used (for example the style of images) is also a problem when concerning the differentiation, which creates direct competition. To differentiate themselves, the selling effect through promoting the products may not be the only suitable approach. The very irreverence that is sought in a brand to stand out, becomes obsolete when we talk about labels that have no design vision and focus only on digital marketing as a tool to generate sales. In these cases, access to diversified campaigns in loco or communication using alternative channels could be some of the strategies to promote the brand name in the social environment. The fact that there is no cult around the brand leads us to question whether the same exists in the internal structure itself, which seems to present imperfections, because as we can see with Wheeler (2009, pp. 22–23) and Olins (2008, pp. 44–53) all the consolidation of an organization or service must be structured primarily by the internal organization, trough the brand architecture, willing all the actors involved in the brand management process to transmit the concepts associated with the brand in a clear and effective way. We believe that they are very new brands in the market and therefore do not have a consolidated position in Portuguese society, given the high degree of intangibility socio-cultural phenomena referred to by Bauman (2000). The lack of internal culture of the brand and the fact that it is still expanding, makes it difficult to convey the idea of identity and worship or even to create them. The word-of-mouth advertising process is the method to convey brand awareness and situations similar to Privet Banking in which sellers have a strong relational component and are holders of a customer portfolio that generates sales. Trademarks should seek a representation that is different from what is being done in their sector in order to make their participation in the market relevant and meaningful. Socioeconomic factors such as the demand for technological liberation,[3] the transparency of the Internet or the veracity of the contents found on the web,[4] influence the audiences and the relations between brands and people. To combat these weaknesses, the content and the capacity to interact with emotions should be coherent points in the discourse and in the communication of the companies. In this sector (tailoring brands), betting on stories and sensitive content can become the key to engage communication of the brand and to attract new audiences. In the view of all these imprecision already promoted, the study of new markets, the intersection with these case studies and the continuity of this work must be fostered so that the structure of the brands can be adapted as well as strategies created, allowing the brands to gain power, to create stronger identities and generate life to solidify brand heritage.

References

Bauman Z (2000) Liquid modernity. Polity Press, Cambridge

Bauer M, Gaskell G (2008) Pesquisa qualitativa com texto, imagem e som: um manual prático. Vozes, Brasil

Erlhoff M, Marshall T (2008) Design dictionary. Prospectives on design termnology. Birkhauser Verlag AG, Germany

Gerhardt T, Silveira D (2009) Métodos de Pesquisa. Editora da UFRGS, Brazil

Heller E (2014) A Psicologia das Cores. Gustavo Gilli, Barcelona

Heitlinger P (2014) Tipos & Fontes. Tipografos.net, Portugal

Lupton E, Phillips J (2008) Graphic design: the new basics. Princeton Architectural Press, New York

Lupton E (2011) Graphic design thinking. Beyond Brainstorming. Princeton Architectural Press, New York

Mateus A, Gomez L (2009) Brands revolution. In: Proceedings das 5as. Conferências Internacionais em design. UNIDCOM/IADE, Lisboa

[3]Relation with the macrotrend "Connection and Convergence" identified by the platform Trends Observer (http://trendsobserver.com/trends/macro-trends/).

[4]Relation with the macrotrend "Trusted Content" identified by the platform Trend One (https://www.trendone.com/en/trend-universe/trends-2018/trusted-content.html).

Menegazzi D, Gomez L (2013) Think, view and use the brand: the TVU brand methodology. In: Proceedings of 3rd international conference on integration of design, engineering & management for innovation. IDEMI. Faculdade de Engenharia, Porto, pp 371–379
Mollerup P (1999) Marks of excellence. Phaidon Press Limited, New York
Oliveira F (2013) Diagramas no processo de design: a esquematização na metodologia e no projeto de sistemas de identidade visual. Convergências - Revista de Investigação e Ensino das Artes VI(12)
Oliveira F (2015) Diagramas e Marcas. Contributo sobre a utilização de diagramas na conceção e análise do discurso visual das marcas. Tese de Doutoramento (Doutoramento em Design). FAUL, Lisboa
Olins W (2008) WALLY OLLINS: the brand handbook. Thames & Hudson, London
Pimentel A (2015). Alfaiates do século XXI. A tradição casou-se com a modernidade [Artigo online]. https://observador.pt/especiais/alfaiates-seculo-xxi-tradicao-casou-se-com-modernidade/ (Verificado em 23/03/2018)
Pradanov C, Freitas E (2013) Metodologia do Trabalho Científico. Universidade Feevale, Rio Grande do Sul
Raposo D (2008) Design de Identidade e Imagem Corporativa. Edições IPCB, Castelo Branco
Raposo D (2012) La Letra como Símbolo de Identidade Visual Corporativa: Codificación y descodificación visual del sistema de identidade. Tese de Doutoramento (Doutorado em Design). FAUL, Lisboa
Rodrigues D (2014) Naming — O nome da marca. 2AB, Brasil
Teixeira M, Neto M (2017) Uma proposta de tipologia para pesquisas de natureza interventiva, Vol. 23, nº. 4. Ciências da Educação, Bauru, pp 1055–1076
Tufte E (2010) Visual explanations: images and quantities, evidence and narrative. Connecticut, USA
Tufte E (2009) The visual display of quantitative information. Graphic Press, Chestire
Tufte E (2008) Envisioning information. Twelfth printing, New York
Tufte E (2007) Beautiful evidence. Graphics Press LLC, New York
Van Nes I (2012) DYNAMIC IDENTITIES: how to create living brands. BIS Publishers
Wheeler A (2009) Designing brand identity: an essential guide for the entire branding team. Willey, New York
Yin R (2001) Estudo de Caso: planejamento e métodos. Bookman, Porto Alegre
https://uoy.me. Accessed during 25/10/2016 to 31/12/2017
https://www.facebook.com/UncoverTheOriginalYOU/. Accessed during 25/10/2016 to 31/12/2017
https://www.instagram.com/uncovertheoriginalyou/. Accessed during 25/10/2016 to 31/12/2017
https://www.facebook.com/cuttersuitarchitect/. Accessed during 25/10/2016 to 31/12/2017
https://www.instagram.com/cutter.suitarchitect/. Accessed during 25/10/2016 to 31/12/2017
https://alphaiate.pt. Accessed during 25/10/2016 to 31/12/2017
https://www.facebook.com/alphaiate. Accessed during 25/10/2016 to 31/12/2017
https://www.instagram.com/alphaiatept/. Accessed during 25/10/2016 to 31/12/2017

From the Genesis to the Project Result: The Success of Design for Place Branding also Depends on the Modality of Contracting

Emílio Ribeiro, Daniel Raposo and Marizilda Menezes

Abstract This chapter results from a research on place branding, within the scope of the Doctoral Program in Design of the School of Architecture of the University of Lisbon, referring to partial results of the research, specifically on contracting models of designers responsible for the design of place branding brands. Therefore, cases are analyzed from the hiring modality and points are presented for hiring improvements. The methodology is qualitative and descriptive, based on the literature review. The chapter assumes that among the determining factors for the success of a brand branding design is the hiring modality. Direct or indirect, open or restricted, they are aspects little considered when defining the demand for a place branding. One observes the importance of a modus operandi of who does and how does the contracting to minimize the failures. As a result, it was observed that the briefing should be associated with other tools for a better understanding of the territory, as well as a parametric analysis of existing brands to avoid plagiarism or similarity. The singularities of the place should be considered to give it authenticity, effectiveness and sustainability. On the other hand, the result of the design of the brand passes through a rigorous research on the territory, its image and its identity that determined the (re) positioning of the place.

Keywords Place branding · Hiring mode · Branding design

E. Ribeiro (✉)
University Ceuma, São Luis, Brazil
e-mail: ribeiro.emilio@gmail.com

Lisbon School of Architecture, Universidade de Lisboa, Lisbon, Portugal

D. Raposo
Polytechnic Institute of Castelo Branco, Av. Pedro Álvares Cabral, nº 12, 6000-084 Castelo Branco, Portugal
e-mail: draposo@ipcb.pt

CIAUD Lisbon School of Architecture, Universidade de Lisboa, Rua Sá Nogueira, 1349-063 Lisbon, Portugal

M. Menezes
University UNESP, São Paulo, Brazil
e-mail: zilmenezes@uol.com.br

D. Raposo et al. (eds.), *Perspective on Design*,
Springer Series in Design and Innovation 1,
https://doi.org/10.1007/978-3-030-32415-5_11

1 Introduction

The identity of a place results from a set of appropriations made over time by diverse experiences and by interaction with others, resulting in a sense of belonging to the territory, a set of dominant beliefs or values and a notion of place, taking into account their history, cultural characteristics and their position in the world.

The identity of the place is conditioned by culture as a process of social construction of one's own identity, both in the appropriation and transformation of meanings.

The cultural characteristics of a place are part of a process of construction that took place in the course of its history. The diversity of the elements that draw their peculiarities are the integral parts that define this cultural stratum and which are represented by the material and immaterial elements of the place. A place branding, part of that context. It must be authentic and therefore must emerge from the attributes that determine the identity of the place. Marketing strategies for a place should consider the place and its material and immaterial elements that characterize it and, in a way, value and identify it. Changing this truth is compromising the essence that, historically, one knows about the place. On the other hand, place branding can be a way to catalyze the competitive participation of the place in a Globalized scenario. However, some care must be taken when the demands for a place branding arise.

In the scope of this study, a Seminar was held in the city of São Luís/Brazil, with the purpose of discussing the models of hiring designers to create brands for place branding.

The competitions, most frequent modality in this type of project, are defined by their amplitude, open or restricted. Openings allow anyone to propose a design for the brand. Restrictions dictate that registrations can only be made by professionals or specialized agencies. On the other hand, it is important to determine how the identity and the place image are interpreted by the managers of the contraction process.

Finally, the sufficiency of the briefing is analyzed as the only source of information on the territory. It should be noted that the briefing tends to be a medium that translates the wishes of the project applicants culminating in a narrow view of the place and, as a consequence, may compromise the results. Therefore, we present the discussions about the modality of contracting projects for place branding.

2 Methodology

A qualitative mixed methodology was selected, based on the literature review and on the Descriptive Case Study (without control over the object of study, used to describe how it happened) (Yin 2005).

According to Yin (2005) the Case Study is an empirical scientific methodology, useful for the collection and analysis of qualitative data that allows to understand subjects or objects of study complex, similar and difficult to delimit clearly. This methodology allows us to study the cases in their context, perceiving "what", "how" and "why".

The Descriptive Case Study permits us to identify common denominators and draw conclusions based on previous processes, which may be indicators for future projects.

In this way, cases of design of brands for place branding were selected in the scales of the country, city and region. These scales help in the best framing of issues related to brand design for the territory. The typologies of the contracting modalities were analyzed for a better composition of the scenario on the *modus operandi* of the competitions for the design of marks of a territory. The types of competitions, open and closed, and direct signings are classified and analyzed. The elements that define a demand for brand design for a place branding were structured into 3 categories: demand, which refers to who does and treats about thinking the place; the process in designing marks for territories that determines how to do and generate the concept for the project and on the design of the identity, which proposes the means for a correct devolutive on how to deliver the results. The cases presented in this research were synthesized in a parametric analysis that demonstrates the scenario about the design of marks for place branding between 2014 and 2019. Based on the results obtained by the Case Studies and their analysis considering the literature review, some principles are presented and recommendations that may contribute to the effectiveness of contraction models for brand design in place branding.

3 Place Branding: The Valorization of the Territory

The monument of Christ with his open arms is not something exclusive of the city of Rio de Janeiro/BR. There are other places. However, Corcovado hill and all surrounding landscape present characteristics that are unique and peculiar to that place. Cristo Redentor, a symbol of the city of Rio de Janeiro/Brazil, was widely socialized in the media since its inauguration in 1931. Photographs, postcards, cinema, television, advertising, miniatures, instagram, among other means that popularized the monument and, consequently, the city of Rio de Janeiro/BR. Similarly, it is observed that the Statue of Liberty is an image associated with the city of New York/USA, just as the Christ Redeemer is a visual synecdoche, or rather a stratum of the city of Rio de Janeiro/BR that represents it (Fig. 1).

Although there are other monuments similar to Cristo Redentor, the correlation between this monument and the city of Rio de Janeiro/BR is most likely. This phenomenon is due to the fact that the image of the monument of Christ the Redeemer has gained notoriety since its inauguration and, consequently, occupies an important presence in the minds of the people. This notoriety refers to the knowledge of an element of the place that, when establishing a parallel with the knowledge of mark proposed by Keller (2009), is observed as to the concept on the image of mark, that is defined by the set of perceptions or mental images about, in this case, the city of

Fig. 1 Monuments: Christ the Redeemer (Brazil) and the Statue of Liberty (USA). *Source* Google Images, 2019

Rio de Janeiro/BR. Therefore, each territory has cultural characteristics that identify it.

Hanna and Rowley (2008) present a synthesis of place branding terminology based on branding practices for the various scales of geographical indications. From a hierarchical structure, the word place is correlated to a range of terms which, by association, hold a relation to the possible dimensions for place (see Fig. 2). Based on this design, Hanna and Rowley (2008) present the brand dimensions from: culture, industry, agriculture, heritage and tourism (see Fig. 3).

The identity elements representing the place are exemplary of man's creative capacity. Its cultural effervescence reveals itself in the material and immaterial manifestations that identify the place and assume a character of its own and peculiar. The diversity of the monuments associated with time (Lefebvre 2008), matter first, for the citizens who, in their own way, see and relate to this or that artifact of the place and then to visitors who, although temporary, elect cultural productions of

Fig. 2 Branding reputation and image management for place branding and the diversity of terms that define your goals. *Source* Authors (2018)

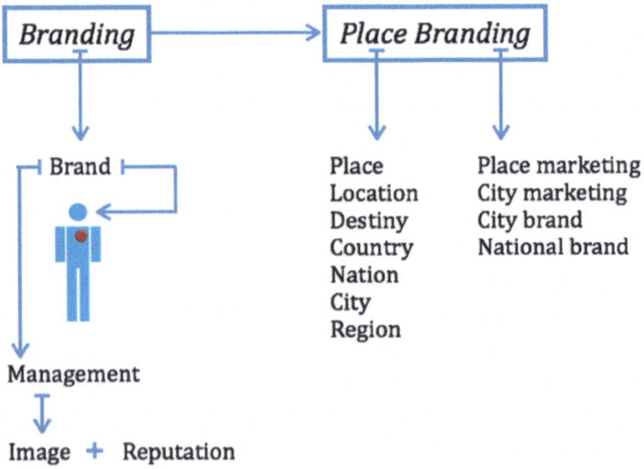

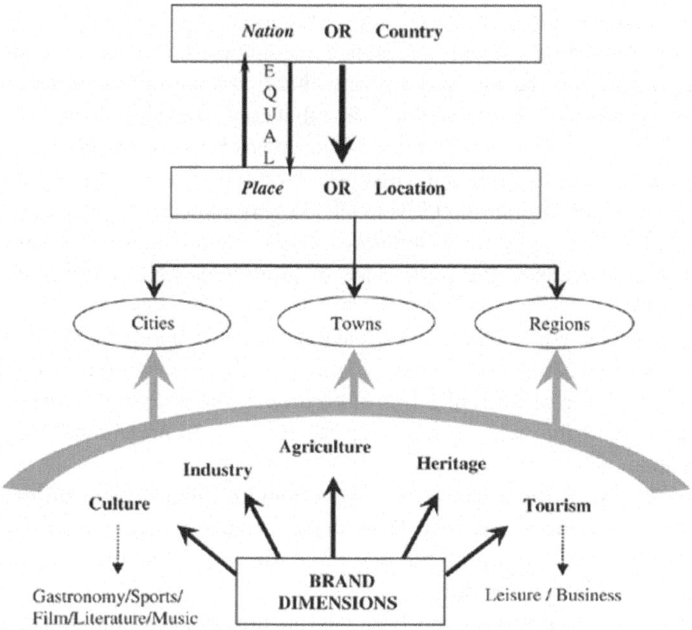

Fig. 3 Terms place and its variants. *Source* Hanna e Rowley (2008)

their preferences. Therefore, associations, even visual, about the place, go through elements that, historically, identify it.

The value of the territory is relevant, however, this perception of value is inherent to its products that go back to its tradition. In this sense, it is fundamental to recognize and highlight local values and qualities (Krucken 2009). It is from this perspective that place branding comes about.

Cities now need their brands materialized and organized by a visual identity that favors their insertion in a scenario of strong competition between places. The reputation of a place, region, city or country, if positive, tends to favor its competitiveness with others (Anholt 2007). Globalization has imputed the urgency of determining a competitive position that strengthens the expertises and singularities of the place. On the other hand, the shortening of the distances between the places in face of the new technologies of mobility offered to the contemporary man better possibilities of coming and going. For tourism, it was important to arouse the interest of this "new nomad" willing to know and live new experiences in different places. This scenario moves the economy and creates impacts that can be positive if, in fact, there is planning that goes beyond the limits of tourism.

The speed of change—social, economic, political, technological, among others—has promoted the rise of new demands in the cities agenda, among them place branding. This dynamic and fluid volatility circumvents the characteristics of contemporary life. Habits and customs that have consolidated in the course of modern culture, share space with the urgency of these days. Everything is ephemeral (Bauman 2007). The image socialized yesterday in social networks, today, will be passed. The new is recurrent. Nothing solidifies. This transience may have affected the perception of the perenniality of artifacts, habits, and customs that go back to the tradition of a place. Being those, the remaining copies of a legacy that conferred singularity to the culture of that people, for the elements that can be an integral part of the place branding.

On the other hand, the applicant's discussion on whether or not the graphic brand is relevant to a place branding (Govers 2013) reveals a bit of this contemporary dynamic where the urgency for a brand design for the territory, for example, is not considered, in some cases, the pertinence of guidelines and methods that favor a promising result.

Public managers, who are eager to leave a frame of their management, do not realize that there is complexity when it comes to the representation of the image of a place. The degree of difficulty lies in predicting the potential interrelationships between the parts of that whole, and the part of each is to understand its part in the whole (Cardoso 2012).

Branding is one of the elements of place branding that, by definition, refers to reputation management (Govers 2013), or, better, brand equity management. A brand for a city should give visibility to the place and, likewise, generate and add value to the territory from its local products.

It is worth mentioning that the brand is an important component for a place branding. It is defined from a double dimension (Raposo 2008): (a) physical image, materialized by a graphic mark composed by symbol, logo and color, articulated with the other elements of the visual identity system (*eikon*) and (b) by a mental image (*imago*) that reflects the set of symbolic, logical, emotional and reputational associations (stored in memory), but associated with the visual image. Brands go beyond images and representations. They reveal our preferences, attitudes and communicate an idea, a behavior to be shared, (Consolo 2015). They are an integral and important part of the social dynamics in which contemporary man lives and defines his choices and options by certain brands, including. Therefore, it is understood that the graphic brand is a fundamental component for the strategies of a place branding in the promotion and valorization of a territory.

In general and particularly in the case of place branding, the graphic brand (symbol and/or logo) constitutes the pinnacle of the brand's visual identity system, not only because it is one of the most commonly used identity signs in communication (just after the name), and because over time the audiences attribute to it new meanings, fruit of the experience with the brand. For this reason, it is often confused brand and visual identity with graphic mark.

4 The Place Branding: Between Being and Being

Place branding is an evolving concept. Academic studies date back to the 1950s (Hankinson 2015). At present, the scope of the term has raised some difficulties as to its nature. On the other hand, place branding originates from branding in the perspective of a management focused on the reputation and image of the place, while the second focuses on the product and the services. As for the diversity of terms—place marketing, place promotion, city branding, destination branding, place branding, among others, for this research the current term will be place branding since it understands that this covers the place in the scales of the nation, city, region and territory (see Fig. 4).

A place branding should emerge from the authentic present in place. However, graphically well-identified marks are observed, but they do not represent the territory or the elements of its identity. A brand is proposed that does not establish any relation with the place and, consequently, it is not possible to establish the necessary connections for its success. One has, then, a brand that is, momentarily, representing what it does not reflect itself, the territory. In this sense, it is a transitory mark, one that is not the legitimate representative of the territory, but it is. Therefore, to represent the territory, one must consider the elements of their identity associated with the notion of belonging of the local people.

Regarding the notion of belonging, Hall (2015) reveals the relation of the subject to what he assimilates as an integral part of his cultural identity. Belonging is an important indication of the affective bonds that city people have with their place. It is the recollection of experiences with the place, which awakens the sense of location in the context in which this man inserted himself. The place and its cultural tradition focused on the habits and customs historically constructed:

> It is in the past of the subject - his cultural tradition - that he governs the perception of the present space, which makes him recognize the places. Not a historical past (or the space of the past), but the past that tells us of the everyday experiences that we live, in space: lived space. The space in which our emotions, good and bad, are impregnated from the events in which we take part, both as agents and receivers. (Malard 2006, p. 29).

This tradition has in Architecture and in the remaining monuments the configuration of the cultural substratum that characterize the city and give it value. Two aspects can be highlighted in the question of value in Riegl (1858–1905): reminiscent value and

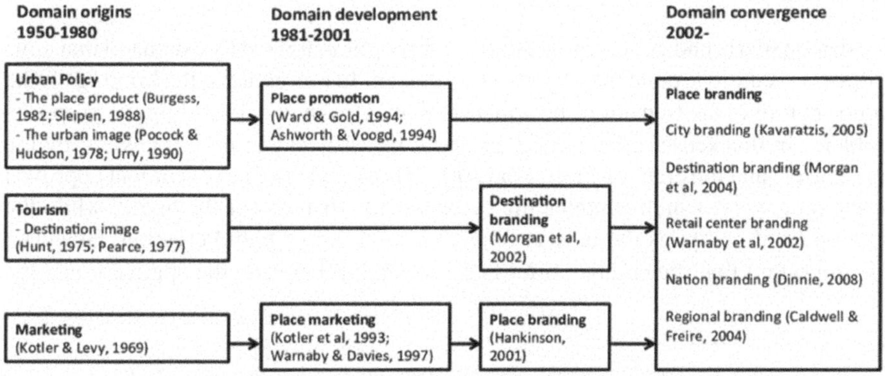

Fig. 4 Time line on the evolution of place branding. *Source* Hankinson (2015)

contemporary value. It is the relation between the old pair and the new pair that must live in harmony in favor of the preservation of the aspects that confer singularity for the territory.

The idea of a hegemonic identity leads to the question of the probable existence of universal standards that leads to the alleged legitimation of those standards (Bonsiepe 2011). Therefore, if one wants to intuit the existence of a static and universal identity, the feeling of the question arises orbiting around the domain and the power, not prevailing the quality of the artifacts, the culture and the identity of a place in the margin of the Center.

In the context of place branding, there are cases of failure of trademarks that did not meet the objectives sought by the applicants. The urgency for innovative, surprising design, complex as the territory itself or for a dynamic identity in its visual manifestations, perhaps, has induced the actors of the whole process, designers and plaintiffs, to fall into a smear. They forget, in some cases, the territory and the elements that, historically, define it. The brand is in the territory. What must be done is to extract, with some skill, the fundamental elements for the representation of that territory. The places themselves are marks (Anholt 2015). Perhaps, due to this universe of possible identities, it has expanded and generated a broad spectrum of cultural signification and representation (Hall 2015). In this sense, Hall (2015) identifies different conceptions of identity from the subject: subject of the Enlightenment that maintains its identity unchanged; a sociological subject that modifies its inner core from the dialogue with other identities that the cultural world offers and the postmodern subject, fragmented and composed of diverse identities. These conceptions intersect with the perspective of modernity having paved the way for the flexible and changing, and therefore for an identity is not fixed or lasting. Therefore, a place branding should translate the territory itself, or rather be the territory and not be, as a transitory and transient element.

5 Hiring: The Genesis of the Solution or the Way to Failure

The design of a brand comes from a need. This is characterized by a demand that aims to draw a scenario around the object of this design. In this context, the briefing[1] is the recurrent tool in the transfer of information that helps in understanding the proposed problem. In this sense, the instructions about the project can be synthesized by the brand brief[2] and the creation brief. For (2008), the brand brief is a document approved by the managers, which contains the essence and attributes of the brand, while the creation brief compiles the project objectives in favor of brand creation. However, it is important that the creative brief is a partnership between the applicant and the

[1]Briefing, series of references provided, which contains information about the product or object to be designed (ADG 2012).

[2]Brief, refers to the sum of knowledge about the project object (ADG 2012).

designer. Ultimately, it can be said that the briefing is an important tool for developing a brand design. But is the brief enough to understand the problem?

The unfolding of the briefing in the brand brief and the creation brief, result in a synthesis that comes from the efforts among the actors in the process. According to Wheeler (2008), the brand brief consists of: vision, mission, brand essence or grand idea, brand attributes, value proposition, guiding principles, target audience, key markings, competitive advantage and stalkeholders; and, for the creation brief, are: team goals, communication goals of all elements of brand identity, list of the most important applications, functional and performance criteria, mental map or SWOT,[3] positioning, protocols, confidentiality statement, documentation systems and benchmarks and presentation dates. It is worth noting that the brief, the brand and the creation, are an integral part of the method proposed by Wheeler (2008). It is a process that unites research, strategic thinking, design and project management. However, it is noted that a branding for a place requires the understanding and dimensioning of the complexity that lies there. It is worth emphasizing that, in terms of territory, its interpretation goes through a multidisciplinary perspective in terms of management, sustainability and competitiveness (Gaio and Gouveia 2007). For Kotler (1999), the presence of multidisciplinary groups is important to compose a diagnosis about the reality of the place through tools, among them a SWOT, so as to have a broader view of reality concomitant with a ψstrategic plan that attributes value and competitiveness to the territory.

In response to the demands for design of brands for a place branding, it is observed that the briefing is the recurring tool in proposing the problem. The contests, especially those that use online platforms, use the briefing to mediate the transfer of the project objectives. It is a document focused on the understanding of the actors involved in the management of the demand process that designers tend to use as the only resource in response to the proposed problem. However, there are cases in which the briefing is the result of an in-depth analysis of the problem being associated with data and information that are made available online and in other cases a simple and direct proposition about the problem and the objectives to be achieved.

In the method proposed by Wheeler (2008), Phase 1 refers to conducting research that aims to clarify the strategy, goals and values, where it is also sought to perform an analysis of the brands and architectures of existing brands, besides of a language audit. These components of phase 1 are synthesized in an audit report for the knowledge of the key decision-makers. It is a tool that can be used throughout the entire process with the objective of building a strategic and complete brand identity system (Wheeler 2008). In conducting the research reveals itself as a tool for understanding the company which, in the case of place branding, refers to understanding about the place.

[3]Created by management consultant Albert Humphrey in 1960, S (Strengths), W (Weaknesses), O (Opportunities) and T (Threats) is a tool to assess the current context and to imagine future possibilities. In its realization, SWOT makes it possible to gather the opinion of the people involved, directly and indirectly, in the context of the analysis. The more opinions are collected, the deeper the analysis (Pazmino 2015).

Brand is understood as the crux of place branding. Without it, without image, without contact, without reference. But where and how to extract it? If for Anholt (2015) the territorial brand already exists in the territory itself, why some projects fail to materialize it into a graphic brand? Or why some projects were objects of repudiation by the city? The answer to these questions matter for a better reflection on process of obtaining a brand design for the territory, from the hiring. It is worth mentioning that the image of the territory reflects its identity and there are places that have a positive image in people's minds. However, there are images that do not reflect the reality of the place or differ from the pretentiousness of place branding to the (re) positioning necessary to meet the desired competitiveness and to improve its reputation. Therefore, only the briefing may not be enough to answer the questions related to the territory, and among them, perhaps, a new image that emerges from an authentic identity.

The image arises from the interaction with a place that derives from its territorial identity, which is the result of the media spread through the territory (Holloway and Hubbard 2001). On the other hand, identity, the physical and psychological assets of the place carry with it all the complexity present there that converge to a dialectic between the emitting and receiving elements of the territory (Gaio and Gouveia 2007). In this sense, identity is the transmitting element and the image the receiver. Therefore, place branding refers to the strategic vision of the place for the place, or rather, for its unique and competitive elements that must be, in addition to being evidenced, strategically positioned. These are the positive aspects where the determinant of the brand essence of the place reflects a cross between the people of the place and the things that are produced and realized there (Anholt 2015).

As for the image of the place, the hexagon of the places branding developed by Simon Anholt assesses the image and reputation of a nation (see Fig. 5). This model presents six components: tourism, exported brands, internal and external politics,

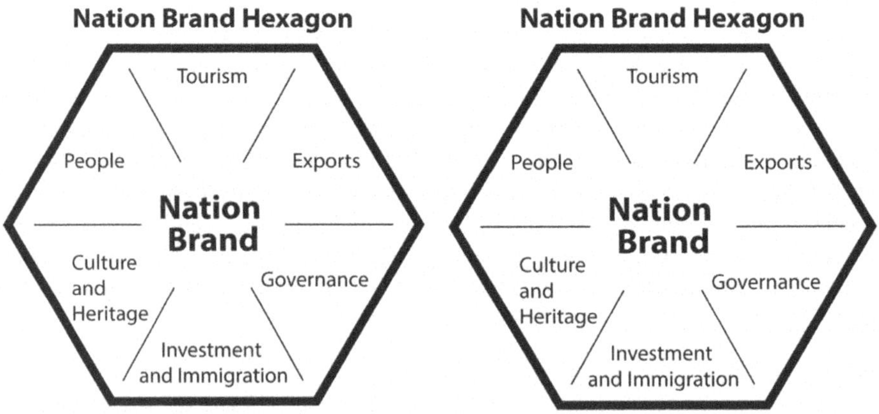

Fig. 5 Model for assessing the image and reputation of the Nation and the city. *Source* Anholt (2007)

investment and immigration, culture and heritage and people. These components are criteria that help in the capture of information that allows an evaluation on the reputation and, consequently, the image of the country. Based on these components, the Anholt-GfK Roper Nation Index ranks among the top 10 ranked countries. The perception of the reality of the place, based on data, is important to position itself in the face of the fragilities and potentialities identified. On the other hand, when the hexagon is adapted to a city the elements become: presence, which refers to comparing the city to the others at international level; place, is defined by the tangible and intangible aspects that characterize it and competitive opportunities emerge from its singularities; potential, rests on the opportunities evident there in the economic, social and cultural spheres; rhythm, it is the lifestyle of its citizens; people, refers to what we think about those who live there, and, prerequisites, is the mental image about the place that must be verified and (re) positioned.

In relation to identity, the model proposed by Cai (2002), focuses on the identity of the place and part of the mix of brand elements composed of 3 components: attributes, affectivity and attitudes. In the construction of the brand image, the model proposes: a brand program, marketing communication and management of the secondary associations. The model emphasizes the importance of the role of brand identity but does not specify how to build and develop it (see Fig. 6).

The communication about the territory in all its manifestations, including visual, must emerge from a process that starts from the identity of the territory. Thus, the most viable path to a place branding is through the understanding of identity, because how this place identity is defined impacts conceptual decisions for place branding design and therefore affects its execution (Kavaratzis 2004). Therefore, the identity of the place should be understood as a complex process of identity construction, which emerges from the dialogue between the stakeholders and the place branding (Kavaratzis 2004). Formulating a strategic vision for the place should involve managers, people and all potential partners. In this way, the model proposed by Kavaratzis (2004) is a process that leads to new actions that culminate in the infrastructure, landscape and incentives and opportunities for the various publics. Therefore, when the four stages of the model are fulfilled, the actions can be communicated (see Fig. 7).

The proposal for an analysis of the reputation and the image of a place, suggested by Anholt (2007), is important so that one can know the fragilities and the potential of the place. The question is how to think the place from these data. On the other hand, the identity of the place is the central element, in the perspective of Cai (2002), so that a place branding can reveal a consistent brand identity. On the other hand, Kavaratzis (2004) presents a model of communication of the image of the city that starts from a strategic vision that arises from the perspective of the social actors of the place. However, regarding the brand design for a place branding, it is observed that in phase 1, of the process proposed by Wheeler (2008), a research culminating in a report that compiles the data serve as the basis for the emergence of the brand and creative briefs, which can be consistent, depending on the rigor for which they were designed, from the stages proposed in stage 1. In summary, what is proposed is the appearance of a brand that reveals the essence of the place from of their identity. However, it is observed that, when identifying the need for a place branding, the management of

Fig. 6 Model focused on identity. *Source* Cai (2002)

the place, main demander, seeks alternatives for the design of their brands through a tender or direct contracting. The designs of a few places have culminated with brand launches that do not dialogue with the territory and, as a result, have frustrated the wishes of those who identify in place branding a way to face the strong competition in the process of Globalization.

The contracting orbits around two fundamental aspects: who does and how does (see Fig. 8). Public managers and institutions linked to the spheres of government are the drivers of the hiring process. The model of how to do tends to 2 alternatives, through open competition or direct contracting. The idea of open competition starts from the reasoning that, the more comprehensive, the more people contribute

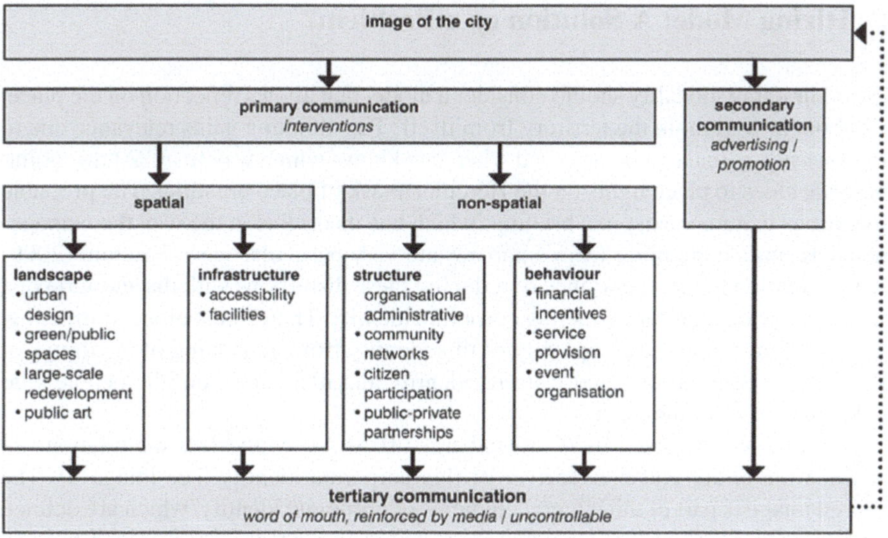

Fig. 7 Communication of the city image. *Source* Kavaratzis (2004)

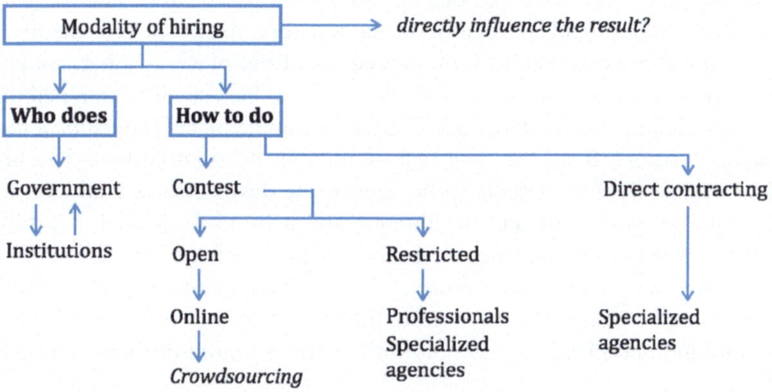

Fig. 8 Hiring mode for place branding projects. *Source* Authors (2018)

and participate. The pretense of involving everyone, designers and non-designers, weakens the model. Involving all the social actors of the place is important, however, one must know how and when to involve them. When the competition is limited to design experts, it is noted that the problem is solved in part. With regard to direct contracting, the transparency of contracting should be evident.

Regardless of the contracting modality, it is observed that the briefing is a recurring tool. It is a compilation of information that, hypothetically, comes from a deep analysis of the local specificities in relation to its place branding.

6 Hiring Mode: A Solution or a Problem?

The contracting modality should consider a model that allows reflection on the place. It's about how to think the territory from itself. This thinking gains relevance due to the consistency that can be obtained when one knows which way to go. At this point, there are clues to place lights on the Herculean task of place branding. The probable distortions that may exist in a briefing, which has its origins in the way the managements assimilate the place from a narrow and very particular view. Fascioni (2009) draws attention to the importance of reducing these distortions with the adoption of a participatory method for extracting corporate identity. Therefore, before contracting, one must dimension the complexity of this identity from a mapping of the things of the place that can be revealed from its identity and its image from the people who make and who live the place.

In the model proposed by (Cauwenberge 2015), he synthesizes the relationship between the issuer and the receiver of this corporate identity (see image 9). The desired image is part of the integral elements of corporate identity, which are defined in the course of its history. It is the cross between what it says it is and how its attitudes reveal, in fact, how it is perceived by those who somehow connect to that information. The image is not a static representation of corporate reality, it is dynamic and therefore there may be disparities between what is said to be and the way it is viewed. On the other hand, in relation to the territory, the perceived identity is the image that must be considered by the managers of the place as a parameter to be crossed with the identity that starts from the attributes historically contained therein.

A place branding should start from a reflection on the place. It must be a moment to think the territory from the manifest identity in its singularities. The briefing should be one of the instruments to be sent to the designers, as well as a SWOT that maps the opportunities and the threats, and a Butterfly Model to establish a comparison between what is valuable in this territory and the problems existing in other territories and, from that crossing can identify competitive opportunities. This information can be augmented by a mind map that branches the heritage elements, identity, culture, and place symbols, as well as the potential of their industries and

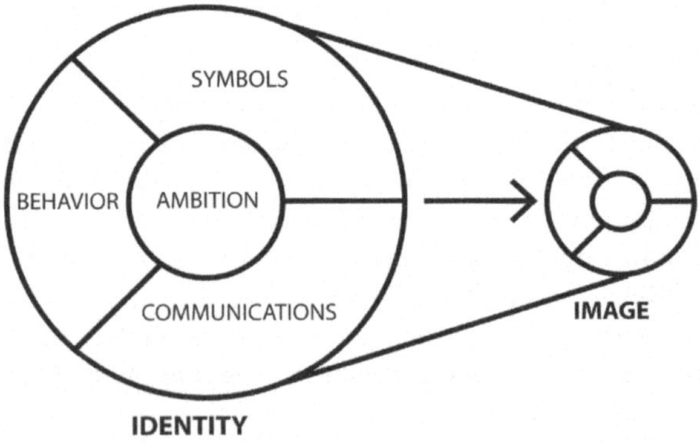

Fig. 9 Image as a reflection of the mix of corporate identity. Source: Cauwenberge (2015)

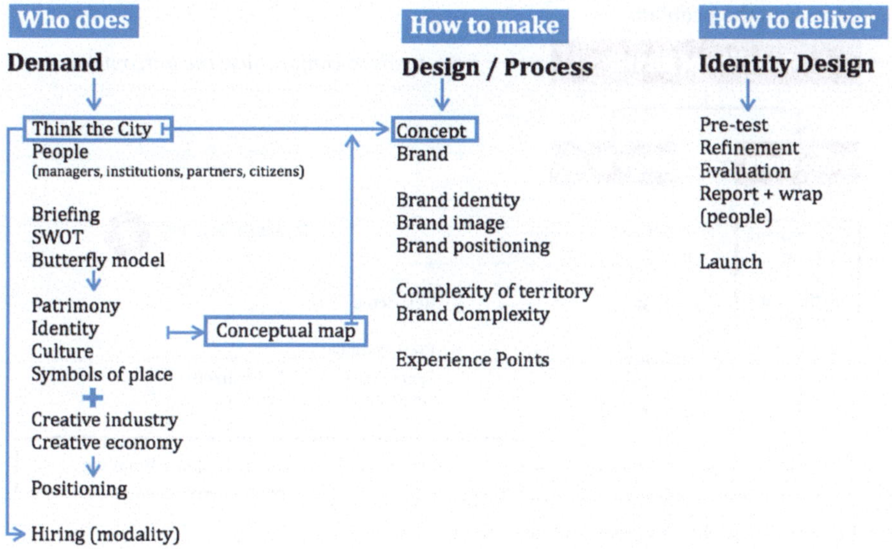

Fig. 10 elements for a design demand for place branding. *Source* Authors (2018)

their creative economy. The positioning should be presented clearly and be in tune with the results of the diagnosis. In summary, there are 3 phases to highlight: who does, which corresponds to demand; as it does, which refers to the design process and how to deliver that determines how the project should be presented as a result fulfilling all the prerequisites proposed (see Fig. 10).

Associated with the diagnosis, there is the modality of demand, which can be through a competition restricted to professionals and specialized agencies (see image 11). There may be cases where the best route is the direct hiring of a specialized agency. However, this process should be transparent and justified to all actors, including local people. It is understood that large and unrestricted competitions are not relevant. Those where any person, qualified or not to the field of design, can participate. These cases generate a large number of proposals that overwhelm jurors, in addition to the results presented, which are mostly amateurs and have no relation to the assumptions of what is intended for the place branding of that place. You lose a lot of time.

7 Cases

In order to demonstrate a brief scenario on the *modus operandi* of the design of marks for place branding, the following cases were chosen: Firenze/IT, due to the strong repudiation by the citizens when it was launched; Bologna/IT, for conducting a wide and unrestricted competition; Porto/PT, for the important success of the result;

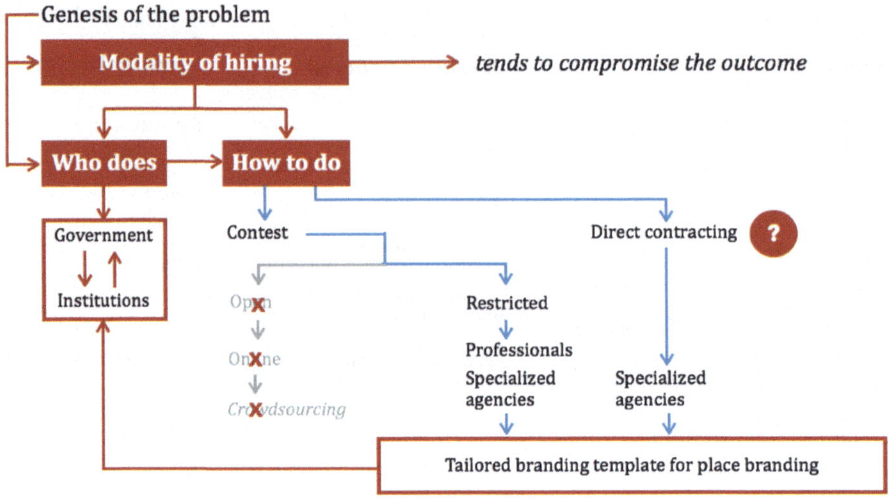

Fig. 11 hiring modality. Source: Authors (2018)

Alicante/ES, for resulting in a mark surrounded by criticism; Bucharest/RO, for drawing us the attention of how not to make a contest; Paraguay, for the short time between one brand and the other; Argentina, through direct contracting and, Swiss Wine, because it is a region that intends to strengthen its wine brands.

In 2014, the proposed brand for the City of Florence/IT was the subject of much criticism. The model of hiring was the crowdsourcing[4] that gave a briefing to the participants who, from him they created their proposals that culminated with 5 thousand inscriptions. Few were relevant to the proposed objectives. The absence of the lily in the winning proposal, generated an important difficulty for the citizen to accept the result that was much questioned at the time. The newspapers Il Corriere Fiorentino (2014) and La Nazione (2014) presented the results of the dissatisfaction, 89 and 91% of respondents, respectively.

The city of Bologna/IT, through the Urban Center Bologna,[5] institution that conducts the whole process to create the place branding, part of a research on the perception of the image of the city at the local and international levels. This research, carried out in the first phase of the process, sought to extract the image of the city, from the people's perspective, to a desired identity, or rather, from the perceived image to the identity that defined the position proposed by the briefing (Grandi 2015). Based on the results of the research, a briefing was generated that guided the International Competition for the creation of the brand (Grandi 2015). The model adopted for the

[4]It consists in the act of outsourcing, in an open and collaborative manner, a work traditionally performed by a hired employee of a company. This outsourcing takes place in the form of an open invitation to a large group of people, and is usually conducted from the web (Dickie et al. 2014).

[5]Urban Center Bologna is the communication center with which the city of Bologna presents and discusses territorial transformations and urban policies (Grandi 2015).

contest was crowdsourcing via the Zooppa website.[6] There were 534 projects from 17 countries. The diagnosis about the image and the identity of the City has rescued information about the place from the people. This strategy is the point that defines the difference between the model adopted by Florence/IT and the care proposed by Bologna/IT in the conduct of the hiring process.

From a restricted process, the City of Porto invited 3 design studios to present a brand proposal to the city. The contest starts with a briefing that was presented to the participants in a face-to-face meeting. The White Studio was the winner when proposing a brand that was is not restricted to tourism or the mere promotion of the City, it is a brand that encompasses all services linked to the City. The logo PORTO, Is the main brand of a brand architecture that proposes a visual identity for the Municipality of Porto, for municipal companies, for municipal services and for municipal projects. The visual identity is completed with a family of minimalist icons that can be joined by a mesh that forms a large panel.

The brand for the City of Alicante/ES was the result of an open competition. This contest did not present important criteria for a consistent brand proposal and that it met the proposed objectives (Zamora 2017). The absence of strategic objectives favored subjectivity, which for a brand, including a territory, is a problem. Finally, next to the delivery of the result came the information that the brand would be part of a "premium" strategy (Zamora 2017). So, soon after the conclusion of the Contest arise new elements for the branding strategy. With this result does the proposed solution apply that did not consider these strategies?

The case of Bucharest/RO is an important example to cite due to the sequence of obvious problems at the time of publication of the outcome of the competition (see Fig. 12). The City of Bucharest/RO promoted a competition for brand of the city and the winner presented a proposal with similarities to the City of Taipei/TW (Faura 2017). In identifying the problem, the organizers of the Contest chose the second place winner. However, in the proposal adopted similarities with the graphic marks of the Cities of Prague/CH and Florence/IT, respectively. The question then arises: how is it possible to achieve identical results for different territories? This finding leads us to reflect on how to proceed when we need to think about a territory from place branding. It is the choice of the most appropriate procedures for extracting the identity of the place, and this requires a method.

Fig. 12 winning mark of the competition for Bucharest's place branding and its similarity with logo for Taipei/TW. *Source* Faura (2017)

[6]Zooppa is a creative platform for the production of videos, graphics and advertising ideas for major brands around the world (Zooppa 2019).

A recent interesting case study is the place branding for Paraguay. In 2017, the Government of Paraguay launched its brand. A brand, whose design did not even consider the colors of the Country. A design, which proposed to represent the growth, riches and opportunities of the place from a symbol that was composed of 3 elements: the flower, the sun and the gear. Finally, a color palette based on green and not the colors of the flag of Paraguay. The result was a brand that could not keep up. In April 2019, through a contest, a new Paraguay brand appears. In this, the colors and elements of Paraguay are present in the mark establishes an important distance between the previous and the current mark that did not despise the elements of the identity of the place (see Fig. 13).

From the title: updating of the Country Brand logo, the bases and conditions for participation of the competition for the new Paraguay Brand designated the redesign of the logo launched in 2016. The previous contest was the result of an international public bid that resulted in a visual proposal that abdicated of the flag colors of Paraguay. For this new Competition, the organizing committee made available all the information on the Contest website. Brief, bases and conditions and presentation model were the documents presented. A workshop on strategy and briefing was held to define the meaning of the Country Brand and counted on the participation of Norberto Chaves. The winning brand of the Contest, rescues the origins of the name Paraguay and highlights the letter "Y" in Guarani means water, land of waters (BrandNews 2019).

In 2018, the Government of Argentina announced its new country brand. From a direct contracting, the redesign presents a graphic mark composed by a symbol in the circle format that contains the letter "A" in the lower part of the circle making an allusion to South America (Foro Alfa 2018). The logo is composed by the letter Gotham Rounded. The simplicity of the mark allows, from the shape of the circle, variations in the background with the insertion of images that establish a direct relation with the object of communication.

Swiss Wine Promotion aims to promote the region of the Swiss vineyards (BrandNews 2014). Composed of regional offices, the Association sought to create a graphic brand, a means to represent the territorial identity of that place. The design of the brand focuses on a graphic that represents the region of those vineyards. The monochromatic proposal has in red color and white background a visual relation

Fig. 13 Paraguay mark, left to front and right to current mark. *Source* BrandNews (2019)

with Switzerland that completes with a bar composed in red and with the cross, symbol of Switzerland, that is applied in the part superior of the bottles. The project was a direct contracting.

8 Discussions

Place branding emerges from the relationship between the identity and image of a territory and, consequently, its competitive singularities. The form for extracting those singularities depends on a set of factors that permeate the interests, especially of the contractor. On the other hand, it is possible to observe that the citizen is an integral part of the process and this is added to the other actors and factors that must be considered from the moment in which it is predisposed to think the city from a place branding. It was observed that the form of this thinking should be composed of propositional guidelines for a model appropriate to the context in which this place branding was inserted.

The cases selected for this study—Firenze/IT, Bologna/IT, Porto/PT, Alicante/ES, Bucaresti/RO, Paraguay, Argentina and Swiss Wine, show points in common and at the same time, different about how to conceive a place branding. Among the problems identified in those cases, it was observed that the contracting modality should be considered as an important and strategically thought-out requirement, in order to avoid the introduction of a place branding where a plagiarism is identified, for example. This leads to another relevant question—the uniqueness of the place. In the case of Bucaresti/RO, the similarity with the Taipei/TW graphic brand demonstrates that it starts from a concept that distances itself from the characteristic features of the place. It is worth mentioning that the briefing is a peculiar element in all competitions, however, it is perceived its importance in the complication of the information to understand the assumptions for the design of the brand, but, it is observed that when that briefing is added to a judicious research on the place, its image, its identity, its expertises, finally, on the way one thinks the place and the expectations for its future, being this information presented in a workshop, for example, where this thinking the place can be socialized in a very tenuous perspective on the place and its singularities.

To synthesize the information about the brands analyzed, a parametric table was assembled based on the following criteria: contracting modality, type of place branding from the scales—Country, City or Region, type of brand—modern or postmodern and the amount of colors present in the mark. Regarding the type of brand, it was considered a modern brand whose design is static and formalistic and, for postmodern brand, the one whose design suggests a dynamism in its visual manifestations (Table 1).

Brand design for a place branding should emerge from a contraction model that draws on a set of guidelines composed of recommendations in addition to a briefing. The contraction, direct or indirect, should be based on transparency. The criteria for such hiring should be excellent, and for this must come from a rigorous process of

Table 1 Parametric analysis of graphic marks.

		BRANDS							
		1	2	3	4	5	6	7	8
ANO		2014	2014	2014	2017	2017	2019	2018	2014
HIRING	Contest Open		●	●	●	●	●		
	Contest restricted	●							
	Hiring direct							●	●
SCALE	Country						●	●	
	City	●	●	●	●	●			
	Region								●
BRAND	Modern	●	●		●	●	●	●	●
	Postmodern			●					
COLOR	Monochrome	●	●			●		●	●
	Duotone				●				
	Polychromy			●			●		

Source Authors (2019) based on Wheeler (2008)

investigation into the place, its identity and its image. The singularities of the place should be considered to give it authenticity, effectiveness and sustainability.

9 Conclusions

Territory is composed of tangible and intangible properties. In them, we find the aspects that, in the course of its history, have built up its identity. The image, the product of that identity, reveals how the territory is perceived by people who directly or indirectly know it. Place branding has been the tool used in favor of a competitive positioning for the territory. However, for strategies to promote their success, the hiring of designers and specialized agencies must start from a model that considers the use of tools beyond the brief itself.

Open competitions should be discarded. They tend to have too many proposals that, in addition to not attending the briefing, generate a significant volume of ideas without the desired quality. Direct contracting is an option, however, because it is a territory, it is pertinent to be transparent throughout the process, including the justification for adopting this modality.

The briefing should emerge from a survey on the image and identity of the place as well as being the result of a discussion on the (re) positioning of the territory. The use of tools other than briefing should help in thinking. Local society should be aware of the relevance of place branding and at some point it should feel part of the process. Regarding the type of brand, modern or postmodern, this will depend on the degree of complexity associated with the concept proposed by the identity design, the narrative that is intended to be adopted in the process of brand communication with the public.

A parametric analysis of the design of existing brands can create a scenario that will reduce the likelihood of similarities or plagiarism. As well as the symbols that give it evidence can guide a way to the solution, both formal and chromatic. A place branding, through restricted tender or direct contracting, should reflect the best investment level of public resources. A brand design for the bad territory conceived due to fragile criteria in the genesis of the process and, consequently to its course, should be summarily avoided. Public managers should surround themselves with specialists so that the thinking of the territory does not sum up to a narrow view of the place. Finally, to obtain a model for place branding that considers the before, the during and the later, or rather, of the genesis about thinking the territory passing through the design of the brand until its implantation, will be a contribution to avoid the problems reported here.

References

ADG (2012) ABC da ADG (Associação dos Designers Gráficos do Brasil). Blucher, São Paulo

Anholt S (2007) Competitive Identity: the new brand management for nations, cities and regions hardcover. Palgrave MacMillan, New York

Anholt S (2015) Branding de locais e de países. In: Clifton R, Simmons J (Org). O mundo das marcas. Actual Editora, Lisboa/PT

Bauman Z (2007) Tempos Líquidos. Trad. Carlos Alberto Medeiros. Jorge Zahar, Rio de Janeiro

Bonsiepe G (2011) Design, cultura e sociedade. Blucher, São Paulo

Brandnews. (2014) New logo for Swiss Wine promotion by Winkreative. https://www.underconsideration.com/brandnew/archives/new_logo_for_swiss_wine_promotion_by_winkreative.php. Accessed 20 Mar 2019

Brandnews (2019) New country brand for Paraguay. https://www.underconsideration.com/brandnew/archives/new_country_brand_for_paraguay_by_alejandro_rebull.php. Accessed 20 Mar 2019

Cai LA (2002) Cooperative branding for rural destinations. Ann Tour Res 29(3):720–742
Cardoso R (2012) Design para um mundo complexo. Cosac Naify, São Paulo
Cauwenberge R (2015) 10 models for identity & branding: schematic tools to assist the understanding of complex problems and the identification of solutions. Total Identity, Amsterdam
Consolo C (2015) Marcas: design estratégico, do símbolo à gestão da identidade corporativa. Blucher, São Paulo
Corriere Fiorentino (2014). Ecco il nuovo brand di Firenze, ma la Rete lo contesta: Si legge Renzi. https://corrierefiorentino.corriere.it/firenze/notizie/cronaca/2014/10-marzo-2014/ecco-brand-firenze-2224190697488.shtml. Accessed 23 Mar 2019
Dickie IB et al (2014) O Processo de Desenvolvimento De Produtos Via Crowd- Design: oportunidades para a sustentabilidade. In: Congresso Brasileiro de Pesquisa e Desenvolvimento em Design, 11, 2014, Universidade Federal do Rio Grande do Sul, Gramado. Anais... Gramado, pp 1268–1279. Accessed em 20 May 2019
Fascioni L (2009) Método participativo de definição da identidade corporativa. Estudos em Design, Rio de Janeiro, pp 18–38
Faura R (2017) Bucarest elige marca-ciudad tras una descalificación por plagio. http://www.brandemia.org/bucarest-elige-marca-ciudad-tras-una-descalificacion-por-plagio. Accessed em 5 May 2019
Foro Alfa (2018) Qué dicen los creadores de la nueva Marca Argentina¿Por qué es tan simple?, ¿tiene relación con el gobierno?, ¿es una copia de Mashable? https://foroalfa.org/articulos/que-dicen-los-creadores-de-la-nueva-marca-argentina. Accessed 20 Mar 2019
Gaio S, Gouveia L (2007) O branding territorial: uma abordagem mercadológica à Cidade. Edições UFP, Revista a Obra Nasce, p 2007
Govers R (2013) Why place branding is not about logos and slogans. Place Branding Public Diplomacy 9:71–76. Retrieved 23 janeiro 2017 from http://link.springer.com/article/10.1057/pb.2013.11
Grandi R (2015). È Bologna giallo progetto: city branding. i Quaderni: Urban Center Bologna. Retrieved from http://www.urbancenterbologna.it/images/Bolognacitybranding/Q2_Bologna_City_Branding_WEB_bassa.pdf
Hall S (2015) A identidade cultural na pós-modernidade. 12ed. Lamparina, Rio de Janeiro
Hankinson G (2015) Rethinking the place branding construct. Rethinking place branding: comprehensive brand development for cities and regions, pp 13–31. https://doi.org/10.1007/978-3-319-12424-7_2
Hanna S, Rowley J (2008) An analysis of terminology use in place branding, vol 4, no 1, pp 61–75. Palgrave Macmillan Ltd. Retrieved from http://www.palgrave-journals.com/pb/journal/v4/n1/full/6000084a.html
Holloway L, Hubbard P (2001) People and place: the extraordinary geographies of everyday life. Pearson Education, Harlow
Kavaratzis M (2004) From city marketing to city branding: Towards a theoretical framework for developing city brands. Place Branding Public Diplomacy 1:58–73
Keller, KL (2009) Building strong brands in modern marketing communications environment. J Mark Commun 15(2–3):139–155
Kotler P et al (1999) Marketing places Europe. Pearson Education, London
Krucken L (2009) Design e território: valorização de identidades e produtos locais, 1st edn. Studio Nobel, São Paulo
La Nazione (2014) Il brand di Firenze non piace. E su Facebook spunta un logo alternativo. https://www.lanazione.it/firenze/cronaca/2014/04/10/1051038-nuovo-logo-brand-firenze.shtml. Accessed 23 Mar 2019
Lefebvre H (2008) O direito à cidade. Documentos, São Paulo
Malard ML (2006) As aparências em arquiteturas. UFMG, Belo Horizonte
PAZMINO A (2015). Como se Cria: 40 Métodos Para Design de Produtos. Editora Blucher, São Paulo

Raposo D (2008) Design de identidade e imagem corporativa: branding, história da marca, gestão de marca, identidade visual corporativa. UCB, Castelo Bran- co

Wheeler A (2008) Design de identidade da marca. Tradução: Joaquim da Fonseca. Bookman, Porto Alegre

Yin R (2005) Estudo de Casos: Planejamento e Métodos. Editora Bookman, São Paulo

Zamora I (2017) El nuevo logo de Alicante es un fracaso como ciudad que trabaja para profesionalizarse. http://www.brandemia.org/el-nuevo-logo-de-alicante-es-un-fracaso-como-ciudad-que-trabaja-paraprofesionalizarse. Accessed em 5 May 2019

ZOOPPA (2019) Retrieved 20 march 2015, from https://community.zooppa.com/

The Creation of Brands in the Online Experience: A Study About the Brand Image of Children's Clothing

Vera Barradas, María Victoria Carrillo Durán and Daniel Raposo

Abstract Nowadays, digital social networks represent a new way of being connected. These powerful tools widely present in our lives are a form of communication, information and social experience transversal to several areas of knowledge. What transposes to the outside of the virtual network for "real life", are also business models that represent challenges for new forms of communication of the brands, highly based on the relation between the brand and the interested parts. With a strong focus on people and their individual choices. Based on an exploratory analysis of a brand of children's clothing created online and a subsequent questionnaire survey of brands from the same sector of the Iberian market, this study made it possible to clarify the steps for the proper creation of a brand based on online experience. Allowing us to understand how the behavior of people in social networks allows and also values the emergence of business forms directed to defined groups and highly personalized preferences.

Keywords Digital identity design · Brand image · Brand experience · Branded content

V. Barradas (✉)
Instituto Politécnico de Portalegre, Campus Politécnico, 10, 7300-069 Portalegre, Portugal
e-mail: verab@ipportalegre.pt

M. V. C. Durán
Universidad de Extremadura, Facultad de Ciencias de la Documentación y la Comunicación, Plaza Ibn Marwan, s/n. 06001, Badajoz, Spain
e-mail: vicduran@unex.es

D. Raposo
Polytechnic Institute of Castelo Branco, Av. Pedro Álvares Cabral, nº 12, 6000-084 Castelo Branco, Portugal
e-mail: draposo@ipcb.pt

Lisbon School of Architecture, CIAUD, Universidade de Lisboa, Rua Sá Nogueira, 1349-063 Lisbon, Portugal

D. Raposo et al. (eds.), *Perspective on Design*,
Springer Series in Design and Innovation 1,
https://doi.org/10.1007/978-3-030-32415-5_12

1 Introduction

The brand image and the experiences lived through the brand in the online context is experiencing a really exciting moment. Brands have never lived so close to their publics and never before have individual opinion and choice ever had this level of relevance. We are interested in focusing on the present research the response that the brand image has before this market of emerging needs. Outlining a plan for how a brand creates value online based on the experience it offers to its stakeholders.

In the course of this research process, we worked actively on the construction of a brand of children's clothing and observed a very peculiar environment of relationship between the brands, between the brands and their public and the society's response to these micro-phenomena that touch the areas of communication, technology, sociology, management, image and design.

The specific choice of brands of children's clothing is also a result of the phenomenon called smart mom—this universe of maternity strongly linked to the new technologies through which mothers share their doubts, needs and experiences, creates in the online environment a community of mothers of which many are influencers and to which the brands of the infant universe is clearly attentive. With a strong unemployment rate, the online environment and social networks have become the most direct response to the creation of new markets and to the creation of professional activity in a fusion of this and personal life. From that moment comes a market centered on the person and his experience. This experience is shared on social networks and is able to create effective return.

For this reason the changes that the online context has had in people's lives and in the way these changes alter the view of social behavior, merge with business issues. There is new behavior towards this online market.

The Spanish and Portuguese markets have common denominators, not only in the high unemployment rate, as mentioned above, but also due to the similarity of online phenomena observed. The blogs of mothers, the importance of children's fashion in the business sector, brands of children's clothing created around social networks and activities outside the network, such as markets and fairs formed and disseminated on social networks.

Doing research in this context also leads to a new reading of the applied research methods. Data and numbers are likely to contain weaknesses inherent in this environment, such as the constant emergence of new data, new users and even new tools that will change behaviors.

The objectives resulting from the proposed theme arise from the need to adapt the methodologies to be applied at the level of the brand image on social network.

The present study has as main object of study the birth of children's clothing brands in the online experience, focusing on digital social network. It aims to analyze brand image and brand experience values as the basis of online reputation and as effective value generators.

Based on the general objectives of the research, we have the following specific objectives:

- Investigate the steps for the design of a brand of children's clothing on social network.
- Analyse the role of the brand's visual identity in the effective communication of the brand on the network.
- Know results of brand activities from online relationships.
- Find within social network the tools that best fit the creation of a brand on digital social network.

In order to reach the objectives set, and resulting from the reading and verification of the state of the art within the object of study, the following hypotheses of research were formulated:

H1 The construction of a brand of children's clothing in the online environment corresponds to defined and concrete steps.

H2 The corporate identity, represents in the online environment, added value in the identification of brands.

H3 In the online context, relationships are established that result in profitable activities for the development and growth of brands.

H4 Social networks have tools and techniques that allow the construction of a brand of children's clothing in the online environment.

2 Methods and Structure of Research

The review of the literature inherent in the research process focused on three main themes:

1. The brand image of experience
2. Digital social network
3. The brands of children's clothing online.

The studies found allowed us to deduce hypotheses as well as to recognize the difficulties inherent in a research with a strong connection with digital social network and social phenomena related to the universe of brands. Applied research allows us to corroborate or reject the hypotheses found through the theoretical survey of the themes mentioned.

The constant mutation and difficult accounting of the online environment led us to resort to two different research methods so that the conclusive data were as close as possible to the reality researched. For the reasons presented, during this research we searched for the main sources of information about the themes within the network. And in this context blogs specialized in the topic, represent a safe and effective source of updated information. The number of users and the daily dynamics of these sources becomes the main meter. Professional experience and approach to the subject has become the main ally in the planning of research that relies on research methodology with an exploratory study basis. This is intended to have a living knowledge of the subject that allows to advance to the second phase of research in a more secure and prepared way. Applied research thus divided into two distinct phases:

1. Exploratory study, active participant observation
2. Non-interventionist research: inquiry (Carmo and Ferreira 2008).

Researching online requires an analysis and reading of the techniques to be addressed. Access to people, networks and groups in an immediate way can be misleading and can be seen as a facilitated means of research. However, research in the online environment poses specific challenges and if collection techniques are not properly used they may be seen as unwanted or even intrusive.

When the questionnaire survey was established as a technique for collecting quantitative data, its design showed the need to find a qualitative basis for work.

Thus, before proceeding to the process of inquiry of brands, we analyzed the main guidelines (guidelines that were established through the development of the state of the art) of a brand of children's clothing, built in the online environment, for children between the 3 months and 7 years old.

This exploratory study and the consequent active participant observation with the brand, besides the data and the analysis that we present, allowed us a dip in the sector that we propose to analyze, thus establishing parameters and marking criteria of the brands to be inquired.

We then entered the planning phase of the questionnaire survey and sample selection. We selected 300 brands (150 Spain and 150 Portugal) through the observation of the medium, with a high presence in the online environment, considering this the universe and starting point of the investigation.

The data processing of the online questionnaire surveys was developed through graphs and their essential reading in the comparison of data and contrast of the hypotheses placed.

3 The Brand Image Experience

In this approach to the theme of brand image experience, and being the term brand, connoisseur of various contexts and consequent meanings, we approach the main dimensions and meanings of the brand. Directing our bibliographic research for the brand in its emotional aspect and able to attribute value to relationships with its public in a reciprocal way. We intend to understand the main concepts inherent to its evolution, making clear the differences between the concepts that are part of the theme. Thus, we indicate as main objectives:

- Define brand and its dimensions (Coelho and Rocha 2007);
- know the most important moments in the history of the brand (Raposo 2008);
- identify brand image (Costa 2004);

- differentiate corporate identity from corporate visual identity through the definition of both concepts (Carrillo et al. 2009);
- define the brand experience and the value of emotion in building rich identities (Martins 2007).

In order to reach the proposed objectives, we begin by defining the brand, using definitions of revised authors. Stressing the brand ages, brand types and brand image. In the theme are still defined the concepts: corporate reputation management, full communication: corporate identity, brand design: the corporate visual identity, typology of the graphic brands and the brand experience and its emotional nature (Villafañe 2004).

We finish the approach proposed in these points concluding that the experience and management of the brand depends on the alignment of the tangibles and intangibles that constitute it. Since the brand is always the result of the alignment of these variables. It is of utmost importance the coherence between these and the experience lived by the client, by the person (Alloza 2004).

If the brands are made of people it is important that these find themselves in and are in a common experience. People are online in organized groups and new ways of relating. In the following themes it was intended to find and describe the tools made available for effective living in the digital context, beginning with searching the root of the social network for the logic of group behavior that leads to phenomena of influence, tendency and micro tendency. Fundamental to the creation of brand experience online.

4 Digital Social Network

After the bibliographic review of the main concepts related to the brand experience, in the previous points, in the present theme we approach the social networks, a fundamental element of the present research. What is wanted in this exact point of the research is to know the origin of social networks looking for analogies that allow us to bridge the phenomenon of the current digital social networks, both from a personal and business point of view. Having as main objectives the theme:

- Search the origin of social networks (Buezas and Camarero 2006);
- identify and analyze the main social software available (Shih 2011);
- understand the role of the blog, corporate web and product web in the online experience (Ros 2008);
- analyze the challenges of interactive design as an online experience generator (Pratt and Nunes 2012);
- study the actors in the current digital landscape and how they influence behavior (Gladwell 2007).

In order to fulfill the stated objectives, we cover the following themes: the origin of personal social networks, digital social networks—facebook, twitter, linkedin,

Google +, pinterest, instagram. As well as the blog as a privileged tool and with an increasing value in content sharing and trend creation (Weber 2009).

The corporate and product web, as a key element in online presence, interactive design and users. We conclude this topic with the focus in the present research, which is related to examples of phenomena generated in the environment previously exposed: influencers, trends and micro-trends. Exposed and tools and the context of the users we enter what we call as the synthesis and contextualization of the reviewed themes, centered on the specific object of the investigation.

5 The Brands of Children´S Clothing Online

The brands of children's clothing represent the case study of the present research, which is why it is fundamental to contextualize the theme, as well as an integration of the previous concepts with the specific case of this universe. Children's clothing in the online context is associated with phenomena that do not only concern the purchase of products. We are not so much interested in children's clothing in the market sense. But rather try to understand what leads the Mothers of these children to become bloggers and how closed groups of Mothers on the facebook social network reach more than 13,000 members. Mothers buy online, but most of all they share. They generate content for brands and live experiences inside and outside the network, with their tribe. In short, the main objective of this specific point of the study is to synthesize, contextualize and relate points and themes already reviewed bringing them to the more specific context of the object of study.

- Contextualize children's clothing in the brand experience market (Cardoso 1994);
- analyze the parallel phenomena of the public linked to that market (Shapiro 2001);
- define the brand experience in the online context (Castells 2004);
- identify consumer behaviors online and analyze the tools of digital markets (Ron et al. 2014).

In this sense, the following themes were addressed: the children's clothing brand and smart moms, as a key player in the online experience of children's clothing brands, the creation of the online experience brand, the consumer society and electronic commerce, and branded content in the image of brand experience online. This last crucial point of connection between the brands and their audiences.

It is concluded that it is fundamental in the management of the online experience brand always to have the attention (from part to part), identification (of the public before the actions of the brand), the emotion implied to the experiences provided and the relation of permanence and fidelity. The brand has to be today, a "being" always available, always attentive, always connected.

6 Exploratory Study: Active Participant Observation of the Brand

As we said, the examination of the brands led to a preliminary observation of the medium (Scott 2013).

Experience brands in the online environment are not necessarily registered as a company and occur spontaneously from the will of the author and owner of those brands. Which makes the measurement process cautious and challenging.

Social networks are user territory and brands live a time in which their image has passed into the hands of the consumer. These two factors are decisive when planning our observation, taking us to a dive in the experience and the construction of a brand in the online environment.

The study was based on the Cherry Papaya kids brand, firstly because it appeared in a timely correspondence with the planning of the research but also because of the proximity to the members and managers of the brand that allowed to explore its development in a dimension of diving observation more accentuated.

Despite this proximity, which allowed us to work very closely and actively follow the strategies adopted by the brand, the data presented here are public. Any user has access to them through the visit to the corporate website/online store and social networks of the brand, which we describe below, in this point (in, III).

The observed items result from the subjects under study in the state of the art, so we found it relevant to observe the brand with regard to:

- Brand typology and definition;
- The corporate visual identity;
- Digital social networks and content branded;
- Other online communication tools: website;
- Characterization of the product;
- Brand activities outside social networks.

The observation and analysis of these concrete points of the Cherry Papaya Kids brand, as well as the proximity to the actions and activities of the brand, since its inception, enabled the present research to know the main sources of information related to the object of study and thus to formulate the universe and consequent sample, for the application of the active research methods previously exposed (Fig. 1)

7 Non-interventional Research: The Questionary Survey

A questionnaire survey was selected for the present study. The fact that the respondents and the researcher were not in a physical situation affects the way in which the number of answers is mastered, but it allows to fill the question of physical distance. Considering that the respondents are in Portuguese and Spanish territories and that

Fig. 1 Visual Identitiy I Digital content

these are connected to the internet (by the nature of the universe) electronic mail was the chosen medium of distribution.

The questionnaire consists of 19 questions, 15 of which closed questions. All questions are straightforward and the sequence of the questionnaire is intended to be logical so that it is the most comfortable for the respondent. In the elaboration of the questionnaire the questions were formalized considering the estimated degree of comprehension of the respondents, resorting to the use of control questions to verify and confirm the veracity of the answers.

We used the survey model by self-administered questionnaire, without the presence of the researcher sent by electronic mail. The emails were sent in three phases, for the first contact a newsletter format was developed aiming at a previous clarification of the study, objectives and later phases, after this email was sent a link through which the brand could access and complete the questionnaire and a third stage provided for in the absence of sufficient answers to recall the response to the survey and in which cooperation was appreciated.

The questionnaire was developed using the type form tool. In Portuguese and Spanish. This allowed us to target respondents and at the same time keep a clearer comparative record. The questions are open and closed questions. The former, although more difficult to quantify, can cover a greater number of information, the latter facilitates the quantification of data and allows faster response.

The online tool allows a result with high levels of usability which was proven by the speed of response. Formally the questionnaire is clear, enlightening the steps to follow and enjoyable.

In addition to the exposed sources (sector markets and blogs of the theme) it was also useful in the definition of the universe the personal network in the facebook social network of the researchers involved. Based on these sources, 300 brands were collected from Spain and Portugal.

This collection was established between April 8 and May 15, 2015, and it was found that the data that has undergone the most changes to the current date was the number of fans of the facebook pages.

The products of the universe chosen, are located in children's clothing from 0 to 12 years. 52.3% of the brands have an online store hosted on a website. The social networks used by these brands are: facebook, pinterest, instagram, youtube, twitter, google+, vimeo, linkedin, tumblr and bloglovin.

Although the comparison between the Portuguese and Spanish markets is not an objective of the research, we can not fail to note that of the brands analyzed in Spain, 24.6% of them have an active blog, whereas in Portugal this figure drops to 9.3%. Similarly, in relation to the presence in the social network instagram in Spain is registered in 74% and in Portugal the percentage is 65%.

Based on the 300 Iberian brands found in the exploratory study, the sample size was established in the simple random sampling model by randomly selecting the number of tags to be surveyed.

The definition of the sample comes from an error margin of 5% and a confidence level of 95%. 50% is the margin of distribution of the answers. From this estimate, we obtain the value of 169, which is the value of the sample considered. Considering this the minimum value for the validity of the resulting data. The randomness of sampling is facilitated by the size of the universe.

The data collected above allow us to establish comparative relations and fundamental conclusions to contrast hypotheses and final conclusions. This quantitative study with the brands demonstrates a great similarity between the Spanish and Portuguese cases, allowing to deduce that there is a strategic line of the online environment independent of the country of origin of the brand. This helps strategic thinking and allows you to deduce common steps for creating an online brand.

Regarding general data and impressions, facebook continues to lead both in terms of business tool and in terms of direct contact with the customer.

Both cases fully agree on the importance of designing a graphic brand in the initial definition of the brand. Perhaps the point of least concordance among cases is exactly who develops that visual identity. In the case of Spanish, 59% of the brand owner, while in the Portuguese case, 57% of the respondents reported having referred to a communication design professional to carry out this task (Graphic 1). More than half of the brands do not have, however, a manual of norms that allows to help the good use of the identity by the one who develops the contents.

Also, most of the brands consider the use of the graphic brand as important. And more than 80% believe that the public identifies the brand by the components of the visual identity (Graphic 2).

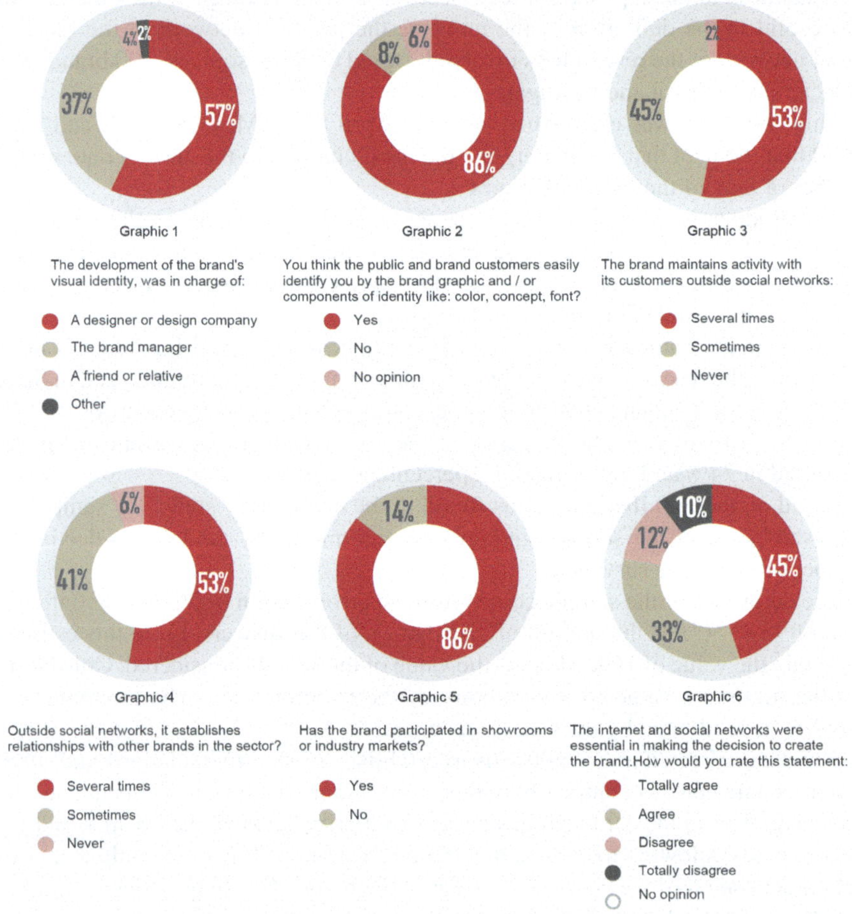

Graphics 1 to 6 Portuguese case

In relationships established outside social networks, only 2% of brands refer to never having personal contact with customers and 6% in relation to other brands (Graphics 3 and 4). This data will be fundamental in the human perception of online relationships. Still in this field more than 80% of the brands participate in events of the sector, such as markets, fairs or showrooms (Graphic 5). It should be noted that this response would be only confirmatory given that the formulation of the universe came largely from data presented by markets and childrenswear fairs. Over 80% agree that they have effective results in this participation.

Regarding the return coming from social networks, facebook continues to play a preponderant role although it is also felt on instagram, pinterest, twitter, blog and website. When talking about the online social tool in which a more profitable interaction is established, facebook is clearly ahead in preferences. With percentages close to 80%.

A fundamental data for the investigation comes from the question in which the brands are inquired about the importance of the social networks in the decision making of the creation of the brand. More than 70% of respondents agree and totally agree that the existence of social networks was essential in this decision-making (Graphic 6).

The brands clarified that the volume of sales of the brand has a direct correspondence with the campaigns of greater interactions in social networks. Being that the photographic images are the type of content with which the brand gets more interaction of its public. A relevant comparative data reveals that in Spain only 3% of the brands consider that personal content is attractive for this interaction, whereas in Portugal this percentage rises to 21%.

Curious and unexpected given the theoretical framework is that for brands the association to a blog sector does not seem to be essential. Since more than half say that connection is not essential. Unlike the website that for the brands is a key tool of online strategy.

Brands made it clear through adherence to the present study that not only are active participants in this context, but also show great interest in the results. Through the e-mail address that we placed at our disposal, we received the votes of success in the development of the investigation as well as reports of interest in the analysis of this new market model. A market closer to that which connects us to the old methods of sale in which the customer had a more significant proximity to the seller, for example, on large commercial surfaces. The facebook network has a preponderant role in the various answers obtained and the care with the brand image is assumed by the actors of this experience that the digital social network and its tools have brought us.

8 Contrast of Hypotheses

H1 *The construction of a brand of children's clothing in the online environment corresponds to defined and concrete steps.*

Data presented reveal that despite the experimental nature inherent in new tools, the fundamental steps for the creation of a brand in this environment are identified. The state of the art digital tools and survey results demonstrate the strategies to be adopted not only for building but also for brand management in this environment.

H2 *The visual corporate identity, represents in the online environment, added value in the identification of brands.*

Taking into account the state of the art and the observation of the brand of children's clothing this hypothesis would be verified. But in surveys, although the vast majority of brands demonstrate the importance of a visual corporate identity and its proper use, we find that it is still in the hands of the owner of the brand and in large percentage the brands show that they do not have access to the graphic norms manual. The design of the visual corporate identity should correspond to technical standards that allow the development of effective identity systems capable of responding to the challenges of online and digital tools. In this way although the importance given in theory is verified and although from our observation results a caution with the contents that include the logo, data of the inquiries do not allow us to confirm in full the truthfulness of the hypothesis.

H3 *In the online context relationships are established that result in profitable activities for the development and growth of brands.*

The importance of activities outside social networks, but derived from this common starting point, has been demonstrated at all stages of the research. Traders not only regularly participate in markets and fairs of the sector, but also show that they have effective results. The partnerships that are established between brands and between brands and blogs are within the network strong example. Corroborating this hypothesis also allows to affect the human character and the power of online relation. Making digital social networks a platform for building personal networks with strong interference in the development and success of the brand.

H4 *Social networks have tools and techniques that allow the construction of a brand of children's clothing in the online environment.*

This hypothesis is linked to hypothesis 01 in the way it is verified. If it is first agreed that it is not only possible to establish steps for a model for the creation of an online children's clothing brand, it is confirmed here that social networks and monitoring applications are fundamental aid in the construction of a brand online.

9 Conclusions

We live in a new world. For the first time we are (almost) all connected. What unites us is no longer the territorial, family and temporal coincidences. What unites us is more intrinsically connected to who we are. We are united by the photographs we like on instagram and the news we share on facebook. We are linked by digital boards and we organize in circles. Brands experience this new world. They are attentive and offer experiences in exchange for sharing. 69% of the Portuguese brands surveyed and 63% of the Spanish brands do not have a physical store and mostly keep in touch with their customers through facebook. They are available at any time of the day, answer questions, help clients in making decisions and discover affinity, become friends. If the digital social network on the one hand entails the lack of commitment of physical distance, the brands work in the opposite direction and create codes of identification with their public. From this identification is born the emotional bond that is worked daily by the brand.

The future comes from the past and mothers faced with a fragile professional background have returned to make their children's clothes, to be at home and to abandon the false ability to be super women. Mothers have been given the freedom to share, they have lost the fear of asking questions and commenting on responses and from doing business. The evolution of blogs and the intelligence of online relationships gave women the power of constant interaction with their peers, and with it also learning methods and business models.

At the same time that the brands claim to have had the existence of the social network as an influence for their creation, they also need a face-to-face contact. 88% have already participated in markets in the sector, markets that reach profit values well above expectations and with a strong growth trend.

On the other hand, the challenges that this new world offers are inexhaustible. Sociology, design, communication and management are bound to remake their size and establish new rules. Capturing the client's constant attention, working relationships daily, responding to expectations through product quality, astonishing, winning customer time and delivering emotion-filled experiences.

But of all the challenges, one should be seen by brands as the primordial of this new relationship. To have a brand today is according to Coelho and Rocha (2007), *tell every day a seductive story that never ends and continues the next day.*

The dynamic of this living relationship between the brand and its public and consumers, leads to a constant change of the paradigm. At the same time that power is on the side of people, these as users are generators of content. These contents are filtered and analyzed by the brands and these brands have in their power more and more means of personalizing their products. The brand responds with what the consumer feels identified and somehow re-assumes a power relationship. Between the paradigm changes and the exchange of roles, it is necessary to constantly capture the attention of the client, daily working the established relationship. A story that engages all stakeholders by crossing and elevating the emotional experience of the brand and the emotional values of consumers.

Acknowledgements This chapter results from the doctoral thesis entitled "*La imagen de las marcas de ropa infantil creadas a através de a experiencia online*." completed on January 4, 2016 at the University of Extremadura, Spain.

References

Alloza A (2004) MK Marketing + Ventas. El caso BBVA. La marca experiencia y la reputación corporativa. N 189. España

Buezas TC, Camarero DB (2006) Antropología, Teorias de la Cultura, Métodos e Técnicas. España: @becedario

Cardoso A (1994) Marketing de vestuário de criança: a percepção das crianças sobre as marcas de vestuário, sobre a moda e sobre o estilo de vestir. A influência dos grupos de pares (grupos de referência e de pertença) na escolha do vestuário das crianças. Tese de doutoramento. Universidade do Minho, Portugal

Carmo H, Ferreira MM (2008) Metodologia da investigação—Guia para a Auto-aprendizagem. Universidade Aberta, Lisboa

Carrillo MV, Núñez de Prado S, Tato JL, Delgado JP, Castillo A (2009) Comunicación integral y resposabilidad social de las organizaciones. Person Educación, México

Castells M (ed) (2004) La sociedad red: una vision global. Alianza Editorial, Madrid

Coelho C, Rocha P (2007) Brand taboos. Booknomics, Lisboa

Costa J (2004) La imagen de marca. Ediciones Paidós Ibérica, Barcelona

Gladwell M (2007) A chave do sucesso. Dom Quixote, Lisboa

Martins JS (2007) A natureza Emocional da Marca: Construção de empresas ricas, 6a edn. Elsevier, Rio de Janeiro

Pratt A, Nunes J (2012) Inter-active design. Rockport, USA

Raposo D (2008) Design de Identidade e Imagem Corporativa. Branding, história da marca, gestão da marca, identidade visual corporativa. Edições IPCB, Castelo Branco

Ron, Álvarez, Núñez (2014) Efectos de los contenidos de marca em niñosy jóvenes bajo la influencia del branded contente. ESIC Editorial, Madrid

Ros V (2008) Posiciona tu marca en La Red: e-Branding. netbiblo, España

Scott J (2013) Social network analysis, 3a edn. SAGE, London

Shapiro AL (2001) El mundo en un clic. Como internet pone el control en tus manos. Grijaldo, Barcelona

Shih C (2011) The facebook era, 2a edn. Pearson Education, Boston

Villafañe J (2004) La buena reputación—claves del valor intangible de las empresas. Ediciones Pirámide, Madrid

Weber L (2009) Marketing to the social web, 2a edn. John Wiley & Sons Inc., New Jersey

A Practice-Based Research Model for Interaction in Print Design

Marco Neves

Abstract Print design is defined by a relationship with technologies that promote static existence of objects. This characterization is usually opposed to interactive possibilities and modification by users. Given its practical nature, the presence of interaction in print design benefits from practice-based methods. To establish a systematic presence of interaction in print design media, we developed a specific method, based on features of design practice. There was an initial planning phase for conception and enhancement of ideas and selection and definition of objects to further improve. We than created models and prototypes and from these, we undertook a design project where we included interaction features in a controlled way to test their application in print production. The method presents a contribution for the inclusion of interaction in print design, modifying objects and their relationship with people.

Keywords Print design · Practice-based research · Interaction

1 Introduction

Almost every daily concern seems to be allied with a technological rising and media broadcast. Design occupies a changing place in face of this, between a professional practice which provides solutions to already established needs and a research field aiming to offer an optimization of material conditions and suitability to people and their time.

In the case of print design, most used processes to conceive and produce visual communication appear to show no interference from research conducted in the area, at least one which may contribute to such media and user adequacy. Print design is historically centered on unilateral elaboration of material for production through printing technologies. Print media are understood as a group of static objects and

M. Neves (✉)
Lisbon School of Architecture, CIAUD, Universidade de Lisboa, Rua Sá Nogueira, 1349-063 Lisbon, Portugal
e-mail: mneves@fa.ulisboa.pt

D. Raposo et al. (eds.), *Perspective on Design*,
Springer Series in Design and Innovation 1,
https://doi.org/10.1007/978-3-030-32415-5_13

information. When compared to its digital counterpart, they lack dynamic information and most of all, interaction with their users. Print media does not support hyperlinks, does not change its behavior or the system by which information is transmitted, if the choice rests on users.

Nevertheless, print media can become more interactive, even without considering an electronic existence or adaptation. Interaction has been defined when it comes to design (Preece, Rogers and Sharp 2002; Moggridge 2007; Saffer 2007) and operationalized (Neves 2013). Previous studies have indicated most common features of digital production of interaction (Cooper et al. 2007; Neves 2016a), how it would relate to print production (Neves 2016b) and how such features would apply to a print produced specific object (Neves 2017; Neves and Caixeiro 2019). These studies have shown interaction can relate to print design and how to translate common interactive concepts to an already defined printable format. However, it is still necessary to explain how to systematically implement interaction on print design processes and how to change the way print media relate to their users.

By interaction, we do not refer to a technological environment, usually constituted by digital systems, which allows updating, modification and contribution to the very same system, through a relationship with hardware and software. We refer to a general concept (Bonsiepe 1999), explained by the possibility placed in artefacts for action-reaction, alternating messages and behavior decision (Bolter and Gromala 2003; Cooper et al. 2007). In other words, to provide a common framework to various media, as a continuous mapping of working hypotheses adaptable to several situations, in conception and development stages. If so, this strategy may be an innovation statement in print design, or at least, an attempt to increase possible solutions during visual communication.

So, the main goal is to implement interaction as a regular procedure in print design, benefiting objects submitted to printing technologies and the relation they have with their users. How to implement, increase or just control the levels of interaction in a print design process.

2 A Practice-Based Method

For this purpose, we followed a qualitative based methodology. We undertook a practice-based research, where we mixed a structure for project development, based on information gathered from previous research (Neves 2013, 2016a) with a print design project approach. Such specific process was outlined to achieve a material outcome in these circumstances.

Having a research method established through project practice is meant to test new information and ideas and to fit the creative practices being worked on (Logan 2006). The correct way to name the inclusion of a design project in a research method might be 'practice-led research' (Rust et al. 2007; Nimkulrat 2007; Haslem 2011) or 'practice-based research' (Pedgley and Wormald 2007) which derives from the action research method (Archer 1995).

Rust et al. (2007) explain it as "research in which the professional and/or creative practices of art, design or architecture play an instrumental part in an inquiry" (Rust et al. 2007: 11). But inclusion of a design project within a research method may lead to misconception, since design activity does not meet all the necessary criteria to be considered research (Pedgley and Wormald 2007). Still, we should distinguish research made about a certain practice, from the one with purposes of practice and yet from the one done through practice (Archer 1995: 11).

Archer (1995) gives us a reasonable description of the method: "there are circumstances where the best or only way to shed light on a proposition, a principle, a material, a process or a function is to attempt to construct something, or to enact something, calculated to explore, embody or test it" (Archer 1995: 11). The association between interaction as a concept and the development of print design is not fully theoretical, and may be better understood with material evidences of such association.

It is although advisable that such material construction can be paired with a description and reflection of the process which enables it (Pedgley and Wormald 2007: 82). So, a part of the entire developed project is presented as a real object, produced and distributed. This will help to make the result authentic, as the method itself is conducted simultaneously through a detailed description. In this sense, design projects included in a research method should be documented (through sketches, technical drawings, audiovisual materials, texts, prototypes, etc.) so they can become information to be processed and for conclusions about the method be taken.

3 Print Design Project

What we name as print design project does not intend to address common problem solving (Johnson 2002; Davis 2008) in well-defined and circumscribed situations. Our project initiative was not only to execute, but also to generate a system from which several print design projects could be carried out. It brings together a set of alternative and broad solutions, given the collection of information during research. In this way, it follows:

> What would happen if we used our knowledge about current practices not to answer certain questions by our design, but to avoid answering them? Or, in other words, if we tried to make our design ask questions about use that were open for its users to answer, rather than thinking of the design as a way of providing well-defined answers from the start. (Redström 2005: 136–137)

Also, the intention of a typical print design project is arguable when related to a research aim, for reasons stated by Scrivener (2000):

> In contrast, in projects where the work is progressed through the creation of and interaction with artefacts, issues, goals, and priorities may change throughout the project resulting in a stream of outcomes, thereby never settling on a specific problem or yielding a'final'solution. Perhaps the main reason for this is that the artefact matters as an object of experience. (Scrivener 2000)

Any purpose to generate print design projects in this situation does not fit with already known results. This opposition between project as a solution for a given problem and project as a creative production does not exclude the existence of a set of rules to both. In the latter case, Scrivener (2000) specifies:

- the result of this production should be original (not derive from the work of others);
- the work should be an artefact as an answer to concerns and interests;
- concerns and interests reflect culture and should manifest themselves through produced artefacts;
- artefact must contribute to human experience, even if with no utility.

An acceptable result from this method would be understood as material existence, which will be the confluence of previous knowledge and the one which we were trying to generate through its production. In a simple way, interaction can be understood and implemented in print media and allow users to play a part in it.

To achieve a material outcome for these requirements, we outlined a structure for project development following the suggestion "a key feature of action research is that it adopts a dynamic, cyclical process which moves through phases of planning, action, observation and reflection" (Bloor and Wood 2006: 10).

3.1 Structure for Project Development

To conceive print media from interaction categories, the following structure was used:

I. Planning

   - listing interaction categories from literature review and previous research;
   - conceiving and developing ideas;
   - selecting and defining objects.

II. Action

   - prototyping.

III. Observation or discussion

   - experimentation;
   - reflection;
   - changes and improvements.

IV. Extra

   - gathered theory;
   - process documentation;
   - process and reflection analysis.

Planning consisted in establishing possible connections between print and digital interaction and how can it be applied to certain objects. Conceiving and selecting ideas for objects were moments of free production. In the sense that it is not intended to change a subjective part of design projects or the individual role of each designer. However, selection may be subject to greater criteria specification, such as proximity to interaction categories or technical restrictions.

After selection, there was a second phase called 'action' where at least one idea should be developed with better detail through prototypes, to enable real object production.

Observation was done on all collected material and reflection about suitability of interaction as a concept, to print media development.

The last phase was mainly used to describe all work carried out and to aid on method synthesis.

4 Project Development

Picking up on our project development structure, all phases were carried out, so that this method would convert into applied research, understood as knowledge acquisition, conversion or extension for use in some situations (Archer 1995).

4.1 Planning

Connections of Interaction Categories

For the planning phase, we established a bridge between interaction categories, identified in digital artifacts, with techniques frequently used in print production. To better visualize it, we built a three-column structure, relating such categories and techniques with a listing of some of the most known printed materials developed by graphic designers. Digital interaction categories are summarized in: point and click, drag and drop, pull down menu, customization, browsing, visualization and text insertion. As print production techniques, we synthesized in: folds, cuts, overlay, transparency, creative participation of users (through drawing, writing or other) and relationship with digital artefacts.

The main materials developed by graphic designers can be frequently named as books, magazines, newspapers, catalogues, posters, fold outs, signage, visual identity, flyers, stationery and business cards.

The purpose of the scheme is to add a digital interaction category to a print technique, or to make an interpretation of digital interaction, through printed means. But always ending in a connection to a known print media. In Fig. 1 we can see all categories and established connections.

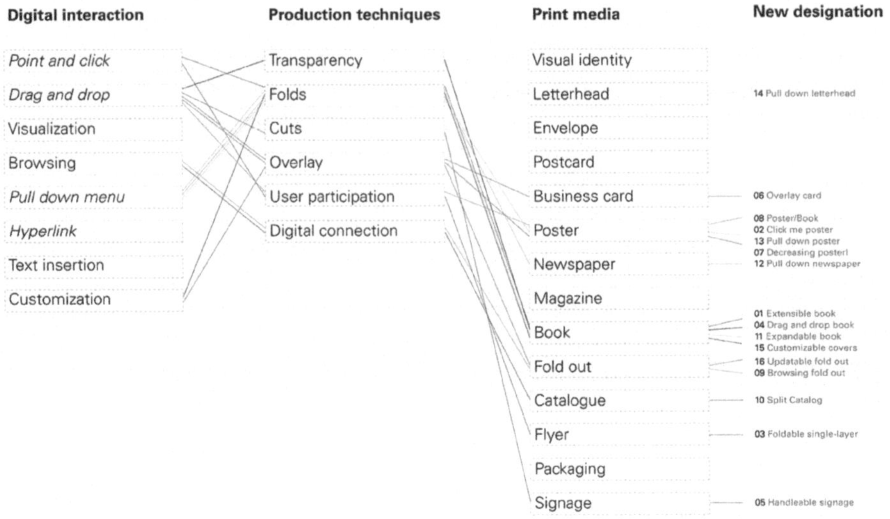

Fig. 1 Interaction categories and print design connections

By performing these connections, media on the last column should endure modifications when challenged by interaction categories. These modifications were translated into sketches and short descriptions, with the aim of demonstrating appropriateness of this new proposal for better user experience. We have doodled some ideas through early sketches, without any order. Figure 2 shows early sketches for some of these connections.

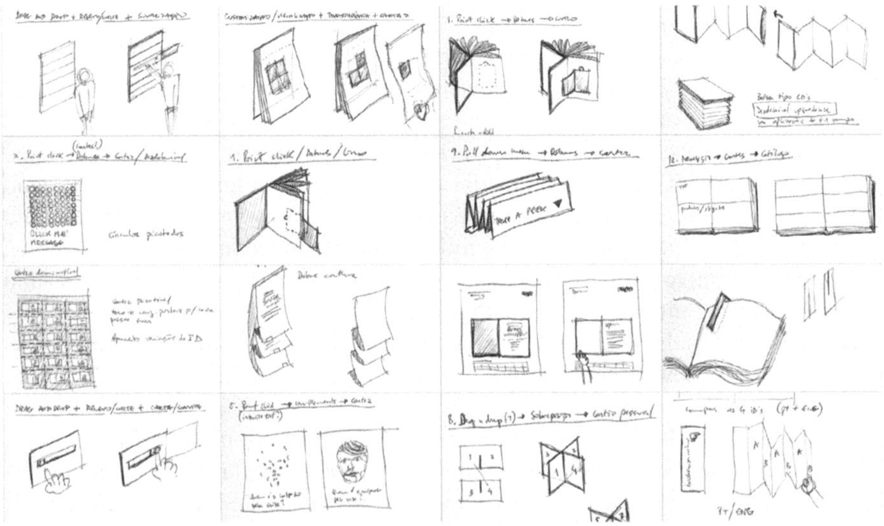

Fig. 2 Early sketches (selection)

These first sketches are brief representations where we tried to advance, through fast and schematic drawing, a visual explanation of the intention of the object, by applying categories of interaction. Each sketch represents an idea, which is the result of an established connection between categories of interaction and which may be developed as a printed medium with interaction.

Poster/Book
Following on early sketches, one of those ideas was further detailed through a diagram, devising a model, which could be applied to a real working context. Such model became the basis of our design project to test interaction features in a print design.

The design project consisted of a poster/small book. As a poster, it is printed front and back; the front is ready to be posted anywhere and the back will serve as book pages. It has two folds which allow it to be placed in an envelope and sent by postal service. In both fold lines, dividing the paper sheet in half, height and width, two perforated lines are produced. Each part divided by these lines can be detachable. Numbers in small type size were placed in the poster to serve as indicators. Users can follow these numbers and sort all detachable parts, fold them all in half, and thus form a small 16-page book. The diagram can be seen in Fig. 3.

This dual function keeps two levels of information. A more immediate and direct one for exhibit and a more detailed reading in book form. Transformation is done by users, which modify the object to access a second level of information.

4.2 Action

The idea for this object was applied to a poster for a congress in Lisbon, Portugal, named 'Books with a view'. This international congress was dedicated to architecture and culture of the eighteenth century and celebrated 300 years of the birth of Portuguese architect and urban planner Eugénio dos Santos. The congress was held in Lisbon between November 23th and 25th, at the Calouste Gulbenkian Foundation.

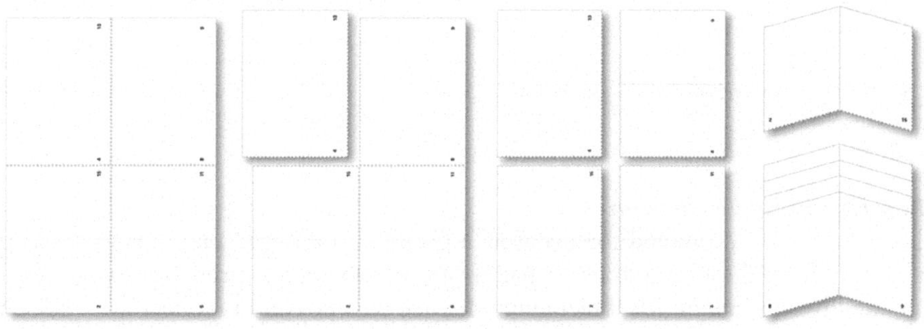

Fig. 3 Poster/small book diagram

The poster for the congress gave the possibility to test previous results in a real context and became a material evidence of what we intended to prove, as well as the carried-out research. We have tried to adapt previously generated ideas to the real conditions of the situation at hand. Therefore, all information which is possible to raise through this project cannot be dissociated from its specific material context.

Prototypes

We initially made low fidelity prototypes to understand what would be the best placement for information. It would also guide us in establishing numeral indications and order of contents, between poster and book design.

We determined an A2 size (420 × 594 mm; c. 16.5 × 23.4 in.) for this poster, so that the final book could be A5 (148 × 210 mm; c. 5.8 × 8.3 in.). We prepared an Adobe Illustrator file with A2 measures, where a division between four zones was done, each corresponding to an A5. We also prepared simultaneously an Adobe InDesign file, with A5 measures, where we developed all book pages. This was a decisive step to see how the poster could be presented and afterwards divided as a book. It also allowed to understand how book pages could be designed in line with the poster or, alternatively, how could they influence poster layout.

The chosen poster layout was overlapped by the set of numerical indications which corresponded to book pages and by the lines for production of folds and puncture. Indication of page numbers was adjusted on the poster, to follow book margins. A layout grid for book pages was placed over the poster as a temporary aid, to determine the exact place for numbers to appear on the poster. The front of the poster was also imported into the Adobe InDesign file and placed in the corresponding book pages to visualize the result.

Both front and back of the poster were prepared as preliminary version to produce a high-fidelity prototype. The poster was printed on colored paper, to approach the remaining objects produced for the congress, without requiring a significant amount of ink. However, during layout stages we used a background color in both files, to replicate the final paper color effect.

After printed, the prototype was folded and its behavior observed, as if it were made available by postal service. It was also observed its behavior when divided and overlapped to form the book.

4.3 Observation

Main Results and Reflection

This poster/book promoted the event as a poster and allowed participants to keep the main information as a book. It had a run of 200 copies, printed in two colors, front and back through offset technique. It gave us a possibility to test the research in a real context as it also constitutes a material evidence which intends to prove the carried-out study.

Final produced object includes attributes related to interaction. It addresses alternating messages, by allowing users to make a request with the purpose of modifying the object. From an initial message (poster), the object replies with another message, related to the first one (making of the book). It also presents users with the possibility of action, for which there will be a reaction. In this way, users are defining the following behavior for the object.

This practice-based research method originates a progressive level of fidelity which can be observed through various results: schemes, sketches, high definition prototypes and actual production of the object, which becomes an integral part of material culture. The obtained group of evidences contributes in two ways to human experience. First, the produced object was used by a group of people. Secondly, both interaction categories scheme and consequent development of ideas and proposals, contribute to the study and practice of print design.

By observing the prototypes, we understood the first layout did not favor the book, where certain pages seemed devoid of content. However, our primary concern, to maintain a layout which allowed interaction, was obtained. Number sequence was easy to understand and transformation from poster to book did not present difficulties.

What we obtained during this step showed us a process through which two objects were simultaneously conceived and developed, with an increased level of difficulty – the relationship between them. Therefore, making a print media convertible by its users adds some effort to the design process and implies a careful consideration on how to prepare it.

Changes and Improvements

Changes were made to the congress poster/book to improve communication and smooth transition from poster to book. We created a pattern with symbols extracted from the event's visual identity which would assist in a better distribution of information. This option was also relevant for book page layout. Thus, a large portion of the poster area was filled with an incomplete pattern. Small drawings made from a mouse cursor used in digital artefacts were added, with intention of directing the eye. Final version is shown in Fig. 4.

Reflection on the Process

According to Pedgley and Wormald (2007: 80) design projects should reflect "two specific indicators of excellence: 'completeness' and 'esteem'" to be considered as research. The first one refers to production of high quality results in design. These results can be artifacts of pre-production, working prototypes or models. Essentially, material to be presented should not allow doubt on previous research. The second indicator concerns the meaning such results may have outside the academic institution in which they were carried out.

This practice-based research method originated a progressive level of fidelity, which can be observed through various results: schematic executions, low and high fidelity prototypes and actual production of the object, which becomes an integral part of material culture and available for several users (Fig. 5). We may also understand

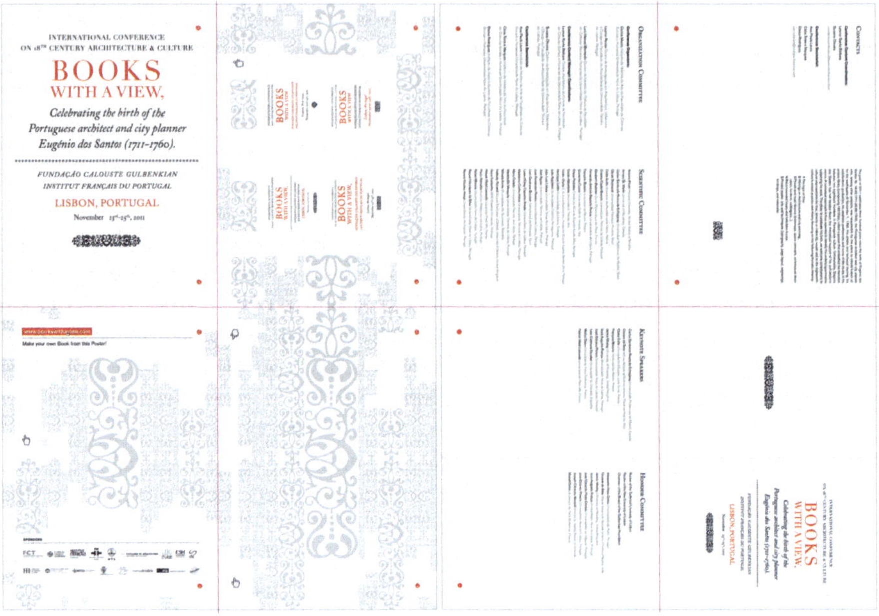

Fig. 4 Final printed poster (front and back)

it as part of a professional practice, carried out by the researcher himself and subject to acceptance of a designer-client relationship.

Scrivener (2000) proposes standards for projects obtained through creative practice, when integrating research. Most significantly, artifacts should be produced and be original in their context, contributing to human experience. They should be made as a response to questions, concerns and interests, expressing them somehow.

When comparing our project results with these standards, we realize there was a significant production of artifacts, all of them original and placed within a specific context. This context refers to the practice of print design and implications that interaction may have in such area.

All materials resulting from this method, manifested at various levels, the issues raised by our research. Indeed, interaction categories scheme and consequent development of ideas and proposals for print media, becomes a contribution to the study and practice of graphic design.

The method is also oriented by a specificity, in which various elements that compose it are related to the possibility of obtaining interaction in printed matter. This starting point has made work an intentional task, through which it sought to generate new knowledge.

Even the results of project development, apart from the produced object, present themselves as proposals of possible application and the method can therefore be generalized to other situations.

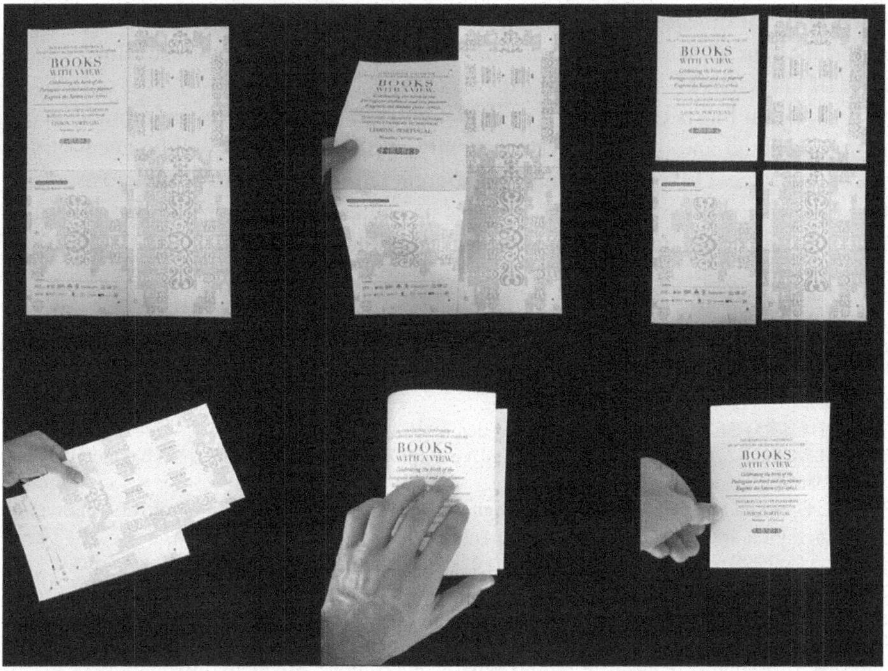

Fig. 5 From poster to a book

5 Conclusions

This practice-based research method is simultaneously a statement of restlessness and a devising of proposals. Both can be understood at two levels: interference in a work process, which will precede a project practice and the possible results from this change.

The conducted work intends to, at first, through a set of generated ideas, add solutions to graphic designers when print media is designed. Secondly, whether these options can be used, incorporated into a professional practice or disseminated in material existence, they show modification in user experience.

Some connections established through the interaction categories scheme originated ideas which could be applied in other print media. Despite the options for their presentation and explanation, each idea must be interpreted as flexible in the face of specific situations.

The presented project development had explicit and defined purposes. It was intended to implement and control interaction in media obtained through printing technologies. Procedures for this to happen were described and explained. Also, the remaining generated ideas and their progressive definition can be applied, replicated or modified. This group represents a work basis, which obtained results according to the outlined objectives. It is transferable knowledge and therefore perhaps more important than what was produced, since the object is just a demonstration of existence of knowledge.

Although it has been applied to a real context, it is not yet proven interaction can be implemented and obtained systematically in print design. It will be necessary to repeat the process several times and in different circumstances to be able to ascertain this. Likewise, after repeating the process or gathering a greater number of results, it will also be necessary to test and collect information about their use.

References

Archer B (1995) The nature of research. Co-des Interdisc J Des 6–13

Bloor M, Wood F (2006) Keywords in qualitative methods: a vocabulary of research concepts. Sage Publications, London

Bolter J, Gromala D (2003) Text rain: the digital experience. In: Windows and mirrors: interaction design, digital art, and the myth of transparency. MIT Press, Cambridge, MA

Bonsiepe G (1999) Interface: an approach to design. Jan van Eyck Akademie, Maastricht

Cooper A, Reimann R, Cronin D (2007) About face 3: the essentials of interaction design. Wiley Publishing, Indianápolis

Davis M (2008) Why do we need doctoral study in design? Int J Des 2(3):71–79

Dubberly H, Haque U, Pangaro P (2009) What is interaction? Are there different types? Access http://www.dubberly.com/articles/what-is-interaction.html

Haslem N (2011) Not an academic type: what do communication designers know about research? Iridescent Icograda J Des Res 1(1):178–191

Johnson M (2002) Problem solved. Phaidon Press, London

Moggridge B (2007) Designing interactions. The MIT Press, Cambridge

Neves M (2013) Printed interactivity towards a new understanding of graphic design. Iridescent Icograda J Des Res 2(2):22–37

Neves M (2016a) Enhancing user experience in graphic design: a study in (unusual) interaction. In: Rebelo F, Soares M (eds) Advances in ergonomics in design. Advances in intelligent systems and computing, vol 485, pp 801–810

Neves M (2016b) Printed matter as an interactive system. In: Marcus A (ed) Design, user experience, and usability: novel user experiences DUXU 2016, part II. LNCS, vol 9747, pp 625–637

Neves M (2017) The future of print design relies on interaction. Des J 20(Suppl 1):S4087–S4100

Neves M, Caixeiro I (2019) Interactive printed book: a design experience. In: Monteiro MR, Kong MS, Neto MJ (eds) Modernity, frontiers and revolutions: proceedings of the 4th international multidisciplinary congress (PHI 2018) 3–6 Oct 2018. CRC Press, São Miguel, pp 209–214

Nimkulrat N (2007) The role of documentation in practice-led research. J Res Pract 3(1)

Pedgley O, Wormald P (2007) Integration of design projects within a Ph.D. design issues 23(3):70–85

Preece J, Rogers Y, Sharp H (2002) Interaction design: beyond human-computer interaction. Wiley, New York

Redström J (2005) Towards user design? On the shift from object to user as the subject of design. Des Stud 27:123–139

Rust C, Mottram J, Till J (2007) Review of practice-led research in art, design & architecture. UK, Arts and Humanities Research Council. Access http://shura.shu.ac.uk/7596/1/PacticeledReviewNov07.pdf

Saffer D (2007) Designing for interaction: creating smart applications and clever devices, 2nd edn. New Riders, Berkeley
Scrivener S (2000) Reflection in and on action and practice in creative-production doctoral projects in art and design. Working papers in art and design 1

Gravitim APP and the Contribution of Digital Media in the Process of Pregnancy

Nuno Martins and Tânia Araújo

Abstract This paper aims to demonstrate the importance that digital media can have in the support to the process of pregnancy, namely in the contribution for a more informed relationship between the pregnant woman and the healthcare professional. This research consists of the development of a mobile application of pre-natal care with the support of the Health Local Unit of Alto Minho (ULSAM), where the app is being studied in a joint work with doctors and patients. With the development of our mobile application the aim is to create a more effective and secure pre-natal care that contributes to an improvement of the levels of transparency and effectiveness in health care services. In recent years, there are several computer systems that have been developed and implemented in the health area in Portugal. Portugal is first in the digital ranking of Health, having been distinguished by the European Observatory of Health Systems as an example of good practices. The development of digital platforms is a strong investment of the National Health System (NHS) for the improvement of the quality of its services (Ministério da Saúde (SNS) in Retrato da Saúde, Portugal, Lisboa, 2018). This paper begins with a contextualization of this problem; a presentation of a critical analysis about the main digital platforms existent in the market; and the proposal of the application under development.

Keywords UI/UX design · Digital design · Prenatal care · Pregnant health bulletin · Digital application

N. Martins (✉)
IPCA and ID+, Vila Frescaínha São Martinho, 4750-810 Barcelos, Portugal
e-mail: nmartins@ipca.pt

T. Araújo
IPCA, Vila Frescaínha São Martinho, 4750-810 Barcelos, Portugal
e-mail: tania.araujobcl@gmail.com

D. Raposo et al. (eds.), *Perspective on Design*,
Springer Series in Design and Innovation 1,
https://doi.org/10.1007/978-3-030-32415-5_14

1 Introduction

The main objective of this research was to contribute, through Design and digital media, to a more informed and monitored prenatal diagnosis and surveillance process.

The solution consisted in the development of a mobile application (app), of public utility, in the area of Obstetrics, in order to contribute to a better communication between the health professional and the pregnant woman. The objective was to design an interface that represented, in a structured and dynamic way, the results intrinsic to prenatal surveillance.

This app intends to be a complementary means of support for the pregnant woman and the health professional, without the pretension of replacing the habitual and necessary direct and personal relationship between the health professional and the pregnant woman.

This project also aimed to respond to a set of specific objectives:

1. **Contribute to social equity, in terms of literacy and access to diagnosis**.
   To raise awareness of the importance of a Digital Culture in the field of Obstetrics.
2. **Centralize and monitor prenatal surveillance information in the mobile application**.
   Through the fieldwork carried out (follow-up of consultations with pregnant women), we have seen the use of various physical supports, from the Pregnancy Health Bulletin (BSG), the Diabetes Bulletin, the Blood Pressure Registry, and the Consultation and Exams.
   It is important to develop a means of communication that facilitates the accomplishment of these registers, between the pregnant and the clinician, which aggregates all the information, replacing the physical supports.
3. **Provide a set of interactive tools that help the expectant and health professional to get better results**.
   All data intrinsic to the state of health of the pregnant woman and the fetus are recorded in the bulletins and in the software that the health professionals operate. However, given the diversity of software used by health professionals (*SClínico, OBSCare, Astraia* e *PEM*) ()the interpretation of data becomes extensive and more complex, leading to a greater delay in understanding them.
4. **Sensitize pregnant women for greater pro-activity**.
   Depending on the clinical cases, not all pregnant women express an interest and full responsibility in the monitoring of pregnancy. Through the mobile application it is possible to sensitize the personal and collective interest on the importance and benefits of an informed pregnancy process.

2 Methods and Techniques Adopted

This study was developed at the Health Local Unit of Alto Minho (ULSAM), in Viana do Castelo. We began to follow the prenatal diagnosis consultations, in order to observe, analyze and understand how the process of prenatal surveillance is developed.

We also developed a survey (90 women) to obtain a generic idea about mobile applications in pregnancy. After the data were collected, the results were analyzed, managing the research objectives and creating a plan.

Later, we developed a moodboard, in order to organize and direct the information architecture of the application. Information architecture is based on competitive analysis; creating people and using cases/flowcharts; and the development of wireframes and workflows, for later high-fidelity prototyping.

In the development of interface design (UI), there were several methodologies associated, namely the identity of the application, the visual design of the workflows and the possible successes and again the prototyping.

To better understand the user's interaction with the application, usability tests have been developed with the target audience (a significant number of people, about 10 pregnant women and 5 health professionals). We used specific metrics to verify and evaluate the usability tests. In this way, we intend to improve the visual experience and the process of conception and development of the application.

At present, the application is being tested as a pilot project, in partnership with ULSAM, and we already count with good results. All the users of this sample, consider the application very important in supporting the process of diagnosis and prenatal surveillance.

After application development, data collection, review and feedback of usability testing, the goal is to finalize the application. And, later, to aspire the application to be implemented throughout the Portuguese National Health Service (NHS), both in the public and in the private system.

3 Reference Mobile Applications

For the design of our App was also important to ascertain the national and international applications in the market. In this chapter we summarize our review of the top Apps: Nove Meses (2015), BabyCenter (2017), A Minha Barriga (2017), Babynote (2017) and *Gravidez+* (2017).

We performed an exploratory analysis by collecting data from these applications, such as: application designation, popularity, average rating, price, features, language and technical specifications (Table 1).

Regarding functionality, the application that meets almost all criteria is *Gravidez+*. The *Babynote* is the application with less functionality.

Table 1 Comparison of existing applications

FUNCTIONALITIES	NOVE MESES	BABYCENTER	A M. BARRIGA	BABYNOTE	GRAVIDEZ+	Total app
Profile		X	X	X	X	2/5
Gestational age	X	X	X	X	X	5/5
Personal diary	X	X			X	3/5
History / Prenatal surveillance records						N
History / appointment consultations	X		X		X	3/5
History of analyzes, ultrasounds, vaccines			X		X	2/5
Weight index	X	X	X	X	X	5/5
Glycemic indexes (diabetes)						N
Blood Pressure indexes (hypertension)						N
Calendar with alerts	X				X	2/5
Multi-channel alerts						N
Monitoring with doctor						N
Information guide / tips / advice	X	X	X		X	4/5
Prenatal surveillance guide			X			1/5
Album / gallery images / videos	X	X	X		X	4/5
Share content (email, social networks)		X		X	X	2/5
SPECIFICATIONS	**NOVE MESES**	**BABYCENTER**	**A M. BARRIGA**	**BABYNOTE**	**GRAVIDEZ+**	**Total app**
Native Android	X	X	X	X	X	5/5
Native IOS	X	X	X	X	X	5/5
Web application/RWD	X	X				2/5
Smartwatch				X		1/5
Popularity	+ 100.000	+ 10 M	+ 1000	+2500 (2016)	+ 17 M	
Average rating	3,8	4,7	4	4,1	4,5	
Price	free	free	free	free	free + v. pro	
Languages	PT; ENG.	PT Brazil.+*	PT; ENG.	EING. +**	PT Brazil. +***	
Product advertising			X			1/5
Promotor	BIAL		BARRAL			2/5

N - None
* SPA; ENG; GERMAN; FR; ARABIC ** GERMAN; POLISH *** SPA; ENG; GERMAN; DUTCH; ITALIAN; RUSSIAN; CHINESE

None of the applications analyzed presents the main prenatal surveillance data, such as blood pressure registers/indices, growth curves, glycemic indexes/indices, multichannel alerts (in the app, by email, by SMS), and such as monitoring the diagnosis in real time with the health professional.

Only one of the five applications analyzed, *A Minha Barriga*, presents the prenatal surveillance guide. This guide consists of informing the pregnant woman about the timing of the respective laboratory tests, ultrasounds and specific analyzes. However, high-risk pregnant women (hypertensive and diabetic) have no indication in the guide about the exams or analyzes specific to their own clinical condition. The guide presents only a general schedule of prenatal surveillance.

All the applications have the gestational age in which the pregnant woman is, and almost all of them allow to create a profile of the user (the only one that does not allow is the *Nove Meses*). Other data intrinsic to prenatal surveillance that are not included in the applications analyzed are: personal history, place of surveillance and health professional, as well as the recording of clinical and analytical data (blood group, etc.).

Only two of the five applications analyzed make it possible to consult and verify the history of exams, ultrasounds and vaccines.

About technical specifications, all applications work on Android and IOS, and only *Babynote* on smartwatch. All apps are free and have the option of the Portuguese language.

The app that has the highest number of downloads is *Gravidez+*, with more than 17 million; followed by *Babycenter* with more than 10 million; the *Nove Meses* with more than 100 thousand; *Babynote* with more than 2500; and, finally, *A Minha Barriga* with more than 1000.

4 The Problem

After analyzing the reference mobile applications, we verified some problems and lack of data intrinsic to the evaluation of pre-natal care. In this way, when analyzing the problems that a product can present, we must answer the question "Why are we developing this product?", Simultaneously with the definition of the target audience "Who are we developing?" And what is the problem "What problems are there?", to find the assertive solution. It is the answer to these questions that will help measure the success of our application:

1. **Why are we developing this product?**
   The development of a digital solution contributes to a more informed and monitored pregnancy process, improving physician-to-pregnant communication, fostering digital culture, centralizing prenatal care information, and providing a set of interactive tools by form to obtain results more effectively, and by sensitizing pregnant women to greater pro-activity.
2. **Who are we developing for?**
   The target audience of this app are mostly female people, framed in low, middle and/or upper cadres, as well as health professionals (obstetricians, nurses, gynecologists).
3. **What problems exist?**
   This study shows a diversity of physical supports essential to the pregnant woman and the clinician, from the BSG, registration of consultations and examinations, the Diabetes Bulletin and the Hypertension Bulletin. Whenever the pregnant woman travels to the National Health Service (health units or hospital) or to the private service (clinics), she must be accompanied by all these recording media. Although the clinician registers all the information in the computer programs, the most relevant information is always present in the BSG (paper format). However, if the pregnant woman forgets the BSG (or another bulletin), the information risks losing.

Thus, the development of a mobile application with adequate monitoring of pregnancy is an advantage in the clinical context, in the context of the interaction between pregnant and health professionals and self-management of pregnancy by the pregnant herself, especially in pregnant women at risk, and may prevent future complications.

5 The App Gravitim (Mobile Application)

5.1 *Value Offer*

The success of a product is primarily to meet the basic needs of people, before meeting the needs of the highest level.

As the authors of the book "Universal Principles of Design" (Lidwell 2010), the principle of the hierarchy of needs specifies that a product must first respond to lower-level needs (e.g. functionality) and then level (e.g. creativity).

According to the authors, based on the Maslow Pyramid, the five major levels of needs are:

1. The Functionality: referring to the analysis of the most basic requirements of design (structuring);
2. Trust: stable and consistent performance lends credibility to the user;
3. Usability: be easy and effective for the user;
4. Competence: empowering users to do better than before;
5. Creativity: it is the level where all needs are met and users begin to interact in an innovative way.

Developing a differentiating product that responds to the needs and objectives of different users requires that the key characteristics, needs and frustrations of a specific segment are observed, resulting in a solution with differentiated characteristics, benefits and that leads the user to an emotional involvement in the digital experience.

5.2 *Personas Model*

To understand and direct the purpose of the application, we use the Personas model, which consists of archetypes based on behavioral patterns, which we observe during the fieldwork conducted at ULSAM (follow-up of prenatal care visits) (Fig. 1).

After verifying and analyzing the personas, we directed the development of the application to two fundamental people: pregnant hypertensive (Laura) and pregnant diabetic (Manuela).

SÓNIA	LAURA	ÂNGELA	ANA	MANUELA
DEMOGRAPHICS: 22 years higher education 17 weeks	DEMOGRAPHICS: 33 years licenciada 36 weeks	DEMOGRAPHICS: 40 years obstetrician doctor 2 children	DADOS DEMOGRÁFICOS: 54 years nurse 3 children	DADOS DEMOGRÁFICOS: 39 anos administrative 29 weeks
KEY CHARACTERISTICS: - gestation period - schedule of exams and ultrasounds - 1st pregnancy	KEY CHARACTERISTICS: - hypertensive - without alerts - uncontrolled - 2nd pregnancy	KEY CHARACTERISTICS: - glycemic values - growth curves - relevant peaks	KEY CHARACTERISTICS: - weight - height - blood pressure	KEY CHARACTERISTICS: - hypertensive - diabetic - 3rd pregnancy
USER CASE: Sónia does't know gestation is found, nor the periods required for the performance of certain exams/ultrasounds.	USER CASE: She is considered to be pregnant with risk because it's hypertensive need to control daily values of hypertension and blood pressure.	USER CASE: Dr. Angela needs evaluate the different peaks of hypertension and diabetes pregnancy and the curves of fetal growth.	USER CASE: Nurse Ana must introduce the indices of weight, height and pressure when the pregnant woman have a medical appointment.	USER CASE: Because she's diabetic, Manuela has to monitor and introduce mandatory every day, in different times of day, the values of glucose.
GOALS: - Daily Pregnancy Guide - history and exam alerts and ultrasounds	GOALS: - dashboard with historical values - control of values	GOALS: - dashboard with values indices - dashboard with history of growth curves	GOALS: - clinical history - control of values	GOALS: - dashboard with clinical history - control of glycemic,hypertension and blood pressure values
FRUSTRATIONS: - doubts about pregnancy - not knowing the schedule key procedures in the prenatal surveillance	FRUSTRATIONS: - have to enter the values manually in the support bulletin - have no alerts to help you control hypertension	FRUSTRATIONS: - introduction of values in different software and in BSG - calculate the values - possible loss of information	FRUSTRATIONS: - have to enter manually all values obtained - fill in different bulletins	FRUSTRATIONS: - have to enter the values manually in the support bulletin - have no alerts to help you control diabetes

Fig. 1 Personas

5.3 *Information Architecture*

After analyzing the competition and the development of the archetypes (personas), we proceed to the elaboration of the information architecture process, in order to determine objectives and organizational stages in the development of the application.

Based on the fieldwork carried out at ULSAM, together with the considerations proposed by the clinical team, for the development of this research, we carried out the survey and analysis of the context, users and their content, based on the diagrams of Louis Rosenfeld and Peter Morville (2007).

We started by schematizing a flowchart, defining the main actions and contents necessary for the user.

After the development of several studies, we came to a final version of the flowchart. For the sake of standardization, ease and monitoring, we decided that registration in the application will be through the number of users, based on access to the Citizen area, of the NHS. In this way, it is possible to migrate the user information available in the SNS to the application. It is also possible to enter new data, completing the fields that are still to be filled, regulating the profile of the user. In this flowchart, the page splash and registration is only for the first use of the application, in later uses the user is directly directed to the dashboard.

5.4 Wireframes

Based on the flowchart previously developed, we developed the wireframes (white board sketches) in order to synthesize and design the functionalities, the flows and the content of the application. Subsequently, we proceeded to the wireframes scheme in the Sketch software.

We start by designing the user input into the application, through the splash page and the "login" area. The splash page is the first screen that the user views, identifying the entity where the pregnant woman is being watched (in this case we chose ULSAM). The user is cleared of the content of the application by means of a short message. The entry in the application is performed through the "START" button, as we can see in Fig. 2. After clicking the "START" button, a new window will open where the user will be able to login, using the user's number and password. The process is the same as the way the user accesses the SNS through the "Citizen's Area" platform. If the user don't already have the registration in the "Citizen's Area" platform, the user can also automatically register by the "REGISTER" link.

After login, the user profile screen opens with a welcome message. In this area of the app, the user can insert a user photo, and put various information, such as: personal data; surveillance data; current gestation and personal history, etc.

After checking and entering data into the profile, the user is directed to the dashboard. In this dashboard is represented the current gestation of the pregnant (number of weeks, days and the respective quarter) and a graphic icon that will vary the visual configuration throughout the gestation. According to the profile of the pregnant woman (normal, diabetic or hypertensive), the dashboard itself will present the menus intrinsic to the surveillance of the pregnant woman herself. If you

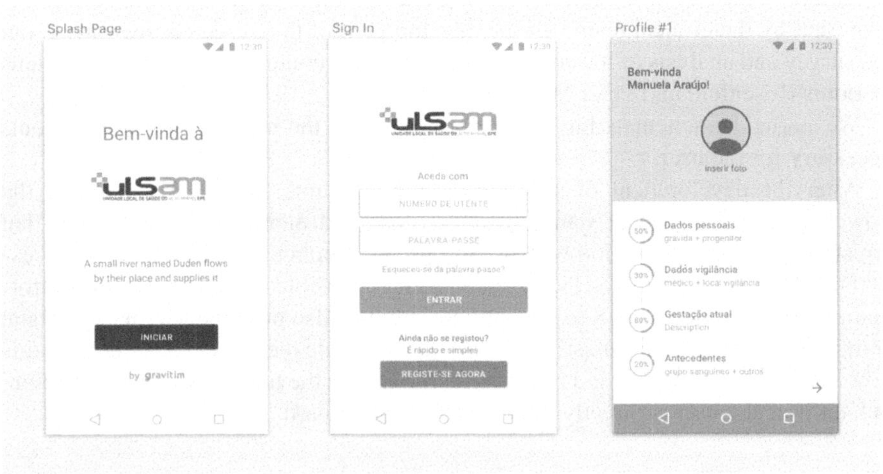

Fig. 2 Wireframe

are a normal pregnant, the menus represented will be: weight; calendar; consultations/examinations. If you are a diabetic pregnant, the menus represented will be: glicemias; Weight; calendar and consultations/examinations. If you are a hypertensive pregnant, the menus represented will be: blood pressure; Weight; calendar and consultations/examinations.

5.5 *Workflows*

For the development of workflows we use Sketch software. We started by identifying two personas: the diabetic pregnant woman and the pregnant hypertensive woman, and then we outline and direct the possible actions and thoughts of the user through the workflows.

In the diabetic pregnant person, one of the frustrations was to have to manually introduce, at different stages of the day, the glycemic values in the support bulletin (physical support). Another pertinent question was that the pregnant woman did not have glycemic index/peaks alerts, since the bulletin does not allow visual and monitored evaluation.

In this sense, as we can see in the workflow of the glycemia (Fig. 3), the purpose of this application is to allow the introduction of the values, in different stages of the day, by the pregnant herself. It also allows the issuing of alerts for the pregnant woman and the health professional.

Another important feature of the application is the daily, weekly and monthly visualization of different glycemic indexes, allowing a more effective and regular diagnosis.

In the hypertensive pregnant persona, we verified in the field work that pregnant hypertensives record the values of blood pressure in small personal notes or notepad. And it is only when the pregnant women go to the clinic, that they inform the doctor

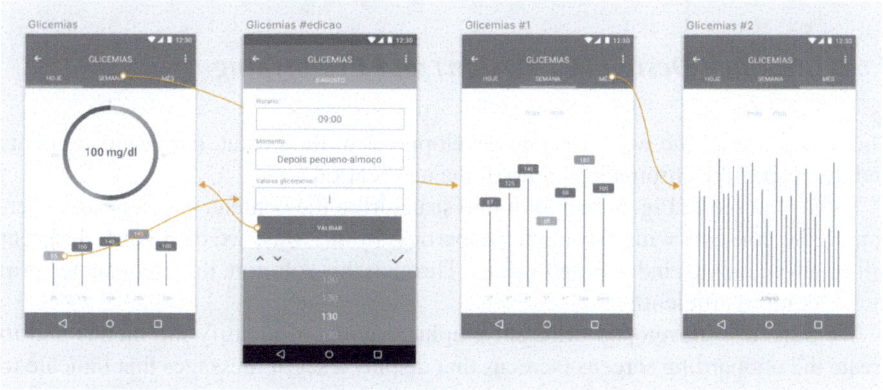

Fig. 3 Glicemias workflow

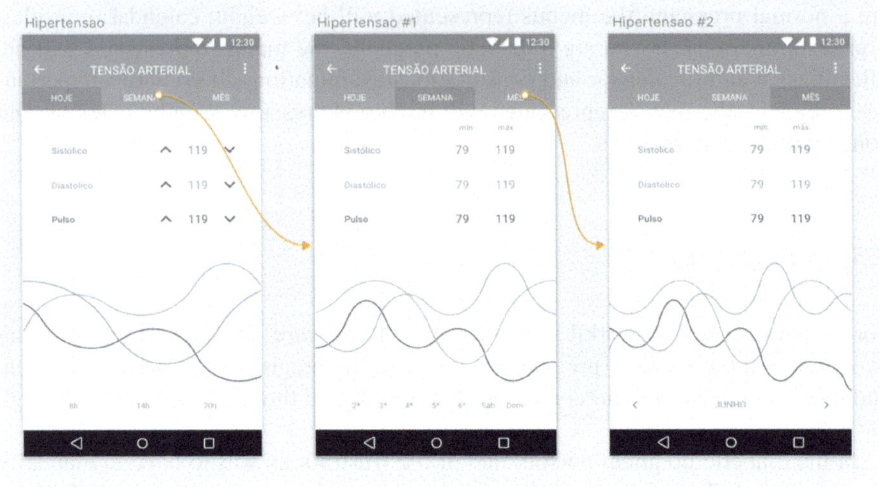

Fig. 4 Workflow arterial tension

of the values recorded during the absence of surveillance. The problem is that they do not always make the records or indicate median values, without the precise values of the indexes.

Through the application, they can record the values more effectively, at different times of the day. And they can also have the daily, weekly and monthly record history through visual and chromatic compositions—which helps a good reading of the information (Fig. 4).

In this way, the application results in a dynamic prenatal surveillance bulletin. And it has the particularity of allowing a more preventive and participatory monitoring by the pregnant woman and the health professional. The app also centralizes the information, obtaining more effective results.

5.6 Interface Design (UI Design) and Prototyping

The next phase of the work was the development of the layout, the visual elements and the respective chromatic palette of the application.

As we can see in Fig. 5, in addition to structuring the content, we chose to differentiate the contents with chromatic properties. In this way, we categorize different information, menus, indexes and values. Through this solution, the user can perform the tasks more efficiently.

We have also developed different graphical icons to identify the menus and to create the onboarding screens (screens that display a set of messages that indicate to the user the resources available in the application).

Fig. 5 Different main screens of the application

To implement the prototype, we used the digital platform Invision, a free platform that allows the development of prototypes of digital products, with the particularity of transferring and updating the wireframes developed automatically in the Sketch software. Through the prototype, we have an application close to reality, allowing perform usability tests with users.

5.7 Usability Tests

There are several methods for studying and evaluating usability. The most commonly used is user testing. According to J. Nielsen (Cooper 2007), gathering a group of 5 users, who may or may not have knowledge about the application, it is possible to evaluate 85% of usability problems. For Robert Virzi, with a group of 10 participants it is possible to detect 90% of the problems. There are more usability evaluation techniques, namely, prospective techniques, predictive techniques and objective techniques.

For app usability testing, we focus on two types of techniques: prospective techniques and objective techniques.

The usability tests were observed by a researcher, UX/UI Designer, at the Gynecology and Obstetrics Service of the Local Health Unit of the Alto Minho (ULSAM), in Viana do Castelo, through direct contact with the users. For the sample, 10 pregnant women (4 hypertensive pregnant women, 3 diabetic women and 3 low risk women), aged between 23 and 41 years, and 5 specialists in prenatal diagnosis were selected.

During this stage, the researcher's role was to observe the reaction of the users, in the interaction with the interface. The observation was made through an online digital system. The researcher asked the users to perform certain tasks and indicate the difficulties, doubts and suggestions for improvement.

After the activation of the prototype, we proceeded to compile a list with different tasks, corresponding to different user profiles:

Tasks list	Number of pregnant women
(a) Validate your personal data	1
(b) Enter the glycemic values of today, from the first moment of the day	2
(c) Complete the task of the day at fault	1
(d) Check the hypertension values for the month of July	3
(e) Validate your vaccination report	2
(f) Consult your surveillance site and its assistants	1

From the tasks proposed, the tasks most tested were (b), (d) and (e), with pregnant women at different gestational ages, different health conditions and different age groups.

According to the observation made to the participants, all were able to carry out the proposed tasks without any restriction. There were only different times in the accomplishment of the task, due to the different levels of digital literacy of the participants.

According to ISO 9241-11, one of the parameters to measure usability is user satisfaction. To evaluate the degree of satisfaction of the participants, we developed a questionnaire, QUIS, with the objective of evaluating the user's interpretation and satisfaction with the application prototype.

In this questionnaire, users were asked about issues such as ease of interaction with the application, the quality of the data generated and presented, the usefulness of the application in the diagnosis, the level of digital literacy and the overall appreciation of the application.

The questionnaire was divided into five parts:

(1) ease of use (related to user interaction);
(2) data quality (related to the interface and interpretation of data);
(3) utility (related to the purpose of the application);
(4) digital literacy (related to the level of knowledge);
(5) overall appraisal of the application (the degree of satisfaction with the application).

By analyzing the answers to the "Ease of Use" questions, 87% of users consider the application to be clear and comprehensible; 93%, which does not require a high mental effort given its simplicity and is intuitive and easy to use; and, 100%, which presents very pleasant visual compositions.

Regarding the issues of "Data Quality", users were unanimous in considering that the quality of understanding, interpretation and management of the data generated is very good.

In the questions regarding the usefulness of the app, all users considered the application an important and useful tool in the management and monitoring of pregnancy, and in the significant improvement of prenatal diagnosis.

About "Digital Literacy", we obtained different responses: 40% of users mentioned that they already use an application about pregnancy; 20% who use more than one; 87% using applications, but from other areas; and 60% replied that they did not use any pregnancy app. Regarding the last part, "Global Assessment", the users were unanimous, considering the application very beneficial.

6 Conclusions

To develop and implement this digital application, it was fundamental to think about the specific problems of the user and to study a solution through a multidisciplinary work. This project involved not only pregnant women, but also health professionals, those responsible for health institutions, designers and computer programmers.

The implementation of a digital prenatal surveillance bulletin aims to contribute to: greater social equity in terms of literacy and access to diagnosis; centralize and monitor prenatal surveillance information; to sensitize the pregnant woman to the self-management of pregnancy; improve understanding and communication of prenatal diagnosis; provide a set of interactive tools that help pregnant and health professionals effectively achieve results; and, finally, to prevent future pathologies and complications.

Due to the potential of this project—not only nationally but also internationally—we believe that the pilot project being developed with ULSAM clinical teams and users can be an important lever for its growth and expansion.

References

Babycenter.com (2017) APP Baby Center [online]. Novemeses.pt. Available at: https://www.babycenter.com/2_babycenter-app-tutorial_10419338.bc. Accessed 23 Oct 2017

Babynoteapp.com (2017) Babynote APP [online] Babynoteapp.com. Available at: https://babynoteapp.com. Accessed 23 Oct 2017

Cooper A, Reimann R, Croni D (2007) About face 3—the essentials of interaction design. Wiley Publishing, Inc., Indianapolis

Direção-Geral da Saúde (2011) Diagnóstico Pré-Natal. Pre-natal diagnosis. E-Konomista. Available at: http://www.saudereprodutiva.dgs.pt/diagnostico-pre-natal

Google Play Store (2017) APP A Minha Barriga [online]. play.google.com. Available at: https://play.google.com/store/apps/details?id=pt.angelini.barriga&hl=pt_PT. Accessed 23 Oct 2017
Health-and-parenting.com (2017) APP Gravidez+ [online]. Health-andparenting.com. Available at: http://www.health-and-parenting.com/pregnancy/pregnancy-app. Accessed 23 Oct 2017
Lidwell W, Holden K, Butler J (2010) Universal principles of design. Digital edition. Rockport Publishers, MA. 978-1-61058-065-6
Ministério da Saúde (SNS) (2018). Retrato da Saúde, Portugal (Portrait of Health, Portugal). Lisboa. ISBN 978-989-99480-1-3
Morville P, Rosenfeld L (2007) Information architecture for the world wide web. O'Reilly Media, Inc
Novemeses.pt (2015) APP Nove Meses [online]. Novemeses.pt. Available at: http://www.novemeses.pt/pt/aplicacao-9meses. Accessed 23 Oct 2017
Salgado C, Espirito Santo R, Valente S, Saldanha J (2017) Preenchimento do Boletim de saúde da Grávida: 18 anos de Evolução. Pregnant Health Bulletin Completion: 18 Years of Evolution. Ata Pediátrica Portuguesa. vol 48, pp 19–27. Departamento de Pediatria. Centro Académico de Medicina de Lisboa
Serviço Nacional de Saúde (2016) #Estratégia Nacional Ecossistemas de Informação de Saúde (ENESIS) 2020. National Strategy for Health Information Ecosystems. http://spms.min-saude.pt/2016/11/estrategia-nacional-ecossistema-informacao-saude-enesis-2020-aprovada-pelo-governo, 8 Nov 2016
SPMS.min-saude.pt (2017) SClínico Hospitalar [online]. Spms.min-saude.pt. Available at: http://spms.min-saude.pt/product/sclinicohospitalar. Accessed 30 Dec 2017
SPMS.min-saude.pt (2017) PEM—Prescrição Eletrónica Médica [online]. Spms.min-saude.pt. Available at: http://spms.min-saude.pt/product/pem. Accessed 30 Dec 2017
Virtualcare.pt (2017) AnesthCare, BreastCare, ObsCare [online]. Virtualcare.pt. Available at: http://virtualcare.pt/index.php/produtos. Accessed 30 Dec 2017

Communication Through Character Design: 'Inside Out' Case Study

António Manuel Rodrigues Ferreira, Pedro Mota Teixeira and Daniel da Cruz Brandão

Abstract According to Teixeira (A Representação Emocional da Personagem Virtual no Contexto da Animação Digital: do Cinema de Animação aos Jogos Digitais. University of Minho, 2013), in the animation field and regarding an animated character, there are seven non-verbal dimensional layers with communicative value. In this sense, a concise relationship between all layers is crucial to allow an unequivocal communication process between the character and the viewer. In this chapter we will be analyzing character design and two other related layers: facial and body expression. Previous studies have shown the significance of non-verbal expression in communication—which is the case—and audiovisual companies are focused on implementing universal symbols, to minimize ambiguity. So in that sense, this study consists on analyzing a paradigmatic example, which is 'Inside Out' by Disney/Pixar partnership, and specifically its main characters—Fear, Anger, Sadness, Joy and Disgust—,which are direct references to primary universal emotions.

Keywords Animation · Character · Design · Emotion · Gestures

1 Introduction

Character design is widely used by the animation industry as reinforcement on character's personification as well as on emotions' transmission. Although several authors

A. M. R. Ferreira (✉) · D. da C. Brandão
Centro de Estudos de Comunicação e Sociedade (Instituto de Ciências Sociais—Universidade do Minho), Campus de Gualtar, 4710-057 Braga, Portugal
e-mail: id7709@alunos.uminho.pt

D. da C. Brandão
e-mail: danielbrandao@ics.uminho.pt

P. M. Teixeira
ID+ Instituto de Investigação em Design, Media e Cultura (Grupo de Investigação: CAOS Comunicação, Arte, Objecto e Sinergia, Instituto Politécnico do Cávado e do Ave), Campus do IPCA, Lugar do Aldão, 4750–810 Vila Frescainha, Barcelos, Portugal
e-mail: pmteixeira@ipca.pt

D. Raposo et al. (eds.), *Perspective on Design*,
Springer Series in Design and Innovation 1,
https://doi.org/10.1007/978-3-030-32415-5_15

still inspire themselves on standard character types (villain, hero, antihero, child and grotesque), as well as in their peculiarities, the fact is that nowadays many of these standards are mixed or even questioned, making room for a randomness margin to expand the boundaries on characters' concept creativeness. In this sense, instead of getting a wide range of unwanted reactions by the audience, since the well-defined Disney revenues were implemented for decades, animated features are getting more and more spectators and fans through the years. The athletic, powerful character with high leveled alertness and resilience may not be the hero. The annoying and sometimes hateful character may not be the villain and it can even be loved by the audience. The cliché of a slender protagonist and a showy hero, in contrast to bad, ugly and rude villains continues to work, but it is no longer universal.

Additionally, the progressive investment on emotions' expressions through body and face, contributed to the communication process enrichment between the character and the viewer.

2 The Importance of Emotions

In an interview for the blog 'Character Design', De'Von Stubblefield (Sly 2012) emphasizes the importance of emotions' representation:

> I learned to ask what the character is and to be sincere when drawing emotions. The only way to be sincere is to draw from life experiences and to identify with what I'm drawing.

Les Pardew (2007, p. 1) has the same opinion, considering that in character animation the ability to express emotion in a convincing way is a task that requires understanding the phenomenon:

> Animation, like acting, is based on life. Life is the foundation, even if the animation is exaggerated. By first learning how to see emotion, you will then be better equipped to portray emotion in your animation. (…).

According to the same author, that is why the animator must give himself to recognize and be able to make great changes, sometimes very subtle, that mirrors certain emotions. For Ekman and Friesen (2003, p. 7) emotions are evidenced by the expression on the face:

> Emotions are shown primarily in the face, not in the body. The body instead shows how people are coping with emotion. There is no specific body movement pattern that always signals anger or fear, but there are facial patterns specific to each emotion.

In this context, the face represents the most important communication instrument of a character, defining his personality and mood.

Facial expression varies enormously from photorealistic or realistic aesthetics to American cartoons and other stylized ones.

McCloud (2006, p. 94) enumerates four visual languages that provide different readings, especially when expressing the characters' actions and emotions: realistic (simplified or stylized), exaggerated or caricatured and symbolic. All of them converge to the same concern: achieving a high degree of expressiveness, emotion and communication with the viewer.

3 The Nature of Emotions

Since *Descartes' Error* that Damásio (1994) argues that emotions are indispensable for the existence of all rational life. This makes us unique, and it is precisely by the emotional behavior that we distinguish ourselves from one another. Damásio also argues that the nature and intensity of emotional reactions depend not only on the brain but on its interaction with the body, and also that our intellectual and moral values are determined by our emotions. Thus, from his laboratory experiences, this author demonstrates that:

> Emotion is an integral part in the processes of reasoning and decision making, for the worst and for the best. (Damásio 1994, p. 61)

Turner (2000, p. 101) follows the same principle:

> The ability of humans to engage in complex decision-making processes does not eclipse, therefore, emotionality; sophisticated decision-making depends on the capacity of human beings to produce broad spectrums of emotional valence.

In an attempt to define an emotion, Keltner et al. (2014, p. 4) make the following statement:

> An emotion is a psychological state or process that mediates between our concerns (or goals) and events of our world.

We can control our expressions through a voluntary control over our facial movements. We can prevent others from understanding our true intentions and we can even send false signals with appropriate facial expressions (Eibl-Eibesfeldt 2017, p. 476). This is why the actor can express himself and rehearse various expressions of emotion. On one hand, if the human being can control their expressions, on the other hand, body posture and mood are less probable to be voluntarily controlled (Ekman 2003). It's precisely because of the human being's ability to recognize these differences, some incredibly subtle, that the animator is obligated to reconstruct these interactions in such a way that they are believable to the observer's eyes, who can look at a character with "true" underlying feelings and goals, instead of watching a simple set of moving pixels (Kundert-Gibbs and Kundert-Gibbs 2009, pp. 56–57). More than that, we create authentic mental bridges searching for facial expressions in abstracter, simplified drawings (McCloud 2006), or even in animals' representations.

> The universally uniform recognition of expression could develop independently and could be acquired anew by each individual, since the expressive movements reliably indicate specific moods of the individual. One argument that our understanding is innate is based on the fact that we also respond clearly to even though it is a very simple form of facial expressions,

> and in particular, it is an example of facial expressions. In fact, the eagle's face is a perfect model for an expression of heroic decisiveness, making it particularly suitable for a heraldic symbol, even though nothing in the eagle's behavior actually corresponds to behavior we consider to be heroic. The eagle looks the same way when it is frightened The camel, on the other hand, appears arrogant, because its demeanor causes the nostrils to be held high relative to the eyes, although this "snooty" looks like nothing with the animal's current mood state at any given time. But since we reject someone by sticking our nose in air, an intention of withdrawing from the other partner whom we dislike, we interpret the camel's disposition as unfriendly. Other animals are responded to more positively, because they appear to be smiling. (Eibl-Eibesfeldt 2017, p. 475)

4 Facial Expression

> Much of life and emotion of a digital character is read through the face, and while facial setup is a mix of motion and deformation systems, the issues intrinsic to this aspect of character development require specific attention. (O'Neill 2016: 4)

The face is one of the most important reference instruments on interpersonal communication (Eibl-Eibesfeldt 2017, p. 443). For O'Neill (2016, p. 3), the aliveness and emotional states of a character are generally read through facial expression and therefore all these aspects, about deformations, movement systems and facial expressions' structure, require special attention from the animator (McCloud, 2006, p. 92). Studies by CH Hjortsjo about facial modifications boosted by various types of emotional experiences motivated authors such as Paul Ekman and Wallace V. Friesen, whom in the mid-70 s set out to research facial expression and emotion, focusing on its universal context rather than cultural context (Kundert-Gibbs and Kundert-Gibbs 2009, pp. 56–57).

> We have conducted cross-cultural comparisons in Great depth, especially with regard to tribal societies, and congruence exists even in very fine details of facial expression. The lively expressions of man signals specific elemental behavioral tendencies (mood states) and needs, beyond cultural barriers and are thus applicable for all humans. (…) It is surprising how much facial movements coincide among the various cultures and races throughout the world, in view of the fact that the differentiation of the facial musculature shows considerable racial variation. (…) And yet the facial expressions of the various human races are so similar that we safely understand them cross-culturally (…). (ibidem)

Ekman and Friesen have discovered evidence that seven basic emotions are expressed similarly in most cultures (Matsumoto and Hwang 2011): happiness, sadness, fear, anger, surprise, disgust and contempt. These expressions may be conditioned by cultural variables, regulating dimensions such as intensity or repression of each expression (Keltner et al. 2014, pp. 75–76).

In this context, these facial expressions do not only represent internal biological self-reactions but also work as easily recognizable and identified signals that help social relations to be established (Eisner 2008, p. 57). This is the reason why facial expressions are triggered most of the time in the presence of other people.

5 Body Expression

McCloud (2006, p. 103) agrees with Ekman and Friesen, in the sense that the body shows how people cope with emotion and suggests that the face reveals the various combinations between eyebrows, eyes and mouth—which results in coded emotions—and that body gestures are contextually variable. McCloud exemplifies three distinct contexts in which the face mirrors the universal emotion of fear and the body reacts variably—facing a pistol, fire and a snake.

Ferreira et al. (2015), in a study about body expressions in animation, propose that expressions of universally recognized body emotions correspond to a group of various types of gestures, that is, a sum of meanings, which often compensate the inevitable facial expression. McCloud (2006, p. 104), from the same point of view, argues that there are relationships between the several body gestures and their messages; but in opposition to the logic of facial expressions, there are no primary body poses from which all others derive.

6 *Disgust* Character

Disgust is a haughty character, looking down her nose upon everyone else. She clearly exposes her rejection stance regarding to almost everything that surrounds her. Although the character's name is *disgust*, the fact is that she expresses the most with contempt emotion or blends between contempt and other emotions (Ferreira et al. 2016, p. 6).

Based on numerous authors, Ferreira et al. (2015, p. 11) organized by categories the different occurrences of contempt emotion and all of them reveal very specific gestures: Signs of disinterest, annoyance, impatience, superiority, invasiveness, dominant, phallic, deformed, simulated discomfort, rejection, mockery, insulting and filthiness. A large part of these gestures can be found in *disgust* character all over the feature film (Ferreira et al. 2015).

Regarding character design, she is a young girl in a sleeveless pompous green dress with V-neck. She wears a lilac scarf around her neck and dark green leggings beneath her dress. The delicate shoes are also lilac, as well as the lipstick that she uses. She has a side swept hairstyle, falling until the shoulders height, eyelashes that stand out due to their considerable size and uses soft make-up on the cheeks and upper eyelids. The mouth is a crucial element on communicating contempt and disgust emotions, so when it goes up it is complemented by the big size of the chin area, exaggerating the human emotion's version.

The predominant color (green) is a key element, since it is highly related with rejected products by the human being. It is no coincidence that green food is not approved by children, since we all suffer from the mammalian distaste response (Rozin and Fallon 1987). It is also usually related to vomit, nasal mucus and insects, that is, elements that provoke nausea and disapproval.

Regarding body gestures, suggests a girl who integrates an aristocratic family that clearly assumes the annoyance and disinterest when socializing with other people. She is not a stereotypical teenager considering that she is not thin and presents a well-composed and healthy shape. This reinforces the concept of indifference to what other people say and the extreme self-esteem and self-confidence that people of this sort possess, considering that nowadays thinness is synonymous to beauty.

7 *Fear* Character

Fear is a highly frightened character, even by the minimum insignificant event. He is immensely confused and does not risk a single decision without some disarray. Fear emotion is mainly caused by the possibility of physical or psychological damage (Ekman and Friesen 2003, p. 47).

On this character, the eyebrows are mostly raised and contracted to the forehead's center, the upper and lower eyelids are raised, the mouth is opened with contracted and retracted corners, forward projected arms, separated and opened hands, the head without rotating oscillation, bent and united legs and a leaning forward spine (Ferreira and Teixeira 2016).

Curiously, it is a character whose head design only works with a blend of the 3/4 view and the profile view. The fact that he has big jiggle eyes, no forehead and no neck creates a weird image on the top, back and profile views. His clothes reveal a mixed concept: a studious well-dressed style (from the waist up) mixed with and a clumsy style (from the waist down). From the waist up he has a pompous purple bow tie, a vest with the pattern of the classic 'Houndstooth' Scottish brand, a buttoned up shirt with light blue vertical lines pattern. From the waist down he wears oversized trousers with folded hems and wide toe shoes. He has a tall and very thin shape, whose elements with greater prominence are the hands, eyes and nose, due to its big size.

Overall, his character design's allusive to an anti-hero with a goofy behavior and body posture. Usually, when someone experiences fear, it is often said that the person is "colorless" or pale. In animation field, lilac color is widely used to represent paleness, reinforcing that fear immensely affects the bloodstream.

8 *Anger* Character

Anger is the most "explosive" character of the entire animated feature film, reinforcing that it is the most intense emotion we can experience (Ekman and Friesen, p. 80). He is never totally peaceful or relaxed, however, when everything leads one to believe otherwise, he reveals himself mocking, being ironic and sarcastic. It is a short duration emotion usually related with violence, possibly involving physical and psychological damage. Blood pressure increases, the face becomes red, the breathing rhythm accelerates, body posture becomes erect, tense muscles, etc. The main causes of anger emotion have to do with moral principles violation, frustration, physical threat, disappointment, rejection, among others (Ekman and Friesen 2003, pp. 79–80).

As far as the character's design is concerned, *Anger* looks like the stereotypical grumpy boss. White shirt, rolled-up sleeves, zigzag patterned tie, brown flannel trousers with folded hems, brown leather belt and black shoes. He is short and has no forehead, which means the upper limit of the head is defined by the eyebrows, and the highlighted parts on the face are the big sized eyes, immensely large mouth and fairly thick eyebrows. This compensates the fact that this character does not have a nose. Although much of the muscular contraction is centered on the eyebrows and wrinkled nose, it did not affect the emotion's communication on this case. The skin's red color is highly related to anger because our blood concentrates on the face when enraged.

9 *Sadness* Character

Sadness is a constantly crestfallen character, negativistic and always with something sad to tell, regardless of context. She is easily bored when some task is channeled to her and rarely smiles during the feature film. As a universal emotion, it is the most enduring one, since it extends for long periods of time. Loss is the main cause of sadness and the innate function of its facial and body expression is "a request for help" (Ekman 2003, pp. 88–89, 91).

The character design follows a social pattern, namely a stereotypical obese girl with several depression symptoms. Her main color is blue, she has big sized glasses, a thick sweatshirt with high-fitting collar, blue pants and shoes, and side swept hairstyle limited by the chin's height. Eyes, eyebrows, mouth and nose are small sized elements.

Overall, *Sadness* shows the "cold side" of human emotions through the intense blue that surrounds her character, but also through the high-fitting collar and the *emo* stylish based haircut. The little given emphasis on her face area—with small eyes, nose and mouth—reinforces the constant low expressive behavior, which reflects a sad, depressing and self-flagellating conduct.

10 *Joy* Character

Joy is the narrative's engine. She tries to happily deal with any kind of context, although she realizes that it is not always possible because we always need room for other emotions. Any kind of sensory pleasure, invoked by our five senses, can cause joy and, as a positive emotion, it is related to fun, contentment, excitement, relief, wonder, ecstasy, high self-esteem, etc. (Ekman and Friesen 2003, pp. 99–103). She has an elegant shape, a light skin tone, a light green dress, large mouth and eyes, blue hair and she is always barefoot. It looks like a young energetic and very happy girl. *Joy* overflows positivism through her highly expressive body and face gestures. Nowadays, elegance, dynamism and bright colors are highly related to happiness and well-being.

11 Conclusions

The dissociation between head and body in animation goes beyond the simple technical issue; they coexist as two interconnected entities, but with sufficient autonomy to express themselves individually. If the face has a natural predisposition to communicate emotions, the body takes advantage of the action. From this perspective, gestures and body movement are also capable of creating strong emotional reactions, reinforcing personality traits and character states of mind. We can say that the fact that these characters of "Inside Out" are self-representative of the primary emotions gives it a double meaning. Firstly, they define a characterization that seeks the potentiality of the "emotion-character" blend, and, secondly, it fits the development of a character design using a thoughtful and planned design methodology carried to its extreme, personified in a triad relationship "emotion-character-visual".

In short, if character design always strives for engagement with the viewer, we can state that in "Inside Out" this reality is not different, as it presents scrutinized evidence of character-emotions that seduce us and lead us through an amazing story, reinforcing its quality of autonomous and imaginative language that exposes the DNA of human nature: its emotions.

References

Damásio A (1994) O Erro de Descartes: Emoção, Razão e o Cérebro Humano (e-book)
Eibl-Eibesfeldt I (2017) Human ethology. Routledge, New York
Eisner W (2008) Expressive anatomy for comics and narrative. W.W. Norton & Comp, New York
Ekman P (2003) Emotions revealed: recognizing faces and feelings to improve communication and emotional life. New York. https://doi.org/10.1007/s11231-005-7891-8
Ekman P, Friesen WV (2003) Unmasking the face (Reprint). Malor Books, Cambridge
Ferreira A, Gomes R, Teixeira PM (2015) Estudos de Expressões Corporais para Animação. In: CONFIA—International Conference on Illustration and Animation. Braga
Ferreira A, Teixeira PM (2016) Análise da longa-metragem "Inside Out" da Disney/Pixar—As Emoções Primárias na Personagem "Fear". In: AVANCA | CINEMA—International Conference Cinema—Art, Technology, Communication
Ferreira A, Teixeira PM, Teixeira JP (2016) Análise da Longa-Metragem "Inside Out" da Disney/Pixar—As Emoções Primárias na personagem "Disgust". In: CONFIA—International Conference on Illustration and Animation
Keltner D, Oatley K, Jenkins JM (2014) Understanding emotions, 3rd edn. Wiley, Hoboken

Kundert-Gibbs J, Kundert-Gibbs K (2009) Acting lessons for CG animators. Wiley Publishing Inc., Indianapolis
Matsumoto D, Hwang HS (2011) Reading facial expressions of emotion. Retrieved 24 June 2019, from https://www.apa.org/science/about/psa/2011/05/facial-expressions
McCloud S (2006) Travers, K, Williams J (eds) Making comic. Harper, New York
O'Neill R (2016) Digital character development: theory and practice. C&C Press, New York
Pardew L (2007) Character emotion in 2D and 3D animation. Boston
Rozin P, Fallon AE (1987) A perspective on disgust. Psychol Rev 94:32–41
Sly R (2012) De'von stubblefield interview. Retrieved 5 Feb 2019, from http://devon-stubblefield-interview.blogspot.com/
Teixeira PM (2013) A Representação Emocional da Personagem Virtual no Contexto da Animação Digital: do Cinema de Animação aos Jogos Digitais. University of Minho
Turner JH (2000) On the origins of human emotions: a sociological inquiry into the evolution of human affect, 1st edn. Stanford University Press, Stanford

Drawing and Image in Design

Chromatic Cognition and Human Behavior

Fernando Moreira da Silva

Abstract For humans, cognition is the processing or interpretation produced by the brain of all information captured by the five senses, based on different capacities, such as perception, imagination, reasoning or memory, and the transformation of that interpretation into fundamental knowledge for our own way of being. However, most of our reactions are still unknown, such as those concerning chromatic cognition. Recent research has been able to identify areas of the brain that are activated during the phenomenon of chromatic cognition, just as we begin to be able to measure human behavior with regard to color issues. The visible brain consists of multiple functionally specialized areas that receive their input largely from two areas of the brain known as V1 and the area around it known as V2. Through these areas the Human Being perceives the Color and these, in turn, can be more or less stimulated when we see different colors. This document presents some results of a quasi-experiment methodology still in development, using the Virtual Reality (VR), trying to verify the human brain reactions, mainly the chromatic cognition, to the different dimensions of the colors. In an earlier phase, in addition to the literature review, other methods were used, such as survey research and direct observation. It is intended to compare the results obtained with the use of these methodologies with those of the quasi-experience. This book chapter focuses on the acquisition of scientific knowledge in the area of chromatic cognition. As future results of this research, we intend to achieve a systematization of scientific knowledge reusable by all within the scope of Color/User interaction and chromatic cognition; and produce guidelines to serve as a projective tool for designers to use and apply color in design projects, as well as a reference to the use of Color for general users.

Keywords Color · Chromatic cognition · Human behavior · Virtual reality

F. Moreira da Silva (✉)
Lisbon School of Architecture, CIAUD, Universidade de Lisboa, Lisbon, Portugal
e-mail: fms.fautl@gmail.com

D. Raposo et al. (eds.), *Perspective on Design*,
Springer Series in Design and Innovation 1,
https://doi.org/10.1007/978-3-030-32415-5_16

1 Introduction

The present book chapter focuses on the cognition of color and human perception, and its importance for design and design development. Although there are some studies in this area, they are still insipid and many of them are not very scientific, so it is considered not only a current theme, but also a pertinent one. After the use of methods such as research by inquiry, direct observation and literature review, the second phase, a quasi-experience with users using the Virtual Reality (VR), is now presented in order to verify the brain reactions to the different color dimensions, in particular the color cognition, comparing the results obtained with the previously used methods.

The waves of light affect us every minute of our lives and enter our energy system, whether we are awake, whether we are sleeping, whether we are visual or blind. Our growth, blood pressure, pulse, temperature, muscle activity, immune system, etc., are all affected by the rays of light. The colored rays affect not only our bodies but also our emotions, dispositions and mental faculties.

As highly colored beings, our forms are made of vibrant, ever-changing colors, and the Human Being responds actively or passively to the chromatic impulses in everything we do. We all have a personal relationship with Color. Often, we give ourselves an instinctive color treatment, just choosing clothes of a certain color or placing certain colors around us, in our homes, gardens or in the work environment. Most of our reactions, however, are unconscious and only when we begin to use color in an informed way can we take advantage of this extraordinary vital force to improve the quality of life and our well-being (Moreira da Silva and 2016). The Human Being, during its evolution, created psychological and physiological reactions to the color, although in many cases they can not be controlled nor explained objectively; however, make color a necessary medium for information, communication and understanding of the environment. Color has the function of attracting attention, transmitting information, adding emotions and stimulating illusions (Gamito et al. 2009).

By understanding the effects of physiological and psychological color, we can select the best colors for our clothes, for our home and work environments. There are not many studies on the subject, and even those are merely descriptive of a period or time. So far, there has never been an attempt to understand the global phenomenon. This article presents a research project, which aims to understand the correlation between human statements when they see different colors, using research methodologies by inquiry and direct observation, and the areas of the brain that are activated when humans see a specific color using artificial cognitive systems.

The main objectives of the research project are the acquisition of scientific knowledge in the area that can serve as a projective tool for designers, as well as contributing, through the dissemination of its results, as a reference to the use of color for general users.

Throughout the project, a qualitative and quantitative, interventionist and non-interventionist methodology was used, with user-centered research, with participatory design, using the survey, direct observation and quasi-experience, supported by mechanical means, in the laboratory, using artificial cognitive systems and virtual reality.

2 Colors Code

> The code is a system of principles that bestows a certain value on certain siamas. Value is mentioned and does not mean, in order to give a more general character, because meaning is used only in relation to communication between human beings; In the case of human communication, the receiver has a voluntary act of comparing message and code and decodes it. (Eco 1967)

This type of message is visual and the signals must correspond to the perceptual needs themselves, which belong to human beings. Therefore, the system of principles must give a certain value to certain signals to become a code and that certain value has a significant basis. In terms of Color, particularly color/space language, the organization of codes follows general rules that are strictly linked to the possibility of manipulation of color language, to the limits and differential thresholds of perceptual ability in general, whether signals or visual field and the categories of meaning that are embraced by the language of color, as analyzed before, according to the color/space signs that are the color communication, and those that communicate by color, or with characters of simple signs, overlapping or supersigned signals. Thus chromatic codes are organized as having the repertoires of color/space signals as the basis, programmed according to categories and shade or shade classes, and depend on the range of signal repertoires of the individuals participating in the communicative process, consideration of the environment and the context of common experience. Primary and primary color/space codes are organized according to requests for harmonic tone types, boundary extension, and levels of differentiation between shadows or tones.

By these codes, which are applicable to the programming of languages for all the fields of the project, one can define the types of chromatic harmonization that, from a group of color/space signals, constitute a new signal indicator of a certain communicative situation characterized as a message by total quality and by the amount of visual vibrations between tones it transmits. Based on the formation of signal repertoires, the overlapping color/space codes for designating communicational intentions of something more than color/space are defined according to the context request. These codes follow the common communicative structure of the languages applied to them, besides the chromatic reference given by the previous code, the qualification established by the communicative structure. The selections of units of interest in the coding are applied to the repertoires of signals and, for the set, a system of rules is established for the transmission of the message (Moreira da Silva and 2016).

3 Neuroscience of Color

In order to approach Color, one must understand the brain and how the nervous system works. Neuroscience studies the nervous system, particularly with regard to Human Beings. This field is significant for the study of Color because it allows an understanding (as much as medical science allows) of the process involved between the arrival of the light wave and the physical reactions that result in the interior of the human body. Chromatic vision results from a very complex process in which the nervous system compares the intensity of light at various wavelengths through specific interactions that are known to be different from those needed to produce the perception of other properties of the universe visual.

It is known that people only see Color when light and chromatic signals that reach the eye reach the brain. The structure of the eye is not at all simple. The very structure of the retina is very complex and acts as the function of the first level of light interpretation. As the color of a given wavelength reaches the retina, all photoreceptors capable of responding do so. The combination of responsive receptors allows the brain to interpret the exact color of light. For example, if light with a wavelength of 600 nm (nonometers) reaches the retina, 30% of the green photoreceptors respond together with 55% or 64% of the red receptors. From this information, the brain will calculate that the color in question is orange. The reason there are two possible red responses is because 60% of the population has the amino acid at position 180 of the opsin protein, while the rest have alanine at that position. Therefore, although they all describe red and their variations in the same way, it is likely that there may be a difference in color perception.

From the retina, visual information passes to the optic nerve, joining the optic chiasm. Here, images are organized so that the information one or both eyes see on the left side of the vision is directed to the left half of the brain and vice versa. With the help of both eyes focused on the same object, the Human Being can perceive depth and distance, while each eye produces a slightly different image of the same object. The visual cortex of the brain completes the task of organizing visual information, which began in the retina. As mentioned earlier, processes in the brain are poorly understood, although the brief description above reflects current medical studies (Moreira da Silva 2016).

> The visual system is adapted to obtain maximum information with a minimum of effort: what is not required immediately, or can be considered as acquired, can be considered redundant, the eye has evolved to see the world in immutable colors regardless of the unpredictable, shifting and uneven lighting. (Lancaster 1996)

Graham (1997) undertook studies in cats to determine the role of various areas of the brain (only a small survey was conducted on humans). Five areas were identified within the cat's brain, which control the perception of: form; movement; color; and coordination of form and movement; and coordination of movement and color. The perception of shape is therefore very important for cats, with movement and color, respectively, in second and third places.

Once visual information has reached the human brain, it is interpreted through the various elements present in this structure. The Human Being has several types of memory that correspond to various brain locations. It is assumed that memories of specific functions, with more or less identified locations, correspond to the cortical location. Although both sides of the brain are involved in mental functions, it is known that each side has its specific functions. The cerebral cortex contains 90% of all nerve cells, and receives and interprets the sensory impulses. The brain contains the voluntary and conscious process, with the right side controlling the imagination and the left side controlling the logic. The thalamus deciphers sight, hearing, taste, and touch; the hypothalamus regulates blood pressure and body temperature, translating emotions into physical (physiological) responses. The pituitary gland is the largest endocrine gland in the body, controlling all other glands; the cerebellum is responsible for muscle coordination; and direct training regulates emotion. All these glands are interrelated and related in the response to visual stimuli. Questions about the role of striated cortex, the first visual area of the superior mammalian brain in color perception, have recently been restarted (Johnson 2001).

Cells in the striatum cortex are excited by light from a distribution of wavelengths and inhibited by light from another distribution, falling into a small region of space. Adjacent or surrounding regions show the opposite pattern of responses. The cells therefore respond optimally to the edges with a specific chromatic contrast that crosses the center and surrounds the regions of their spatial receptive fields. These findings suggest that specific color processing begins earlier in the visual system than previously thought. The first mechanisms of constancy of color, our ability to recognize the colors of objects, irrespective of the color of the light that illuminates them, may occur in the striated cortex. If so, people without striated cortex, like some older people, will lose color contrast processing.

So, the visible brain consists of multiple functionally specialized areas (Fig. 1) that receive their input largely from V1 and the area surrounding it, known as V2. These are currently the most thoroughly charted visual areas, but not the only ones. These are the areas where we perceive color and can be more or less stimulated when we see different colors.

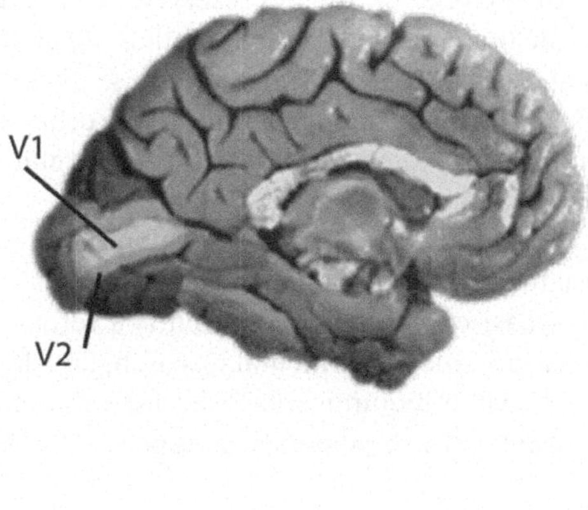

Fig. 1 Brain scheme, with V1 and V2 areas signed

4 Cognition and Artificial Cognitive Systems

Artificial cognitive systems are an emerging field. Cognition anticipates the need for action and develops the ability to predict the outcome of these actions. The field of artificial cognitive systems has the ultimate goal of creating computer systems that can interact with humans. It involves self-nomy, learning and development, memory and prospecting, knowledge and representation, as well as social cognition. This happens on the basis of a characterization of cognitive systems as systems that exhibit adaptive, anticipated and proactive behavior orientated behaviors (Vernon 2014).

Cognition implies an ability to understand how things can be and take this into account in determining how to act (Berthoz 2000). A cognitive system exhibits an effective behavior through perception, action, deliberation, communication and through an individual or social interaction with the environment. The characteristic of a cognitive system is that it can function effectively in circumstances that were not explicitly planned when the system was designed. That is, it has some degree of elasticity and is resilient in the face of the unexpected (Vernon 2006).

> Mental states are always based on real or imagined physical states, and problem space operators always expand into primitive capacities with executable actions. (Langley 2005)

For cognitive systems, cognition is representative in a strong and particular sense: it involves the manipulation of explicit symbolic representations of the state and behavior of the external world to facilitate proper, adaptive, anticipatory, and effective interaction and the storage of the knowledge gained from that experience to argue even more effectively in the future (Hollnagel and Woods 1999).

5 Development of an Experimental Part (Quasi-experience)

This research project has already achieved important results, going beyond the review of the literature on the research topic in question, mainly due to the empirical phase of the project, or applied research. Through recent research, we know which areas of the brain are activated, as well as we can measure the behavior of Human Beings with respect to Color-related issues. Thus, user experimentation is underway with the aim of if the brain reactions to the different dimensions of the color were verified, comparing the results of this experimental phase with those obtained by the other methods previously used.

So, and because we are developing an experiment with humans, we decided to use the quasi-experience methodology.

Quasi-experimental research is a research that resembles experimental research but is not true experimental research, i.e., doesn't corresponds entirely to experimental research requirements. Although the independent variable is manipulated, participants are not randomly assigned to conditions or orders of conditions (Cook and

Campbell 1979). Because the independent variable is manipulated before the dependent variable is measured, quasi-experience eliminates the directionality problem. But because participants are not randomly assigned—making it likely that there are other differences between conditions—quasi-experimental research does not eliminate the problem of confounding variables. In terms of internal validity, therefore, quasi-experiments are generally somewhere between correlational studies and true experiments. Quasi-experience is an empirical interventional study used to estimate the causal impact of an intervention on target population without random assignment, most likely to be conducted in field settings in which random assignment is difficult or impossible, or when we are working with humans or animals. This type of research is often performed in cases where a control group cannot be created or random selection cannot be performed.

> Quasi-experimental research is similar to experimental research in that there is manipulation of an independent variable. It differs from experimental research because either there is no control group, no random selection, no random assignment, and/or no active manipulation. (Abraham and MacDonald 2011)

In a quasi-experience research, the researcher lacks control over the assignment to conditions and/or does not manipulate the causal variable of interest. When we have access to a relatively large number of people who have or will receive a particular type of social work intervention, and we try to figure out what the effects of that intervention may been, then quasi-experience research can be an excellent approach, as it is in the present case.

To be aware of the seen color, one must interact directly with a person's brain, intercepting information at source and translating it into a cognitive color map. In this phase of our research, we are using the Virtual Reality (VR), trying to verify the human brain reactions, mainly the chromatic cognition, to the different dimensions of the colors.

In the present research project, to achieve this, we had to use a brain-computer interface using electroencephalography (EEG) and virtual reality. Our brain absorbs all kind of information, being constantly active, compacting and re-connecting existing data, and integrating everything into a consistent base. It shapes how we see our environment, filters or highlights objects and information most relevant to us. It creates its own knowledge based on our thoughts, emotions, desires and experiences, ultimately driving our behavior. The human brain consists of billions of cells, half of which are neurons, half of which help and facilitate neurons activity. The neurons are densely interconnected via synapses, which act as gateways of inhibitory or excitatory activity. Any synaptic activity generates a subtle electrical impulse referred to as a postsynaptic potential, which can be measured on the head surface.

EEG or Electroencephalography is the physiological method of choice to record the electrical activity generated by the brain via electrodes placed on the scalp surface. It is the process of recording a person's brain waves through electrodes attached directly to the skull. For faster application, electrodes are mounted in elastic caps similar to bathing caps, ensuring that the data can be collected from identical scalp positions across all respondents.

The EEG has the potential to unlock a truly immersive Virtual Reality (VR) by capturing a person's brainwaves while they are in the world of VR, and translating those moving signals from the participants. So, it measures electrical activity generated by the synchronized activity of thousands of neurons, providing excellent time resolution, allowing you to detect activity within cortical areas, even at sub-second timescales.

Therefore, it is relatively easy to detect electrical impulses emanating from a brain. However, the signal is sometimes so cloudy that it is difficult to decode the data. EEG has proven to be effective in detecting general brain phenomena, such as areas of the brain activated when people see a specific color.

6 Conclusions

Until now, during the quasi-experimental phase, we have worked with a group of 43 volunteers aged 23–41 years; and another group of 52 volunteers aged between 45 and 67 years. A helmet, similar to an elastic cap, was used to provide the feeling of Color using immersive virtual reality (Fig. 2). Nowadays, virtual reality helmets have become lighter, more comfortable and with clearer images. However, we had to spend some time to test and adapt the users to the use of the helmet, as well as to the immersive reality, especially the second group. All selected volunteers present normal vision. Some members of the second group evidenced the existence of a more aged vision. However, they clearly distinguished the different used colors. During the experiment, we always used the same color (through the Pantone reference system, for analog and digital support), which we used previously during the research-by-survey phase.

During experimental development, 11% of the first group found that they were color-blind: although they were seeing different colors, they activated the same location of the brain in the striated cortex. In the second group, only 4% were color blind. The results also evidence that regardless of which group they belonged to, 87% of the

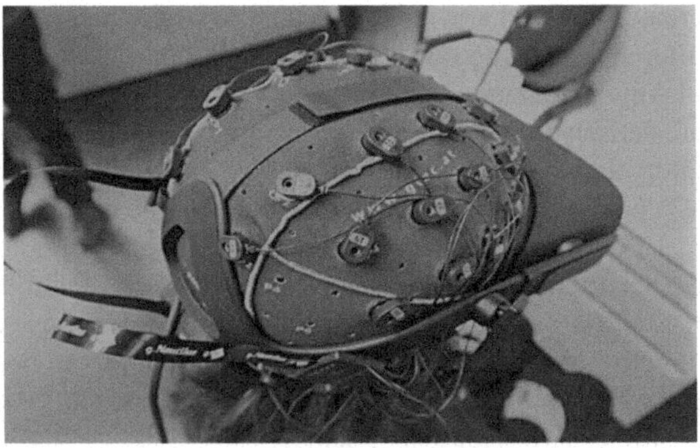

Fig. 2 Helmet used during the quasi-experiment, using immersive virtual reality

volunteers activated the same area of the brain when they saw a specific color. The achieved findings are already very significant data that can contribute to a greater knowledge in the scope of the perception and the chromatic cognition in the Human Beings.

The current project aims the acquisition of scientific knowledge in the area that can be used as a projective tool for designers, besides contributing as a reference for the use of color and for other investigations in areas related to color and the advancement of artificial cognitive systems.

References

Abraham I, MacDonald K (2011) Encyclopedia of nursing research: quasi-experimental research. Springer Publishing Company. Available at: http://0-search.credoreference.com.brum.beds.ac.uk/content/entry/spennurres/quasi_experimental_research/0

Berthoz A (2000) The brain's sense of movement. Harvard University Press, Cambridge

Cook TD, Campbell DT (1979) Quasi-experimentation: design & analysis issues in field settings. Houghton Mifflin, Boston

Eco U (1967) Appunti per una simiologia delle comunicazione visive. Bompiani, Milan

Gamito M, Moreira da Silva F (2009) Cor no Mobiliário Urbano: um factor de Inclusividade, Orientação e Identificação. In: 5º CIPED. Bauru, Brasil

Graham H (1997) Discover color therapy. Ulysses Press, New York

Hollnagel E, Woods DD (1999) Cognitive systems engineering: new wind in new bottles. Int J Hum Comput Stud 51:339–356

Johnson MH (2001) Functional brain development in humans. Nat Rev Neurosci 2:475–483

Lancaster M (1996) Colorscape. Academy Editions, London

Langley P (2005) An adaptive architecture for physical agents. In: Proceedings of IEEE/WIC/ACM international conference on intelligence agent technology, pp 18–25

Moreira da Silva F (2016) Bridging color effects and fashion design: the ColorErg. In: Soares M, Rebelo F (eds) Ergonomics in design—methods and techniques. CRC Press, Taylor & Frances Group, FL, pp 55–72. ISBN: 13-978-1-4987-6070-6

Vernon D (2006) The space of cognitive vision. Cognitive vision systems: sampling the spectrum of approaches. Springer, Heidelberg, pp 7–26

Vernon D (2014) Artificial cognitive systems: a primer. The MIT Press, London. ISBN 978-0-262-02838-7

Sketches Versus New Technologies in Design Creative Process

Ana Moreira da Silva

Abstract As the designer has to conceive and develop solutions for specific problems of different nature, sketches may present themselves as an operative support for conceptual reflection, problem solving and critical analyses within the several phases of the design process. Once, hand drawing and sketching ruled design curricula, because there simply was no alternative. The introduction of CAD, image editing, on-line search engines and other new technology advances have irrevocably changed the way we perceive and map the world around us. As a result, the status and value of hand-drawing has changed and it is important to reflect upon the functions, the teaching and the evaluation of the tools we use to express design. One of the main questions we would like to find answers is if the nowadays wide use of new technologies doesn't invalidate the important role played by hand-drawing as a stimulating instrument when sketching the first ideas and as a critical verification of the several solution hypotheses. Sketching survival facing the artificial intelligence growing importance is this research main issue. Through the study of several statements from various authors we intend to investigate not only the permanence of the sketching important role since mid-last century to present, but also its possible use in the future. From this theoretical approach we also intend to verify the relevance of hand-drawing teaching in the formation of the future designers, despite the paradigm changes that emerge from the new technologies advances.

Keywords Sketches · Technology · Drawing · Design process

1 Introduction

This text stems from a current post-doc research project motivated by the need of producing more knowledge and reflection on Drawing. The present research focus is to analyse nowadays importance of using sketches within design process. We

A. Moreira da Silva (✉)
CIAUD—Lisbon School of Architecture, Universidade de Lisboa, Lisbon, Portugal
e-mail: anamoreiradasilva@gmail.com

CITAD—Faculdade de Arquitetura e Artes, Universidade Lusíada, Lisbon, Portugal

D. Raposo et al. (eds.), *Perspective on Design*,
Springer Series in Design and Innovation 1,
https://doi.org/10.1007/978-3-030-32415-5_17

also want to discuss if sketching may serve as mind stimulus during the creative process phase, using for that a quasi-experience methodology, in laboratory, trying to compare and analyze the brain activity. Based on a survey methodology with two different moments, being the first an inquiry based on a questionnaire applied to undergraduate students, and the second a group of semi-structured interviews applied to designers, to verify the permanence and the importance of sketching in their daily design work, although the constant use and the progress of the new technologies. We intend to get some answers that could help us to define in a more accurate way the future adjustments on drawing teaching methods for Design courses. Here we present the first stage of this research project.

2 Sketches Versus New Technologies

Sketching has been used long ago, especially since Italian Renaissance when drawing started to assume a more important role as the basis of creation. In this period drawing assumed its intellectual character and some authors of those times, like Alberti or Vasari, considered it like something that the hand achieves coming from the human brain (Moreira da Silva 2010) (Fig. 1).

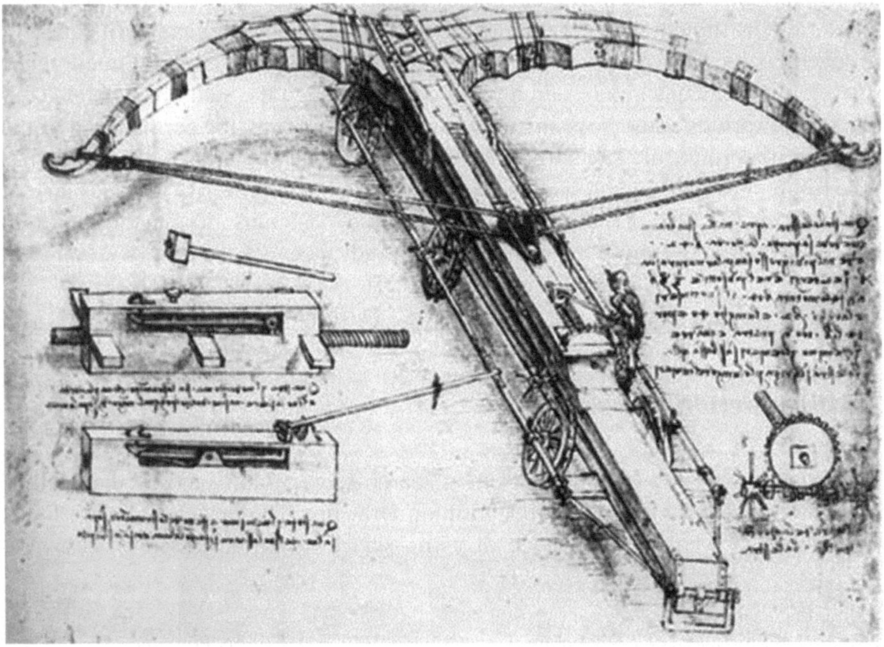

Fig. 1 Leonardo da Vinci (1452–1519) sketches for mechanical inventions. *Source* http://www.leonardodavincisinventions.com/. Accessed April 2019

Since then we can find sketches as the basis of creative production not only in the fine arts but also in design projects. We can consider the relevance of hand-drawing in the act of designing not only as stimulating instrument when sketching the first ideas but also as critical verification of the several hypotheses.

Norman (2010) states that designers are inextricably associated with drawings.

We can find reference to drawing in the very origin of the word design. '*Designare*', is a Latin verbal form, which covers both the meaning of 'to designate something' and of 'to draw something'. In its origins the term already has a double sense: a more abstract aspect of 'to attribute, to conceive, to project' and another one of more concrete nature of 'to register, to configure, to form' (Denis 2000).

Design adds intention to drawing: to imagine, to conceive and to assist in the achievement of solutions for practical problems, which can be of very different nature. During all the complex process which goes from one imagined object to its implementation, designer has one essential media to help him on developing the idea—drawing (Fig. 2).

Drawings have been used for generations of designers. The use of sketches not only helps a great deal in starting the ideas generating process, it also enables high quality concepts to begin with as initial ideas. These ideas can be accessed, combined, selected and developed.

Sketches can be the initial basis for the process of generating ideas, and they also allow to develop the project from these initial ideas. Through this drawings we can easily create, visualize, analyze, combine, cross, compare, criticize, select and develop ideas for future solutions in the several areas of the project.

As the designer must conceive and develop solutions for precise problems which may be of very different nature, sketching can work as operative support for inventive problem-solving during the creative process.

The complexity, characteristic of the transition from the idea to its materialization, is related to the actual extent of projects where a number of factors come into play, such as the ability to understand the context and imagine the solutions, the ability to know and take advantage of the processes and materials as means or vehicles through which the solution can be materialized, the ability to transform ideas into appropriate forms, having in mind the limits and material possibilities. Drawing holds, in itself, the making visible quality of the whole mental process underlying the conceiving stage, from the first sketches of a vague idea of we want to create until the final solution.

Drawing's transcendent importance lies in the capacity it gives to materialize abstract conceptualizations and to create the ideational basis for new forms and objects.

Several authors have written about the importance of drawing in design methodologies. Their general conclusion points out to drawing as an essential tool as it allows ideas to flow and provides a critical investigation of the several alternative solutions in creative processes.

Cross (2005) states that drawings are a key feature of the design process. At the early stages of the process they are communications with oneself, a kind of

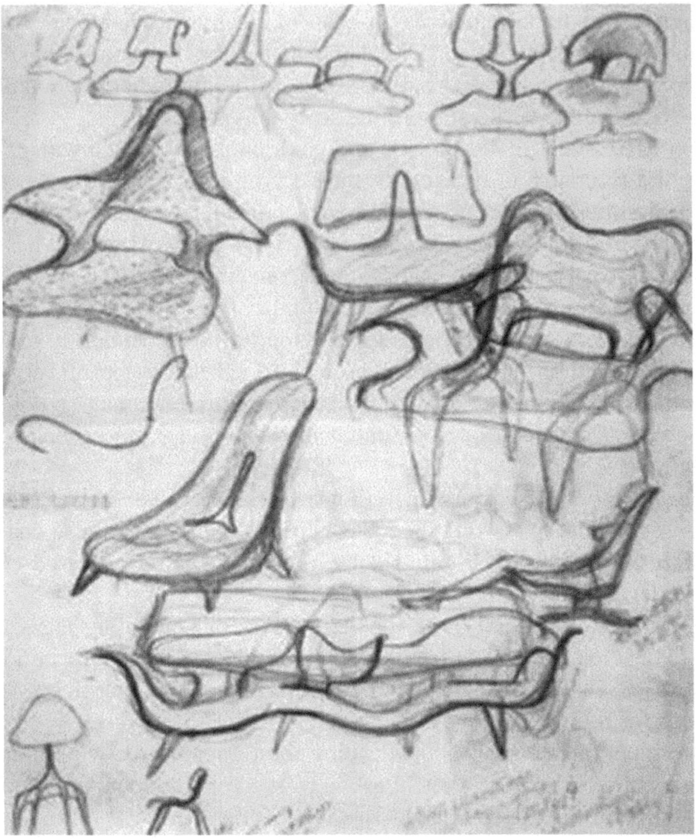

Fig. 2 Charles Eames sketches for several chairs (1946). *Source* https://www.google.pt/search?q=Charles+eames+sketches&tbm. Accessed April 2019

thinking aloud, and concludes that the conceptual thinking processes are based on the development of ideas through their external expression in sketches.

In *The Inexorable Rise of Drawing*, an article that analyses contemporary thinking about drawing, Tilley (2008) tries to identify the reasons why, nowadays, the drawing subject became so relevant. According to Tilley (2008) the interdisciplinary nature of drawing has been acknowledge nowadays and also recognized drawing's facility to allude to or to describe the intangible.

For Schön (2000), the practice of manual drawing goes beyond the simple representation of ideas, encompasses the development of knowledge from the perception and analysis of the drawing itself, considering the sketch as an instrument that even allows reflection in action.

Ortega and Weihermann (2017) state that when students use sketches, new ideas flow, qualities and relationships that were not previously imagined reveal themselves.

According to Baskinger (2008), drawing and sketching are an integral component to the development process for many designers. Sketching tends to be very engaging and invites others in for collaboration. Drawing by hand can enable to think differently about a subject or a design problem and can equip with greater persuasion and impact during collaboration. Hand-generated drawings can also provide a basis for transitioning into digital sketching in a variety of tools. The expediency and impromptu nature of picking up a pencil and letting ideas flow onto paper can be both powerful and compelling. Drawing ideas can serve to clarify, lead, and facilitate collaboration in meaningful ways.

Sketching is fast and easy, allows to explore and refine ideas in a quick, interative and visual manner. No need to learn any fancy design tools, just need to put a pencil or a pen to paper or marker to whiteboard and let the ideas flow. Sketching is a very simple and easy tool used by many creative professionals in any circumstance, because we can draw anywhere, anytime, with any medium (Fig. 3).

According to the architect Jean Nouvel, one of his latest work, the Abu Dhabi Louvre, opened in late 2017, began to take shape from simple sketches made on a paper napkin during a meal in a restaurant (Fig. 4).

The search for solutions, even for simple problems, implies that drawing studies in detail each phase of the process for obtaining the result we seek. The complexity of the required drawings emerges as the designer approaches the solutions which he considers ideal or that are possible. During the conception phase we should not disregard the importance of a trace, of a scribble, of a sketch. They can all contribute to the idea development (Fig. 5).

For a designer, the sketchbook is not just a place to draw but also a place to order thoughts, to gather graphically and visually information and to develop a design response. It emerges as a way of thinking through drawn lines as a creative process, revealing the central role that drawing can play within the development of any project (Fig. 6).

According to Rhode (2011), the real goal of sketching is about generating ideas, solving problems, and communicating ideas more effectively with others.

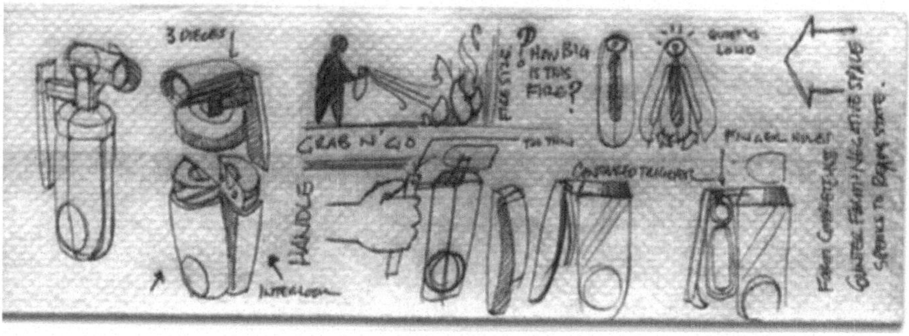

Fig. 3 Sketches on a paper napkin showing the ideas evolution for a fire extinguisher. *Source* https://mv2011.files.wordpress.com/baskinger-pencils_pixels. Accessed April 2019

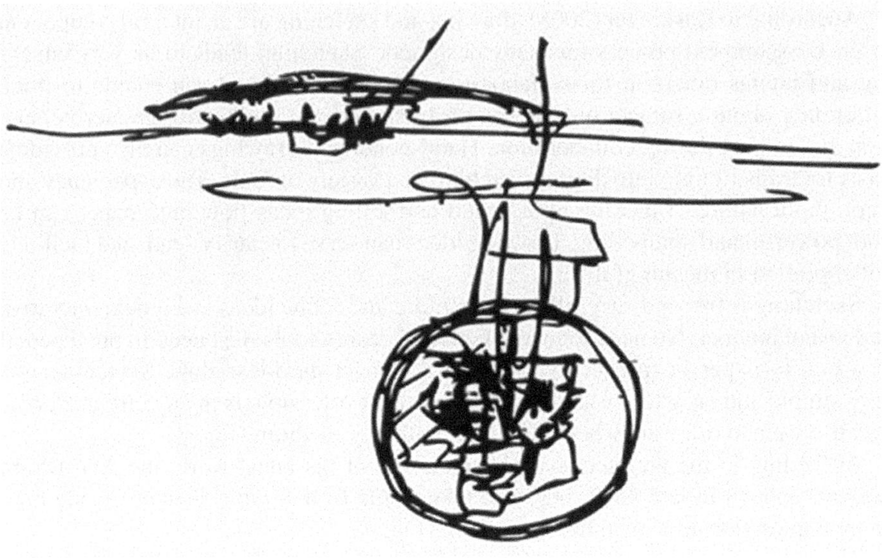

Fig. 4 Sketches by Jean Nouvel for the Louvre Abu Dhabi (2011). *Source* http://www.saadiyat.ae/en/inspiration-details/3/Louvre-Abu-Dhabi. Accessed April 2019

Fig. 5 Graphic design sketchbook layout. *Source* https://www.google.com/search?q=sketchbook+layout+sketches+for+graphic+designers. Accessed April 2019

Fig. 6 Sketching allows to explore a wide variety of ideas all at once. *Source* https://alistapart.com/article/sketching-the-visual-thinking-power-tool. Accessed April 2019

Drawing, within design processes, serves several functions. We can narrow them down to four functions that are indissolubly interrelated: recording, exploration, communication and expression. These functions do not stand on their own, more often one drawing is able to combine several of them (Lawson 2004) (Fig. 7).

The act of drawing allows that the reasoning and thoughts we have developed can be gradually translated and decoded throughout the drawn lines. Somehow, we struggle with our own ideas on the paper space. We scratch, we draw, we overwrite features, we configure, we represent, we visualize, giving physical form to our thinking. There is a direct link between the thought and the hand that performs the drawing: the hand as an extension of the brain, of the reasoning.

For Wilkinson (2016) it can be a simple drawing to start, but something happens when sketching that encourages the creative thinking and ideas begin to flow, something begins to take form and "thinking through drawing" allows the evolution of several solutions and their critical analyses.

According to Norman (2010) the recent profound changes in project processes, due to the contemporary historical context where new technologies are so important, justify an extended discussion on the crucial role played by drawing in design courses, adapted to the new paradigm.

The fact of using new technologies doesn't invalidate the important role played by freehand drawing, both at the initial stage of recording the first ideas and during their subsequent development and in the critical analysis of the different hypotheses.

Rohde (2011) states that sketching is the visual thinking power tool and that adding sketches to the design process is a great way to amplify software and hardware tools. He concludes that the use of sketching during the creative process reveals the central role drawing plays within the development of a project and the importance of freehand drawing, particularly in an age dominated by digital media. For Rohde (2011), sketches can work not only as a visual thinking tool but also as a primary language for capturing thoughts, exploring ideas, and sharing those ideas. So sketching provides a unique space that can help us to think differently, to easily generate a variety of new

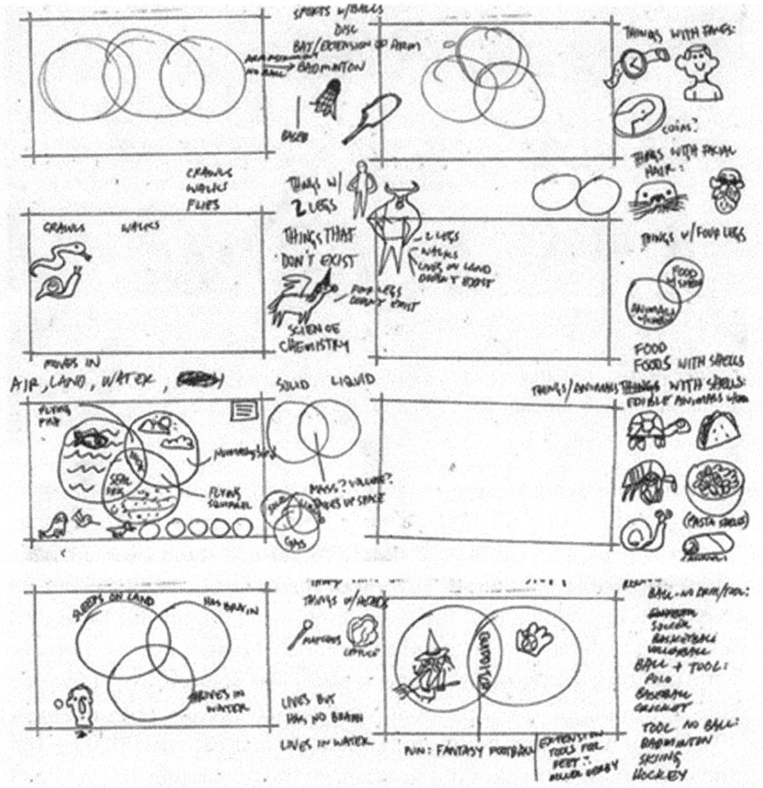

Fig. 7 Mike Dutton's sketches for a Google doodle (2014). *Source* http://www.google.com/doodles/john-venns-180th-birthday. Accessed April 2019

ideas, to explore alternatives with less risk, and to promote constructive discussions with colleagues and clients.

In a previous research approaching this subject, we concluded that drawing accomplishes the most different objectives, such as:

- a way to communicate;
- a means of discovery;
- a process of interiorization;
- a graphic method of study;
- an observation and registering process;
- a research tool;
- a privileged mean for the communication of ideas;
- a link in the mental and creative process.

So, the importance of drawing assumes a broad sense, conferring to the act of drawing the ability to become a means of multiple resources.

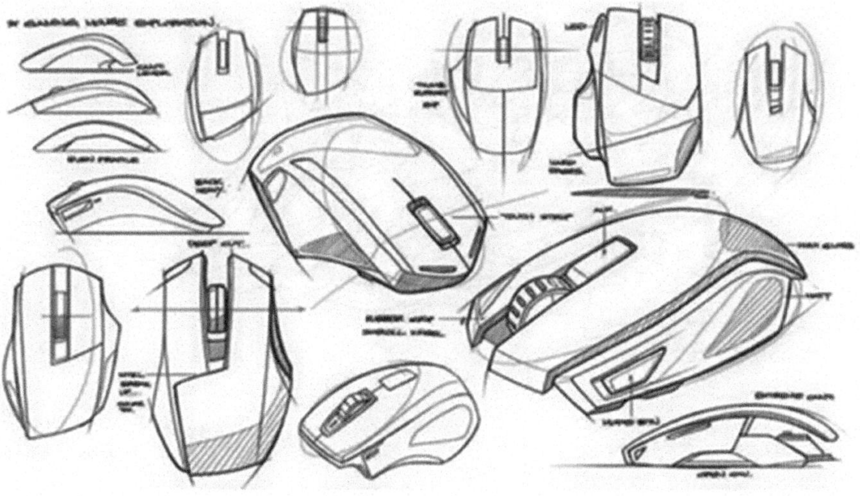

Fig. 8 Sketches for a mouse by Adam Fairless (2016). *Source* https://medium.com/@haydmills/-creative-professions-that-use-sketching-as-a-tool-in-their-design-process-and-why. Accessed April 2019

When analysing and comparing manual drawing and digital based drawing taking into account the creative process, Roome (2011) considers that what is truly interesting and beneficial is the possibility of cross-fertilization between traditional and digital platforms (Fig. 8).

Frank Gehry is one of the contemporary architects who most uses new technologies to achieve his works, however, he always uses sketches to fully develop his creative ideas (Fig. 9).

According to Hamilton (2009) the value of hand-drawing is associated with the knowledge that drawing is the essence of creativity and its value lies in allowing our individual ability to continuously express and create, using self-exploration and creative thinking through the act of drawing, linking together interesting ideas and technical know-how, enhanced by the assistance and the interaction with new technologies.

Domingos (2017) in his recent book, The Master Algorithm, talks about the extraordinary advances of Artificial Intelligence and its promising future, but as he mentioned throughout this book, Artificial Intelligence is developing in the direction of Automatic Learning, but not yet capable of creating something new which is the very essence of Creativity.

Automatic Learning based on Artificial Intelligence can enlarge our data and integrate multiple learning components, and so, can be very useful for generating creativity through knowledge, but still, creativity is something more beyond knowing, it is a special ability that, for now, is an intrinsic part of the human being. Creativity is something that big data does not explain, but, however, machine learning can help us to build more accurate fundaments and research findings for our creative works.

Fig. 9 Sketches by Frank Gehry for the Louis Vuitton Foundation in Paris (2006). *Source* www.google.pt/search?tbm=isch&q=Frank+Gehry+Sketches. Accessed April 2019

Drawing holds, in itself, the making visible quality of the whole mental process underlying the conceiving stage, from the first sketches of a project vague idea to the creation of its final form. Drawing's transcendental importance lies in the capacity that it gives the designer to materialize abstract conceptualizations and to create the ideational basis for new design forms and artifacts.

Drawing today serves as a primary medium for generating, testing and recording a designer own creative and conceptual musings about a project. If drawing is central for conceiving, it is also fundamental for defining how that conception is managed as it moves from its initial stages through its actual development and realization as a material form. Sketches illustrate various aspects of the creative design process. Therefore, the importance of drawing assumes a broad sense, conferring to the act of drawing the ability to become a means of multiple resources to practice the discipline of design.

3 Conclusions

From many authors' statements and according to the professional practice of many designers, we can conclude, from this initial research phases, that drawings are the key to the creative process in design. Drawing presents itself as an operative support for problem-solving and critical analyses in the design creative process.

We can also conclude that new technologies offer the designer new ways, allowing him to save time and facilitating many of the daily practice tasks, even becoming indispensable in many work phases. However, they do not replace sketching, which continues to assume a 'chameleon' shape during the various historical periods, in a constant time adaptation, and, especially by incorporating a critical dimension in design's creative process.

New technologies can be a stimulus for creative capacities, but it doesn't invalidate the important role of sketching in the creative process. When executing a drawing, the one who performs it makes choices, composes, and selects what and how to communicate, both in the representation of reality and in its transformation by invention. Drawings can show appearances of things as experienced by someone who draws, evoke some aspects of human existence, and seize the meaning of the subject under inquiry. The experience of freehand drawing may provide attentive awareness and add reflexive value.

Although we cannot predict what the future may bring in the Artificial Intelligence field, we believe that in the near future thinking machines will hardly replace the creative capacity that sketching can trigger. Drawings are able to trigger ideas while offering the practical tool to express those ideas. Sketches can be the key instrument of creation and control, also playing a critical role within the design creative process.

Despite the paradigm shift required by the changing times, we believe that hand-drawing will remain inseparable from the designers training and professional practice, assuming essential operating support in the project activity which continues being the design basis.

In the future, design education should pass through a systematic approach to hand-drawing in order to highlight and analyze the flexibility with which it adapts itself to various purposes, efficiently fulfilling a wide range of intentions, in a constant and vital adaptation to the continuous changes in the teaching and in the practice of the student and of the designer, given the new techniques and technologies.

To validate these assumptions, the current post-doc research can constitute a contribution for the understanding of the importance of hand-drawings permanence as the basis of the design creative process, although the constant progress of the new technologies.

References

Baskinger M (2008) Pencils before pixels, a primer in hand-generated sketching. In: Interactions, vol XV. https://mv122011.files.wordpress.com/2010/11/baskinger-pencils_pixels.pdf. Accessed Feb 2019

Cross N (2005) Engineering design methods, strategies for design. Wiley, Chichester

Denis RC (2000) Uma Introdução à História do Design. Blucher, São Paulo

Domingos P (2017) A Revolução do Algoritmo Mestre. Manuscrito Editora, Lisboa

Hamilton P (2009) Drawing with printmaking technology in a digital age. http://www.lboro.ac.uk/microsites/sota/tracey/journal/dat/images/paul_hamilton.pdf

Lawson B (2004) What designers know. Elsevier, Oxford

Moreira da Silva A (2010) De Sansedoni a Vasari: um contributo para o estudo do Desenho como fundamento do processo conceptual. Universidade Lusíada Editora, Lisboa

Norman D (2010) Why design education must change. http://www.core77.com/blog/columns/why_design_education_must_change_17993.asp. Accessed Jan 2019

Ortega AR, Weihermann S (2017) Graphic dialogues: the progress of knowledge in architecture studio. In: Kong MSM, Monteiro MR (eds) Progress(es)—theories and practices. CRC Press, Taylor & Francis Group, London, pp 111–114

Rohde M (2011) Sketching: the visual thinking power tool. http://alistapart.com/article/sketching-the-visual-thinking-power-tool. Accessed Apr 2019

Roome J (2011) Digital drawing and the creative process. tracey@lboro.ac.uk. Accessed Feb 2019

Schön D (2000) Educando o profissional reflexivo: um novo design para o ensino e a aprendizagem. Artes Médicas Sul, Porto Alegre

Tilley A (2008) The inexorable rise of drawing. http://www.transitiongallery.co.uk. Accessed Mar 2019

Wilkinson C (2016) Thinking through drawing. https://vimeo.com/15908559. Accessed Feb 2019

The Design-Body Project in Analogue Corpus – Tattooing as a Graphic Expression in Paper and in Portuguese Skin Since the Beginning of the Twentieth Century

Susana Azevedo Cardal

Abstract In this chapter we propose not to think about "the canvas" where it is usually present but change the discourse of its territory and invest in a new identifier space/support, the human skin – the tattooed body as a support of elements for communicational graphic expression. The "analogue *corpus*" generates a new critical process in the history of the tattooed image and the graphic design, as well as in the use of materials, instruments and techniques suitable for the fulfilment of these body marks in paper and on the skin. We intend to examine the evolution of these body marks by the handling of the materials and instruments used in the paper drawings and the skin of the Portuguese throughout the XX century, focusing on three phases of study (INMLCF – 1911/43; Portuguese Colonial War – 1961/74; Post Carnation Revolution, April 25, 1974), as a graphic expression of visual perception on the form of the human body as a means of communication, allowing us to develop a synthesis of these in the development of these body marks as a historical framing. At the same time, we confront "design-project" with "body-project". The tattooed body affirms itself today, as a privileged means of the contemporary representations of the identity of a certain individual, collectively establishing the social portrait of a given generational group, given the qualification and evolution of its professionals, as well as the resources they use to obtain significantly improved results compared to previous years.

Keywords Tattoo · Lettering · Legal Medicine · Portuguese Colonial War · Portugal

---

S. A. Cardal (✉)
FAUL – Faculty of Architecture of the University of Lisbon, CIAUD – Centre for Research in Architecture, Urbanism and Design, Lisbon, Portugal
e-mail: designersusanazevedo@gmail.com

D. Raposo et al. (eds.), *Perspective on Design*,
Springer Series in Design and Innovation 1,
https://doi.org/10.1007/978-3-030-32415-5_18

1 Introduction

> «The original form, prime, is the simple course of the tool. Only hand writing preserves the characteristics of a single line. Hand writing is exercised with a single trace. Lettering is the writing done with built shapes.»[1]
>
> Gerrit Noordzij

Tattooing, in Western culture, and more particularly in Portugal, was cast aside from the artistic milieu for almost a century. From late 1990's onward – and more recently – the tattooed body found itself inserted in the fields of the visual arts, through elements drawn and tattooed, acquiring a biographical, and potentially narrative, and identity connotation. We came across elements of graphic expression, as is the case of letters' tattoos, originating from calligraphy or lettering, who speak directly of the "body marking project" (Ferreira 2008), be it psychological, social, political and/or cultural, that a certain individual adopts. The "denounced confession" by the representations present in these bodies freely expressing themselves, is reinforced by the growing need for affirmation and presence in tattooed individuals, as well as by the recognition of the professionals who execute them. These body marks, given their visibility and projection, become potential tools within the scope of representation, of observation and self-control, as a "body-document/body-file", – a vast graphic portfolio, reflecting the need for difference. Yet like many other bodies, only different in the most intimate details.

Contemporary interest in the drawn tattoo, private in the public eye, highlights the shift in paradigm representing this bodily marking, especially tattooing as a graphical expression, often named "written tattoo", or, as in the past, "inscriptions" (Silva 1926). These marks are associated with character drawing, or lettering reproduced as means of communications. This type of drawing, common nowadays, has been dominating daily Portuguese tattoo consumers' visual communication. The "original" image has a strong impact in urban centers and is an object of desire, linked to fashion trends. Alongside the bodily project (Ferreira 2008) of this social and individual mutation, it points towards a new relation between the consumer and producer, as well as the offer and demand and the outcome in line with the desired quality. We are looking to address the drawing before, during and after the actual execution of these bodily marks, looking at drawings and respective tattoos, gathered at a tattoo studio and given by a graphic designed specialized in lettering for tattoos. Concurrently, we aim to look at different time periods, including archived material from legal medicine and testimonials from the Portuguese Colonial War veterans.

[1] Noordzij (2013, p. 11).

2 Tattooed Characters in the Portuguese Context

2.1 Background

«(…) in a fight *body to body* between the individual and the social.»[2]

Vítor Sérgio Ferreira

To understand the relation between lettering and the tattooed individual, we need to go back until the emergence of tattooing in Western culture,[3] whereby we could find urban social context with some alphabetization, especially in the Portuguese context. During this period, tattooing was mainly comprised of signs with simple lines without any formal concerns. The cursive style mimicked the physical action of writing before being copied onto the human skin. Although this bodily marking was looked at as a "primitive" and "outcast", its development owes, in great part, to the desire of marking the body, and inscribing ink on skin, as a rebellious action of individualized memory, an unchangeable eternal mark. It is worth mentioning that, according to research about tattooing and tattoos, mainly in areas connected with legal medicine, anthropology and sociology, most individuals marked their body out of desire and imitation, "see and do" (Silva 1926). In his taxonomy of tattoo themes and categories, Rodolfo Xavier da Silva, named the ones composed by letters and numbers "inscriptives", and, according to his own data, these were the most common in Portuguese skin.

With time, in Portugal, the "inscriptives" tattoo moves on to the Portuguese Colonial War all the way to present day. However, for the IMLFC[4] Southern Delegation, Lisbon (1911–1943) and during the Portuguese Colonial War, these were also made through sewing needles attached to a stick with sewing thread dipped in India Ink and introduced in the human skin. Composition-wise, until 1974, the type of tattoos remained unchanged, predominantly using block capitals, names and dates and themes and preferred content. Arms and forearms were the privileged anatomical areas for tattoos, without any bodily marking project; they were produced according to the consumer's spontaneous desire. Likewise, there was not an identifiable evolution in its formal configurations and characters' construction. These were written in paper and directly on the skin, keeping its cursive origins, uniquely based on a set of written signs written by a certain individual, without any previous drawing project. Considering the locations where they were produced (bars, ships, prisons and war camps), often without the producer having no drawing experience, or in the tattooing act itself, or even how precarious the materials and tools were, quality was very low. It is worth noting that, especially during the first half of XX century, a large

[2]Ferreira (2008, p. 35).

[3]Whilst sailing for England, Cook discovered Tahiti in the end of XVIII century, and since then "bodily marks were popularized in European Western society" (Ferreira 2008). This bodily change attracted sailors, who started tattooing regularly, both for its originality and the bodily mutation it entailed. Although there are no specific dates, that is how tattooing comes to Europe.

[4]Institute for Legal Medicine and Forensic Sciences.

part of the illiterate population marked their bodies with characters, fully trusting the "markers"[5] (Silva 1926). It is also worth noting that the contexts in which these bodily markings took place had very low hygiene and public health standards; that, allied with the lack of knowledge of how human skin behaves, as well as appropriate materials and sterilized tools. It was "one size fits all" and the tools and materials were the same for all people. There was no post-care, unlike what happens nowadays, whereby there are health and care concerns during and after the tattooing process, as well as considering the body as a living canvas, previously outlined, more, or less, spontaneously.

2.2 *INMLCF (1911–1943)*[6]

Whilst analyzing the photographic archive, looking at the contents and letters' design, both from human skin and iconographic illustration from individual records, we observed that in its majority written tattoos were in block capitals or upper case, accompanied by drawings referring to the female gender, patriotic symbols, religious representations, sports teams' crests, and even criminal or military tools. However, the use of initials, for oneself, family members or even love affairs are predominant in the skin of these individuals. For women, they mainly wrote their husband and/or lover's full name, most of times without any drawing, and usually next to the anatomic place of the human heart.

With regards to the letters themselves, most characters are presented sans serif, falling back on the handwritten characteristics of the cursive style, with simplified shapes to enable faster writing, originating both mechanically from typewriters as well as school-learned handwriting Fig. 1. It is rare to observe more typographic or lettering-based evidence. Generally, "[they] spend their time painting, or allowing their body to be painted" (Silva 1926). These bodily marks were composed by contour lines, very similar to handwriting. There are also cases where a double contour line and no filling are observed, others with filling, but never with shading.

In the bodies with several tattoos we observed a lack of composition and an inadequate tattoo in relation to its placement. These bodies are generally covered by clusters of tattoos with different sizes, often representations or figurative illustrations, with different characters – coeval and random. The tattoo placement followed no specific orientation and followed no formal or compositional guidelines amongst themselves Fig. 2.

---

[5]Designation attributed to tattoo artists at that time.

[6]Azevedo Cardal (2016a).
Azevedo Cardal (2018).

Fig. 1 Human skin fragment (1923), INMLCF collection – South Branch

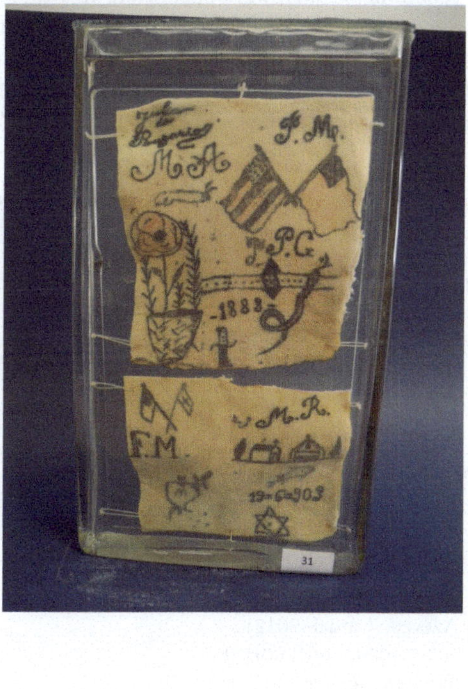

Fig. 2 Human skin fragment (1917), INMLCF collection – South Branch

2.3 Portuguese Colonial War (1961–1974)[7]

The characteristics and the execution of these tattoos are rather simple, with drawings alluding to military symbols, battalions and respective companies, alongside written messages, in block capitals, without any additional ornaments. Beside the military symbols, heart-shaped tattoos are rather frequent in in the typology of these marks. Names and dates are the most usual. As it happens, these marks reflect the identity of a member of this community, that is, unique and individual marks portraying their war and personal experience.

«Amor de mãe»,[8] «Angola 1968», «Esposa e filhos»[9] are amongst many other tattoos executed during Portuguese Colonial War. They identify a specific Portuguese generation. These written records in some Portuguese male bodies are currently being updated, according to each tattoo artist's professionalism and skill: there is demand for maintaining the original record, updating them, according to technique and ability of the tattoo artist. This makes up for the inability, lack of knowledge around the actual letters' design and the actual tattooing technique of the makeshift tattoo artist. These individuals reduced the letter's graphic elements as much as possible, with straight stems combined with bowed segments – shapes without any decoration. Spontaneity is one of the emotions more clearly present in these bodily marks Fig. 3. Formally, they were composed by dry lines, geometrical and rational. However, this formal simplicity amplifies the emotional weight that these texts carry, a "reactionary" writing, appropriate to war contexts. This irrational choice by whom executed the letters formally is essential to the character and credibility of the message. The characters used are typically male, as they reveal a formal weight noted by their robust rectangular forms – we could even suggest that it is "property" of the military context where it is set Fig. 3 (Fig. 4).

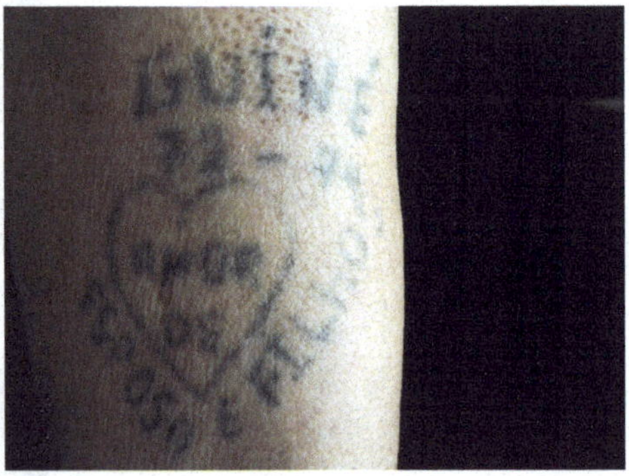

Fig. 3 António Filipe, 62 years old, Guinea 1972/1974 – Mechanical Cable

[7] Azevedo Cardal (2012).
Azevedo Cardal (2018).

[8] TL: «Mother's love».

[9] TL: «Wife and Children».

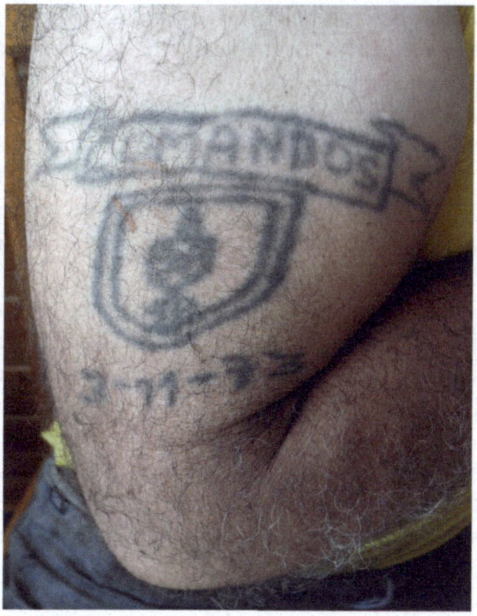

Fig. 4 Armando Ramos, 61 years old, Mozambique 1972/1974 – BCP31/Parachutists

2.4 *Post-Carnation Revolution (April 25, 1974 to Present Day)*

With the change of regime in Portugal, the increase of quality of life of Portuguese people, as well its consuming habits, visual culture acquired a new connotation weaving different artistic areas in a way to match itself with the industrial innovations happening worldwide.

Where design is concerned, calligraphy, during this period, whilst still very much traditional, starts adapting to the new technologies and handwriting moves to a secondary role. The appearance of computers in schools and the focus on digital in detriment to handmade had a deep effect on the Portuguese generation growing during the 1990s. The country itself juggled a balancing act in a way do reign in the information pertaining to what was "new", that rose on the market day after day. In a period where the digital overcame the handmade in the design disciplines, tattooing was still strongly an analogue, handmade activity. Still, a new generation of tattoo artists are using software such as Adobe Photoshop and Adobe Illustrator to achieve better results when it comes to color and luminosity and select and edit digital fonts of their graphic expression tattoos.

Concurrently, in Portugal, during this period, the first tattoo studios open to the public open in Lisbon and Oporto. Later, they disseminate around the country. The human body started being exposed differently, an important factor in the evolution of tattooing, specially when it refers technical aspects, health, safety and hygiene. The representation of the body has always been relevant from a social, historical and artistic point of view. As an art object, the body's perception via tattooing is balanced between the aesthetic experience and its relation to daily life. Unlike the "untouchable" work of art, the human body is in constant contact with other bodies and objects, such as the clothes that cover it daily, therefore in a constant "involuntary mutation", even more when we consider the natural aging process.[10] Furthermore, in Western culture, the concept of beauty has always been wanting of cutaneous props. However, currently, these props are more easily accepted, especially amongst younger people, especially when the notions of being extreme and exuberant are paramount.

The human body is in constant movement, even when perfectly still. Sculpture, painting and photography suspend the movement, unlike other artistic expressions, such as theatre and dance, where movement is necessary for its artistic fulfilment. In theatre, the body reveals a verbal and non-verbal language; in dance, a non-verbal language[11]; body art or body modification, tattooing made with characters, carries a visual-referenced communicational language whereby its understanding differs depending on the culture where it belongs. However, the rise of artists such as Gordo Letters,[12] with its new alphabet branded as «abstract calligraphy», where direct perception is not its main goal. This «abstract calligraphy» is particularly well accepted in bodies whose inscriptions are meant to point to cultures other than Western. However, they still need to capture the reading structure within Western culture, where on the owner and the tattoo artist know what the written content means Figs. 5 and 6. "Their" calligraphic bodies are now an aesthetic and formal propagation vehicle. The bodily image remains as a base for the aesthetic idealism, where each body is shown according to its owner's desire. The tattooed body is

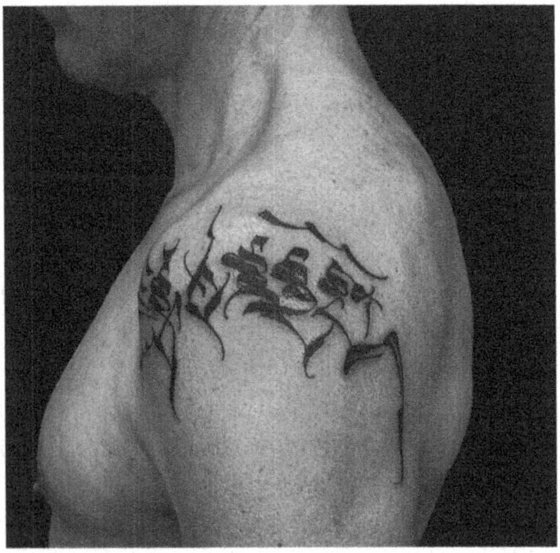

Fig. 5 "Force", male left shoulder (2016) – by tattoo artist Gordo Letters

[10] Wolf (2011).

[11] Idem—Wolf—Id. (2011).

[12] Azevedo Cardal (2016b)

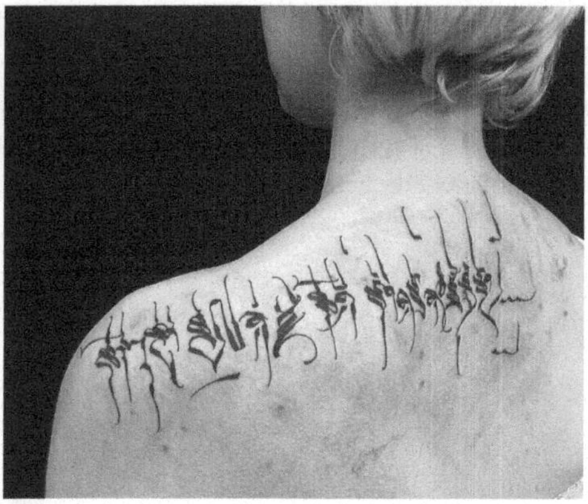

Fig. 6 "the way to paradise", female left shoulder blade (2016) – by tattoo artist Gordo Letters

represented by a variety of symbols that, depending on their placement, will have more or less visibility in public and/or private settings.

3 Lettering Drawn on Paper and on Skin

Lettering is understood "as letters that are drawn and not written",[13] a unique letter drawing, customized and made for a specific application (Flor, 2018). This type of writing is developed through building shapes. These are, however, still subject to editing in order to increase their proportionality and final result (Noordzij 2013).

3.1 The Tattoo Artists

Currently, lettering is the preferred typology amongst professional tattoo artists[14] specialized in developing customized graphic expression tattoos. Generally speaking, they are heavily influenced by the «*chicano*» [15] style, starting the development from one or two letters and conversations with the customer. They first sketch on paper Fig. 7 or directly on a digital device through software such as "Pro Create", developed exclusively for iPad using digital brushes. If the original sketch was paper-based, it is digitalized and then edited with software such as "Adobe Photoshop" and/or "Adobe

[13]Hernestosa and Scaglione (2012, 28)

[14]This new generation of tattoo artists, in its clear majority, holds degrees in Graphic Desgin or Arts.

[15]Fusion between American and Mexican styles. Its origin is traced to the Mexican immigration into USA and is related with gangs. The «*chicano*» style is favors black color and shading, either behind the drawing or inside the actual drawn letters. This type of lettering is usually conscribed by a very thin contour line.

Fig. 7 Male and female personal names, and words "Love" and "Family". Drawings of lettering for tattoos on paper (2015) – by tattoo artist Gordo Letters

Illustrator" and only later placed on the skin, through a decal. The decal is made using a thermos copying machine Fig. 8 or is done freehand, more commonly, with two-tip[16] permanent markers[17] Fig. 9, adjusting the scale to the placement. The paper sketch is developed with pencil or ballpoint pen, and its process is like any sketch developed by a creative professional. The favored color is black, and the preferred placement are the arm—in all its length—, chest and neck. Large lettering is more frequently observed in male bodies Fig. 10, and cursive style is favored on female bodies Figs. 11 and 12. Using upper case in the beginning of words and the remaining letters in lower case is also a very telling characteristic of this type of bodily marks. Texts, sentences, or words that reflect a lifestyle, are in its clear majority in English. Names, family members, love affairs and friendships are tattooed in Portuguese.

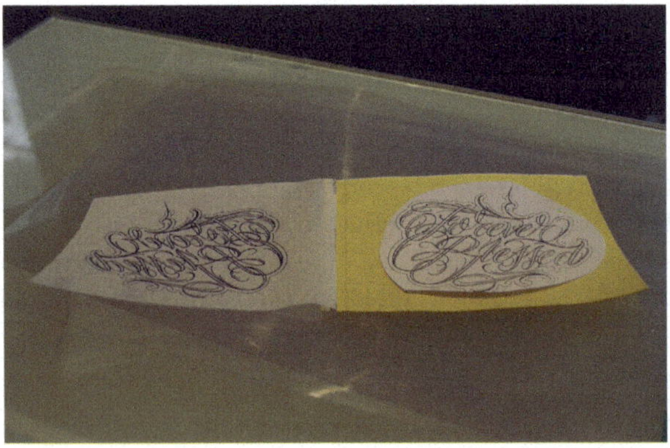

Fig. 8 Decal made in a thermos copying machine and hand-drawn decal (2019) – by tattoo artist Ricardo Frazão, at the studio Just Tattoos

[16]The finer tip is ideal for lines and drawing, and the other one is very similar to a brush. The ink is non-toxic.

[17]Round tipped are preferred. Some tattoo artists also drawn the decal up with ballpoint pen.

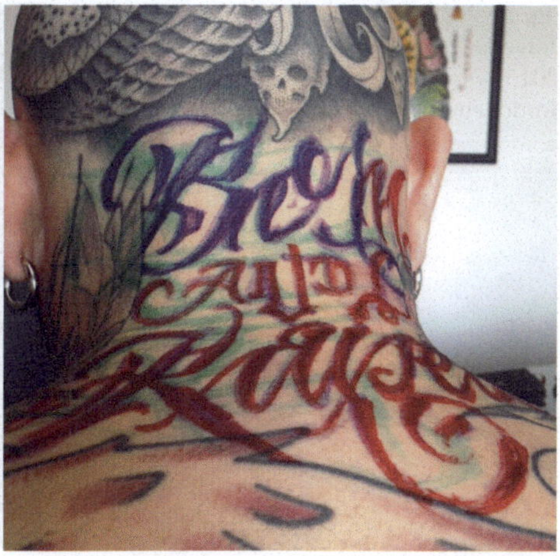

Fig. 9 "Born and raised", preliminary design in freehand (2016) – by tattoo artist Gordo Letters

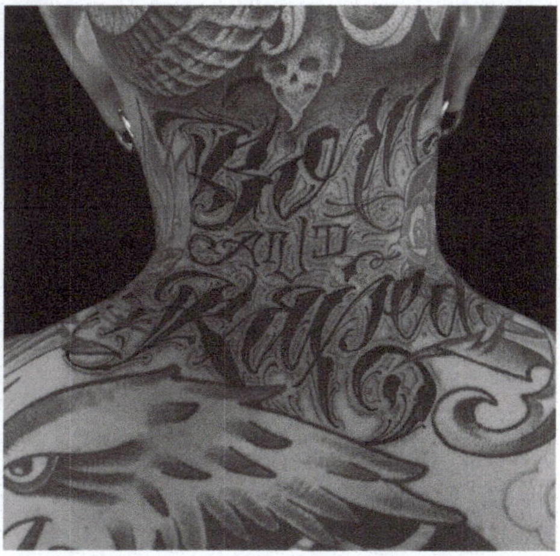

Fig. 10 "Born and raised", tattoo (2016) – by tattoo artist Gordo letters

3.2 The Designers

Nowadays, many designers devote themselves to lettering in several supports such as murals, chalkboards, branding, posters, book covers, packaging, merchandising and clothing. With regards to human skin, there are some designers that work on lettering specifically for this effect, for specific anatomical places, for their own use Figs. 13 and 14 or for others Figs. 15 and 16 who want to have them tattooed. The drawing process is like the ones used by tattoo artists, starting with a paper sketch that gets processed into a digital format, or start working immediately on a digital format. These designers hold degrees in Graphic Design and are well versed in the rules on how to develop lettering, from their terminology and weight, contrast, the use of upper and lower case, spacing and optical adjustment, as well as coherence

Fig. 11 "Child of the Universe", female left arm (2015) – by tattoo artist Gordo letters

Fig. 12 "I refuse to sink", female right rib (2015) – by tattoo artist Gordo Letters

Fig. 13 "Xesta Studio", drawing of lettering on paper by the designer Hugo Moura (unknown date)

Fig. 14 "Xesta Studio", Hugo Moura forearm (unknown date) – by tattoo artist Gordo Letters

Fig. 15 "Bird of Prey…", drawing of lettering on paper by the designer Hugo Moura (2014)

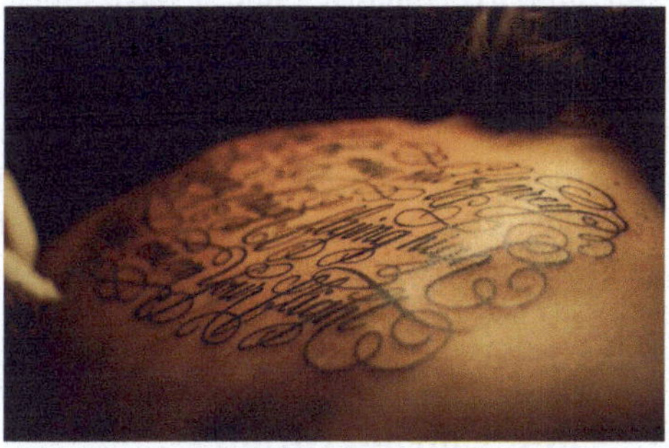

Fig. 16 "Bird of Prey…", male back (2014) – unknown tattoo artist

between width, height, inclination and baseline, serif or sans serif and its origin: calligraphic, cursive or typographic. The selection of the font and style of lettering[18] are very relevant to the "body-design project" and is perfected with the customer at the time of decision.

[18]Serif, or sans serif letters; handwritten; gothic letters; decorative letters, three-dimensional; etc.

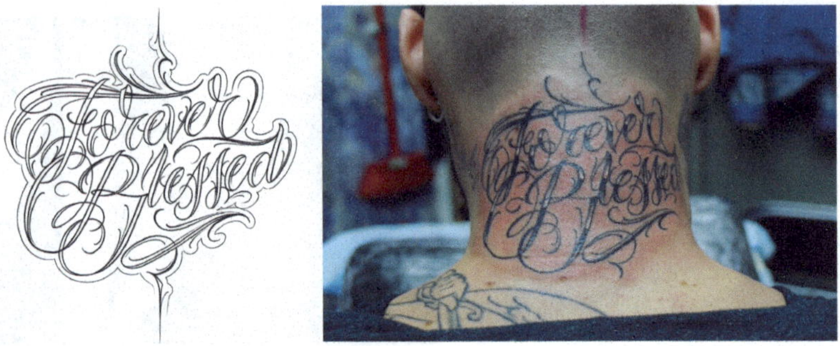

Fig. 17 Drawing on the paper and the tattoo by the tattoo artist Ricardo Frazão

4 Observations on a Tattooed Lettering

The lettering creation in the images below, authored by Ricardo Frazão, is heavily influenced by the «*chicano*» style. The English contents and placement indicate its desire of being widely and universally read (and seen). The drawing is quite decorative and with flowered details, and the letters, upper and lower case, are rather exuberant. *F* and *B* were built similarly, and their endings are alike; the letter *r* rises above the remaining letters on the word "*Forever*". The letter *s* from "*Blessed*" descends from the baseline, the letters *o*, *e* and *v* all have the same height and are built similarly. Finally, *l* and *d* have a deep relation between, as the stems end in the same way. There is a homogenic relation in all the words in this composition: same degree if inclination, the thickness and line extremities are coherent, the thick vertical stems and slim in the remaining. With regards to spacing, this is small but proportional between characters, as their connection is visible. To the final tattoo was an applied a black *dégradé* shading, with white highlights, as small light simulation details (Fig. 17).

5 Final Considerations

It is really on human skin, on the surface of the body, that these individuals personalize and differentiate themselves. In order to achieve the purpose of the unique body image itself, the selection of criteria such as letter design (notorious recurrent use of lettering), content, anatomical localization and language are of high importance in constructing the identity of each individual, in view of their projection towards the society in which they move, that is, the visibility of these body marks, marks the largest or smallest projection towards the observers and the way their bearer connects with the world. The ability to make contact with other individuals is enhanced by the differences and originality of each human being, through a deep understanding given to the body. In each historical period there were different conceptions, from the point of view, of identity construction, from technological advances (materials and instruments used) to project, compared to the initial drawing (on paper or digitally designed manually) and in the human skin itself, during the act of tattooing. Since the body is one of the privileged supports of the ornament, it becomes a field for study and application of new concepts in the area of graphic design. The body has always been a support and vehicle of communication, of sharing visual narratives, a "space" that allows the other to observe coded messages that presuppose an initiation into the reading of symbolic discourse introduced into the skin through these body inscriptions.

Acknowledgements A special thanks to Miguel Fernandes Ceia, for the translation of this book chapter.

References

da Silva RX (1926) Crime e Prisões. Depositaria Livraria Moraes, Lisboa

Ferreira VS (2008) Marcas que demarcam – Tatuagem, body piercing e culturas juvenis. Imprensa de Ciências Sociais (ICS), Iª Edição. Lisboa

Hernestosa M, Scaglione (2012) In: Dias, Ruben e Félix, Ana. *Uma perspetiva sobre letras – escrita e caligrafia/lettering/tipos de letra e tipografia*. Article published at the III Meeting of Typography held from October 5 to 6, 2012, at DAI.ESMAE.IPP, with the general theme "Convergences", p 28

Bibliography

Azevedo Cardal S (2012) Marcas es(ins)critas na pele dos portugueses – Tatuagem e Tipografia na Guerra do Ultramar (1961–1974). In: Book of proceedings UD12, 1st meeting of Ph.D. in design. UA Publisher, University of Aveiro

Azevedo Cardal S (2016a) Marcas "Sobre-Viventes" em Outros Corpos – A tatuagem escrita na primeira metade do século XX em Portugal. In: Book of proceedings UD16, 5th meeting of Ph.D. in design. UA Publisher, University of Aveiro

Azevedo Cardal S (2016b) Pedro Cerqueira, "Gordo Letters" – The first Portuguese [tattoo]grapher?. In: Livro de Atas 7ET, 7th edition of the typography meeting. CIAUD – Research Center in Architecture, Urbanism and Design, Faculty of Architecture, University of Lisbon

Azevedo Cardal S (2018) O«Corpo-Cons(umo)truído» no século XX em Portugal – A tatuagem como expressão gráfica em três fases distintas. Convergences – J Res Teaching Arts XI(22). Retrieved from http://convergencias.ipcb.pt

Castro Ivan (2016) The ABC of Custom lettering – a practical guide to drawing letters. Korero Press, London

Costa J, Raposo D (2010) A rebelião dos signos: a alma da letra. 1ª ed. Dinalivro, Lisboa

Flor M (2018[2016]) Os segredos de ouro do lettering: design de letreiros, do esboço à arte final. Gustavo Gili, São Paulo

Harvey M (1980[1975]) Lettering design – form & skill in the design & use of letters. Bonanza Books, New York

Haslam A (2011) Lettering. Manual de producción y diseño. Gustavo Gili, Barcelona

Kane J (2012) Manual dos Tipos. Gustavo Gili, Barcelona
Noordzij G (2013[2008]) O Traço: Teoria da Escrita. Blücher, São Paulo
Willen B, Strals N (2009) Lettering and type. Princeton Architectural Press, New York
Wolf AS (2011) O Corpo fotografado por Spence Tunick: um retrato contemporâneo do eu. In: Paper presented at the GT of visual and visual history, integral of the viii national meeting of media history

Graphic-Semantic Expression Map: A New Approach in Design Teaching Methodologies

Cátia Rijo

Abstract This paper aims to contribute to the discussion in design teaching and its consequent research context, considering participatory practices and methodologies. In the context of curricular units related to design and teaching experience, the problematic question arises: in the teaching of graphic design, what methodologies and instruments can we apply that associate semantics with visual thinking? This problem led to the need to better-evaluate the methodological instruments currently used and to perceive the feasibility of readapting them in order to overcome the difficulties experienced. Based on the traditional concept map, we create a graphic-semantic expressions map that intends to make the relationship between the various evaluable elements more understandable.

Keywords Design teaching · Teaching methodologies · Graphic design · Expression map

1 Introduction

In design teaching the intention is to simulate, as far as possible, real projects. Using active methodologies, the student is the main agent of their learning. This method encourages criticism and reflection, in which, although accompanied by the teacher, the student is the center of this process. In this way it is possible to orient learning in a more participative way, since the involvement of the student brings fluidity and the essence of active methodology. This method enhances the student's autonomy, developing it as a whole, so that they are able to understand aspects within the various areas of knowledge: knowing-knowing, knowing-how and being or being.

This methodology requires the student to take on an active role. It encourages the student to be autonomous, responsible and capable of fulfilling their potential, to work in a creative and inquiring way and with the knowledge of a reality that is

C. Rijo (✉)
ESELx IPL—Higher Education School from Lisbon Polytechnic, Lisbon, Portugal
e-mail: catia.rijo@gmail.com

CIAUD—University of Lisbon, Lisbon, Portugal

D. Raposo et al. (eds.), *Perspective on Design*,
Springer Series in Design and Innovation 1,
https://doi.org/10.1007/978-3-030-32415-5_19

changeable (Scallon 2015). Active methodology involves research, analysis, reflection and development of ideas individually or in pairs.

In fact, it is generally agreed that active learning is especially important in university education because it helps enhance higher order thinking and learning, such as synthesized analysis and problem-solving skills.

There are a number of tools that must be applied in the classroom in order to apply the active method (Dreeben 1976), although it has been observed that in some situations the students demonstrate difficulty in understanding some of these instruments and their applicability in the creation of a design project.

Many studies have been done in the field of visual or formal language, which I provide as references for the project, considering categories, relations and basic elements of form and visuality (Lessa 2006).

The project was designed within a Polytechnic Higher Education institution, which is built on the affirmation, among others, that importance is given to experimentation and observation, knowledge, techniques and technologies. Considering the practices and methodologies normally used in the teaching of graphic design, the research was developed in search of alternative tools to those commonly used, contributing to research in design teaching methodology.

In the practice of teaching in the area of graphic design, the problem is systematic; there is great difficulty (on the students' part) in the process of transposition of design delimitation. Here it is interpreted as expression design delimitation, or as the design phase for the investigation of context, problem identification, definition of objectives and concept. This designation is the nomenclature used in the educational environment for the graphic composition of the various elements constituting a design basic element of visual communication Such difficulty led to questioning and looking closer at the methodological instruments that are currently used. A little research explores the application of semantic contours in graphic elements.

2 Graphic-Semantic Expression Map—A New Tool in Design Teaching

During the development of a design project, it is necessary to articulate abstract or metaphorical concepts in images, in order to give visual meaning to the concepts. The semantic panel provides a visual form capable of stimulating and inspiring the process of design projects. This is a technique that helps the active method, which aims to represent meanings through the visualization of images, i.e., it is a technique that translates verbal language into visual signs.

Concept maps provide a visual and sensory channel of communication and inspiration for design research and development, which could be considered more logical and empathic within a context, than traditional verb-centric approaches. Concept maps are means of visual or multi-sensory communication (texture, movement, sound) that can be useful in the construction of communication and design process (McDonagh and Denton 2005, p. 36).

This is a tool that relies on communication through visual metaphors and is usually built by collages (manual or digital) with clippings of pictures, photographs or drawings.

This type of semantic panel offers a visual and sensorial channel of communication and inspiration for design research and development, which could be considered more logical and empathic within a context design than traditional verb-centric approaches.

Primarily, mood boards provide a mechanism for students and practicing designers to respond to perceptions about the brief, the problem as it emerges and the ideas as they develop. The construction of mood boards potentially stimulates the perception and interpretation of more ephemeral phenomena such as color, texture, form, image and status (Garner and McDonagh-Philp 2001). Through this it's possible to communicate in a more tangible way abstract concepts.

Ultimately, through the construction of mood boards, it is possible to stimulate the perception and interpretation of more ephemeral phenomena such as color, texture, form, image and status, i.e. a semantic panel can extract references such as color, shape, textures, typography as well as subjective concepts, such as emotions.

Based on the traditional semantic panel, we create a Graphic-Semantic Expressions Map (GSEM) that intends to make the relationship between the elements previously mentioned more understandable.

This map was tested and evaluated as an instrument to facilitate the creation and development of projects in the field of graphic design (Fig. 1).

This tool proposes crossing the collection made in the research phase, the design concept and the elements of inspiration with the typography, color and shape.

It's a methodological map, which will assist the process of synthesis expressive code and also boost the relationship between graphic expression and semantics in design practice.

3 Workshop

In order to test the aforementioned map, a set of workshops to be taught in Higher Education Institutions was conceived to be taught on courses which had curricular units related to graphic design, both in the first and second cycle of studies.

It was a requirement that the curricular unit in question demanded students to develop a graphic brand for a particular product or service. Although the proposed map can be adapted to other projects in terms of graphic design, for a first validation, it was decided to restrict its area of activity to the universe of graphic brand creation, in order to better validate the results.

In the present article we will analyze the results obtained from the master's in digital Identity Design at the Polytechnic of Portalegre, where the briefing consisted of the development of a graphic mark constituted by symbol and/or type, or the visual basis of the Identity of this master's degree. For this, it was necessary to contact the

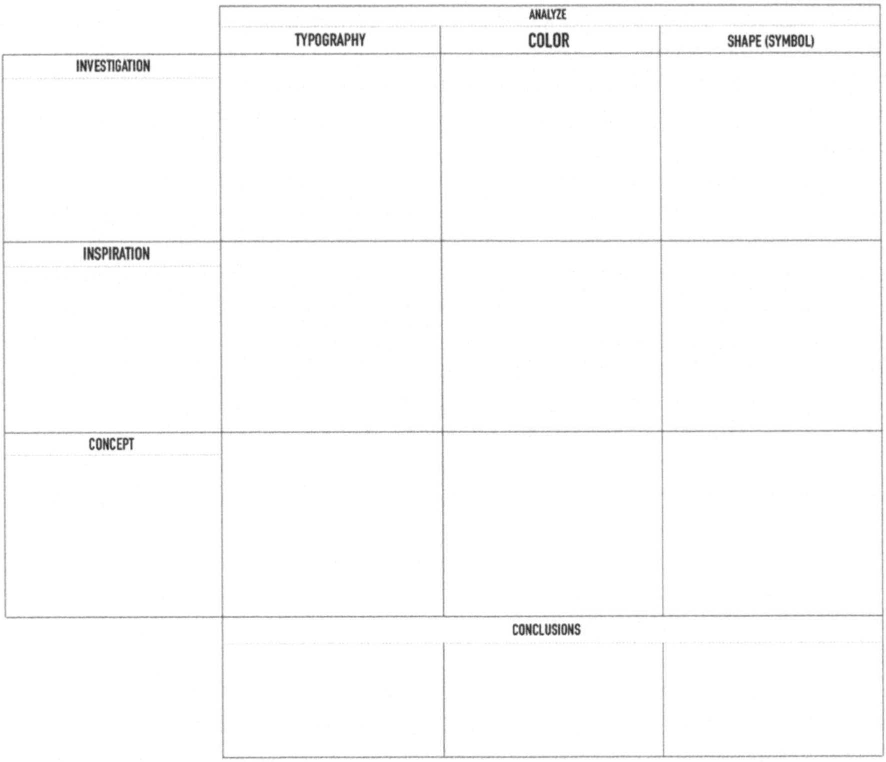

Fig. 1 Structure of the graphic semantic-expression map proposed. Author's image

coordinator of the curricular unit in order to be able to plan the best timing for the accomplishment of the same.

Considering the parameters in which the students should respond to the matrix of graphic-semantic expression, it was established that the ideal time would be after the teacher gave the briefing and the students had already carried out their research for the project and after the definition of the concept of the same.

3.1 Workshop Implementation

The Workshop entitled "From concept to design—Graphic-Semantic Expression Map" took place on the premises of the Higher School of Technology and Management of the Polytechnic Institute of Portalegre on December 14 and 15, 2018.

The workshop was structured in two distinct moments: the first moment characterized by a more expository and exemplifying part (2 h) and the second time characterized by the individual work of the students (6 h).

The first moment was structured in four parts: (1) Exposition of the concrete objectives of the workshop; (2) contextualization of the research, its objectives and presentation of relevant information on the subject to be addressed; (3) presentation of the matrices as tools in the practice of graphic design, which are the objectives and their use and (4) a presentation, with an example, to better explain the problem and the correct use of the tools presented. The second moment was characterized by the application of the students working individually, trying to fill the matrices according to their individual work, whose briefing, collection and investigation had previously been done.

The workshop was attended by 17 students, aged between 21 and 45 years old and lasted 8 h, 4 h on the first day and 4 h on the second. In order to obtain information that could be analyzed to validate if the use of the Graphic-Semantic Expression Map (GSEM) facilitated the process of creating graphic marks, after the implementation of the workshop, the students had three weeks to complete their projects. These were developed in the curricular unit where the workshop was given, and the evolution of the work was accompanied by the teacher of the subject. The teacher, with the permission of the students, shared the final results of the graphic brands at the end, in order to make the comparative analysis of the results of the matrices with the project.

3.2 *Workshop Results*

As it is not possible to evaluate all the results obtained in this article (Fig. 2), of the 17 students who participated in the workshop and who are part of the master's degree, we selected four for analysis.

These were chosen according to the criterion of divergence from each other, that is, it is intended to bring a sample of four projects, which although using the same brief, produced final results quite different from each other. In this way we can analyze the many possibilities for the same project.

In the first example (Fig. 3) the student, after completing the GSEM, concluded that for the creation of the graphic mark, the guidelines were: Typography should be sans serif, with straight rods and could be in regular, medium or bold; the predominant colors are red, blue-gray, and black; the shapes could be curved or straight, with a static composition but giving the sensation of visual uneasiness.

With these guidelines the student created a graphic mark where the symbol is three straight lines with the round terminals, where the middle line is read like an "i", alluding to the name of the master "Digital Identity". The "i" circle changes color among the brand's many applications. The source used was Din Bold and Roboto Condensed Light, both fonts without serif, DIN 1451 is a typeface without serif, which is widely used in signage, due to its readability and its design of straight and straight stems. It is an uncomplicated and unadorned source. From the Grotesks family, the Roboto fountain has a double nature, on the one hand it has a mechanical skeleton and the shapes are largely geometric, on the other hand it has friendly and open curves.

AUTHOR (STUDENTS)	CONCLUSIONS OF GRAPHIC SEMANTIC-EXPRESSION MAP			LOGO CREATED AND VARIATIONS
	TYPE	COLORS	FORMS	
ANA SOFIA BATISTA CARVALHO	HUMANIST SANS-SERIF TYPE STRAIGHT STEM REGULAR OR BOLD		STRAIGHT FORMS GEOMETRIC FORMS MOVEMENT	
ANDRÉ ANTONIO RAMALHO SURRA	HUMANIST SANS-SERIF TYPE LIGTH, REGULAR OR BOLD		ROUND SHAPES DINAMIC FORM MOVEMENT	
ARMANDO CORREIA ALVES SILVA	SANS-SERIF TYPE STRAIGHT STEM LIGTH AND REGULAR		ROUND SHAPES STRAIGHT LINES MOVEMENT	
BEARTIZ JOSÉ RODRIGUES MARTINS	FONT SANS SERIF STRAIGTH STEM REGULAR, MEDIUM OR BOLD		CURVED AND RIGHT FORMS MORE STATIC COMPOSITION BUT A SENSATION OF VISUAL INQUETATION	
CARLA MARIA GONÇALVES RUFINO	FONT SANS SERIF STRAIGTH STEM REGULAR OR MEDIUM		ROUND SHAPES STRAIGHT LINES STATIC SHPAES	
CAROLINA MARTINS MARMELO	GEOMETRIC TYPE REGULAR AND BOLD		ORGANIC FORMS ROUND SHAPES IRREGULAR FORMS	
DÉBORA MARIA SERRALHA TRIGUEIRO	FONT SANS SERIF STRAIGTH STEM LIGHT, REGULAR OR BOLD		CURVED FORMS MOVEMENT SOMETHING WITH STRAIGTH LINE	
JOÃO PEDRO VIEIRA MELO	FONT SANS SERIF STRAIGTH STEM REGULAR OR MEDIUM		CURVED FORMS MOVEMENT SOMETHING WITH STRAIGTH LINE	
LUCIA MARIA ALVES COELHO NUNES	FONT SANS SERIF STRAIGTH STEM REGULAR		STRAIGHT FORMS GEOMETRIC FORMS MOVEMENT	
MARIA FÁTIMA RESTOLHO VERÍSSIMO	FONT SANS SERIF STRAIGTH STEM REGULAR AND BOLD		STRAIGHT FORMS GEOMETRIC FORMS CIRCLES	
MARIANA NUNES GARÇÃO DA SILVA	SANS-SERIF TYPE STRAIGHT STEM LIGTH AND REGULAR		CURVED FORMS MOVEMENT DINAMIC	
PATRÍCIA ALEXANDRA NUNES BRANCO	SANS-SERIF TYPE ESTATIC REGULAR		ORGANIC FORMS CURVED FORMS	
PEDRO MIGUEL FARIA HENRIQUES	SANS-SERIF TYPE BOLD OR CONDENSED		GEOMETRIC FORMS TEXTURES DINAMIC	
RAFAEL MATEUS MOUTA	HUMANIST SANS-SERIF TYPE STRAIGHT STEM REGULAR OR BOLD		GEOMETRIC FORMS CURVED FORMS	
TIAGO MIGUEL CANAVERDE BORGES	FONT SANS SERIF STRAIGTH STEM REGULAR OR BOLD		GEOMETRIC FORMS CURVED FORMS MOVEMENT DINAMIC	

Fig. 2 Resume map with the results of the GSEM and the respective logo created of all students of the master's degree. Author's image

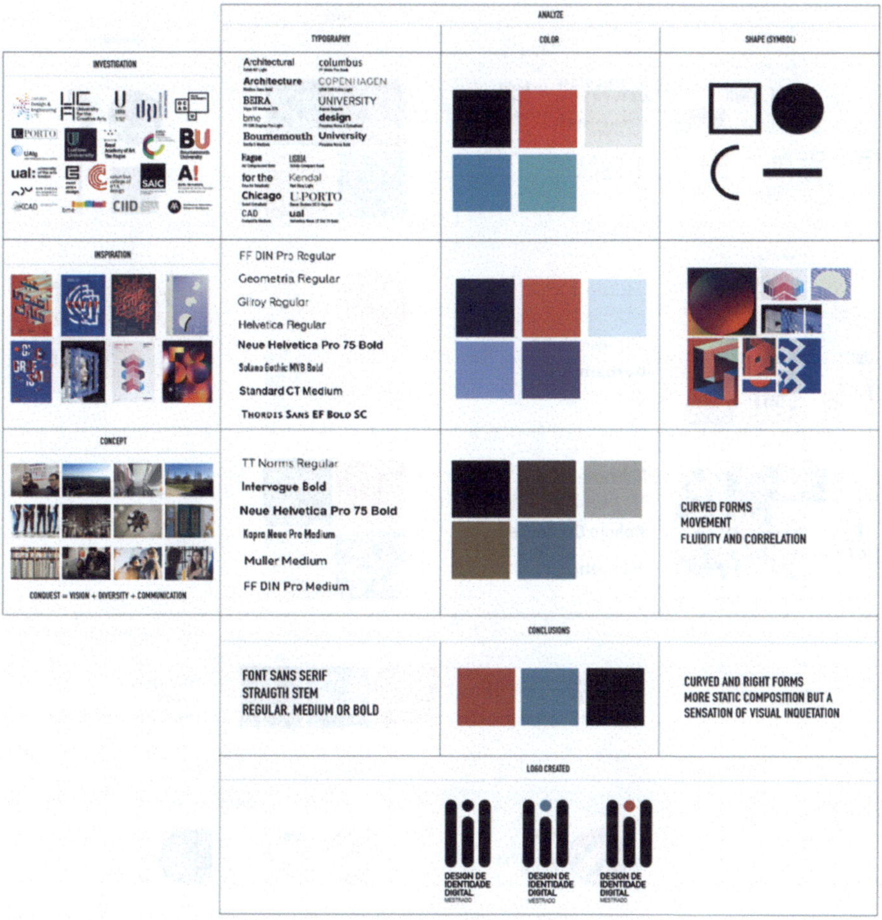

Fig. 3 Graphic-Semantic Expressions Map of the student Beatriz Martins. Adaptation of the author

In the second example (Fig. 4) presented after completing the GSEM, the student obtained the following conclusions: sans serif typography, right light stems, regular and/or bold; orange, blue and black appear as predominant colors and for the symbol, curved shapes can be represented as straight lines.

Through the previous conclusions, the student develops a logo where the "D" of "Design" and "Digital" presents at the same time as type and symbol.

By means of the color contrast used (orange and blue) and the combination of these with the forms created, the symbol presents a sensation of movement. From the main logo the student predicted two secondary decompositions of the him that could be used, through the various supports, where it will be applied.

In the third example (Fig. 5) the student obtained as conclusions Humanist Typography without serif, with straight stems, light, regular or bold; black, gray, turquoise

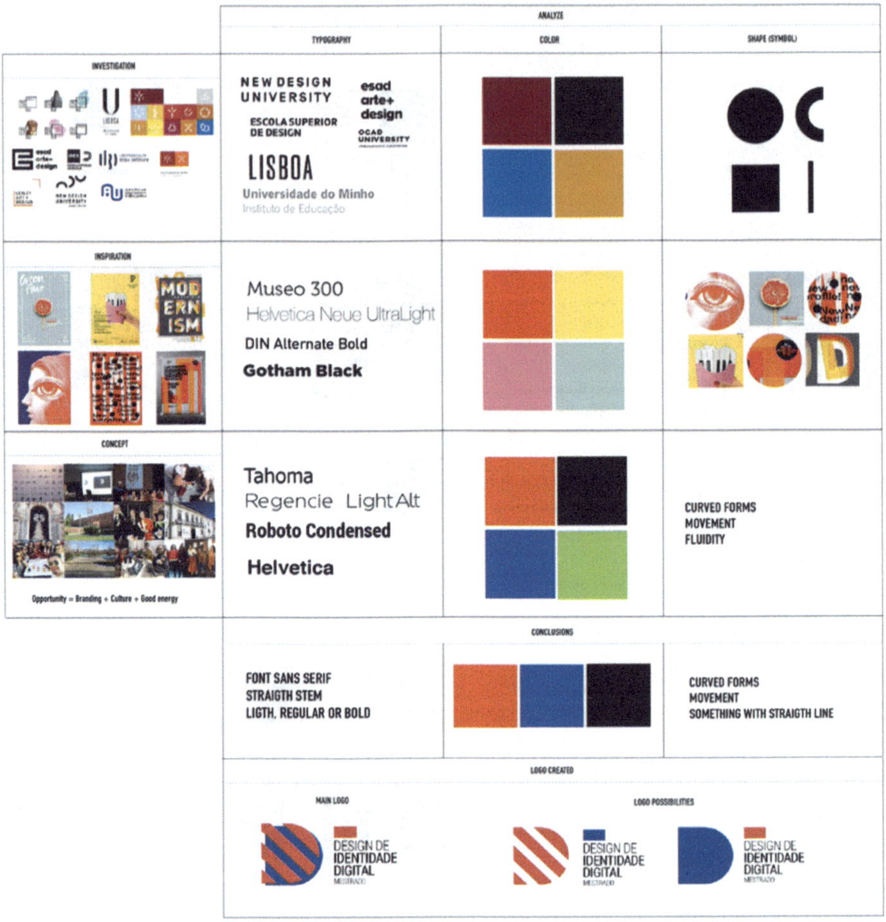

Fig. 4 Graphic-Semantic Expressions Map of the student Débora Trigueiro. Adaptation of the author

blue and gold are the predominant colors and geometric shapes, with straight lines with movement as guidelines for the construction of the symbol.

With these guidelines the student created a logo based on patterns constructed from geometric shapes (Fig. 6), these representations are a simplification of the combination of architectural elements of the ESTG building and elements representative of the formative area in which the master's degree belongs.

The source used is the Tenth Regular Pro Bold in order to establish a relationship with the graphic image of the Polytechnic of Portalegre, since this is the source used in the graphic brand of the mother institution, where this master's degree course takes place.

In the last example presented, the student concluded after completing the GSEM (Fig. 7) that for the creation of the graphic brand, the guidelines are: sans typography

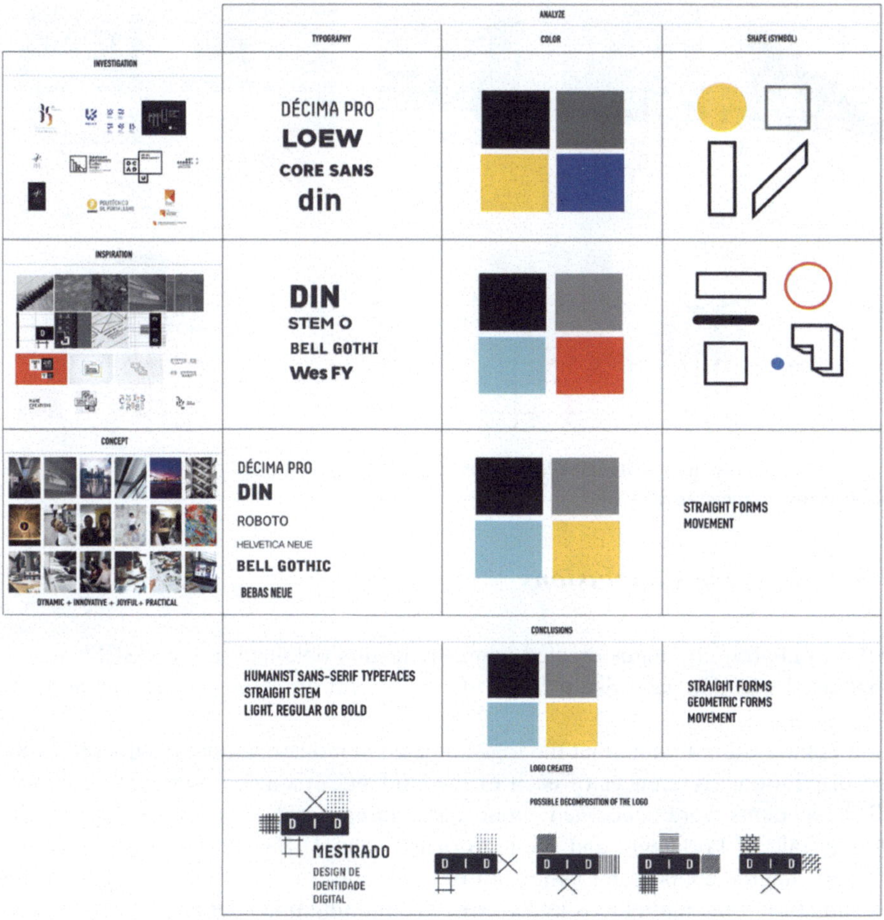

Fig. 5 Graphic-Semantic Expressions Map of the student Ana Carvalho. Adaptation of the author

with a more geometric, regular and/or bold design; the predominant colors are dry green, yellow, orange, beige and two different shades of pink. For the drawing of the symbol the conclusions are organic, round and irregular shapes.

Based on the findings, the student developed a logo where the symbol is an organic form that visually refers to the letter "D", design and digital, but also the design of the inner lines in the symbol allude to the landscape of the town. The various spots of color used meet the various tonalities that can be observed in the Alentejo landscape and internal shapes characteristic of this place. The complete designation of the Master of Digital Identity Design comes in the logo, using the Decima Pro (Bold and Regular) typeface making the link between the Identity of the Polytechnic of Portalegre and the Master's in Digital Identity Design.

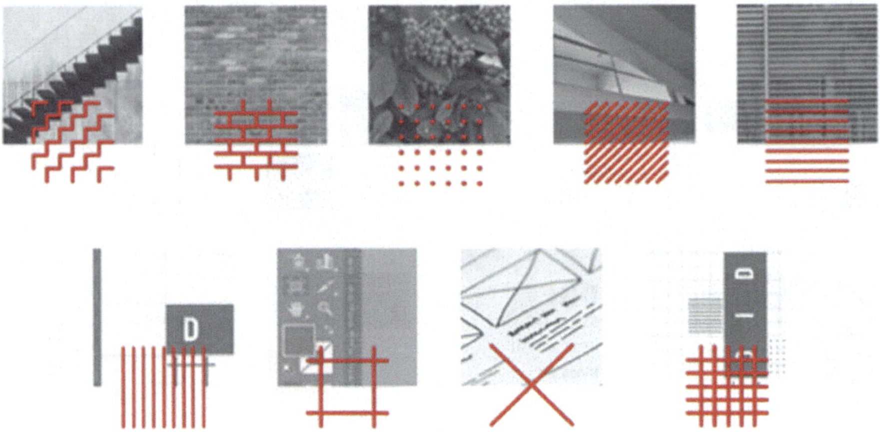

Fig. 6 Photos presented in the GSEM that the student used as a basis for creating the patterns associated with the graphic brand

3.3 *Workshop Conclusions*

After evaluating the logos created from the results obtained in the GSEM, it was concluded that this gave rise to a set of very useful guidelines for the creation of graphic brands.

It is interesting to note that the logos created demonstrate final results reflecting the inspirational references of each of the students, It's also observed that almost all the students were concerned about maintaining a link between the logo of the Polytechnic of Portalegre and the logo of the school (Fig. 8), the location of the master's degree course, and almost all of them chose to use the same typography Decima Pro of the main logos as a way of maintaining this relation.

Another interesting factor to analyze is the fact that the students identify themselves in the brands that they created, these being a reflection of a certain style already adopted by the students. In this way, the tool used was not an inhibiting element of the students' creativity or their own style.

From the feedback received by the students, the GSEM helped fundamentally in the choice of colors and typography, these being the elements causing most difficulty for students when constructing a logo. There is generally a difficulty in combining colors and choosing the type font to use.

It is also interesting to note that all students chose to create a logo composed of symbol and typography, since the briefing gave students the freedom to choose the elements that make up the graphic brand. However, almost all of them felt the need for the symbol to somehow refer to the letter "D", often associated with Design, Digital and the acronym DID (Digital Identity Design).

Fig. 7 Graphic-Semantic Expressions Map of the student Carolina Marmelo. Adaptation of the author

Fig. 8 On the left logo of the Polytechnic of Portalegre, at right the logo of the school ESTG that belongs to the Polytechnic

4 Conclusions

It is noticed that, although all the students had the same briefing, and all of them had the objective of creating a logo for the same institution, all the results that emerged were quite different from each other.

In this way, we can prove what was initially intended: a tool that helps in the research, analysis and systematization of information, giving rise to guidelines for possible paths of action. In design there are no recipes, the results are only guidelines that may help in the creation of a logo. These guidelines are a sum of the different steps inherent in creating a project in the field of graphic design.

With these guidelines, the possibilities are endless, without compromising the creativity or the uniqueness of the project, as we show in Fig. 2, where it is possible to see 17 different results for the same briefing and using the same tool.

In the classroom process, there are many exploratory tools that help students in the research phase, in designing the project and which help in achieving the various design phases.

At the end, an attempt is made to evaluate whether the association of semantic elements with graphic elements promotes the convergence between the project goals and the synthesis of expressive codes, while facilitating the interpretation and creation of new graphic products.

References

Dreeben R (1976) The unwritten curriculum and its relation to values. Gimeno Editora J. & Perez, Tradução

Garner S, McDonagh-Philp D (2001) Problem interpretation and resolution via visual stimuli: the use of 'mood boards' in design education. Int J Art Des Educ 20(1):2–129

Lessa WD (2006) Linguagem visual: algumas considerações. In: Textos selecionados em design 1. PPD/ESDI UERJ, Rio de Janeiro, p 9

McDonagh D, Denton H (2005) Exploring the degree to which individual students share a common perception of specific mood boards: observations relating to teaching, learning and team-based design. Des Stud 26(1):35–53

Scallon G (2015) L'évaluation des apprentissages dans uneapproche par compétences. De Boeck, Bruxelles

Fashion Design, Interior and Product Design

The Experiential and Trans-Aesthetic Substance of Fashion and Design. Culture and Creative-Based Models and Processes

Maria Antonietta Sbordone

Abstract The occurrence of phenomena characterized by 'actuality' is ceaseless, they are produced with trajectories that often escape the attempt to trace a general sense, even the identification of the characters that bring them together, the dynamics that pervades all activities and determines the capacity of absorption remain unknown. The essay reflects on the contemporary scenery of the activities aimed at the production of goods that are based on assumptions oriented to processes of diffused aestheticization of dailylife. It deals with production of tangible and intangible goods at the center of phenomena that cause aesthetic inflation favored by the current economic model that imposes an unquestionable change of paradigm; from a type of strong industrial capitalism to an aesthetic and emotional one define artist (Lipovetsky and Serroy in L'esthétisation due monde. Vivre à l'âge due capitalisme artiste. Editions Gallimard, Paris, 2013). It's a paradoxical combination of the product from the double opposing nature that is played between the industrial production of mass consumption and the goods that communicates with the market, to adhere perfectly to the aesthetic-emotional needs of every consumer/user.

Keywords Contemporary scenery · Experiential and trans-aesthetic · Fashion · Design

1 Foreword

The occurrence of phenomena characterized by 'actuality' is incessant, they are produced with trajectories that often escape the attempt to trace a general sense, even though the recognition of the characters that link them together, the dynamics that pervade all activities and determine their absorption capacity remain unknown. A thin layer of demarcation between what is current and what is no longer, defines the meaning and therefore the degree of actuality of one production compared to another.

M. A. Sbordone (✉)
Department of Architecture and Industrial Design (DADI), Università Della Campania Luigi Vanvitelli, via San Lorenzo, 81031 Aversa (CE), Italy
e-mail: mariaantonietta.sbordone@unicampania.it

D. Raposo et al. (eds.), *Perspective on Design*,
Springer Series in Design and Innovation 1,
https://doi.org/10.1007/978-3-030-32415-5_20

The layer becomes a demarcation level, a sort of algorithm, when finally it is formed it contains and is animates by trajectories and nodal points that open to further layers; a multidimensional universe of languages, of completely renewed contents, of meanings that, though evoking be distinguished traits from the past, transform the common sense and the culture of time.

Translations, recoveries, quotations, time leaps are all mechanisms that contribute to make the different levels supportive each other, they are the explicitation of creative operativeness that substantiates the visual and aesthetic identity of everyday life. Complex construction that contributes to the setting-up of articulated surfaces, where ripples, contact points and discontinuities are highlighted, Enzensberger (2000) defines them as "non-contemporaneity".

Episodes that bounce back from the past or anticipate the future, express the ability to redefine the directions to take in the production of content and consequently in the influence they will exert on the identity of goods.

Fashion and Design take on a central role in the creation of that reservoir of "genres" from which to draw to renew and renew oneself; the contaminative and anticipatory nature acts as a propeller with respect to those immanent processes, succeeding in invading other fields of knowledge and extrapolating practical knowledge to be used.

Goods that arise from the hybridization of concepts, methods, models, practices, shaped within their project areas and practical application, are available for continuous rearrangements, suggesting a substantial interaction with other production sectors.

Fashion and Design are considered "creative and cultural industries" they take on the role of manufacturing industries with production capacity and distribution in large numbers, while holding a privileged role as a thermometer of consumer taste and the cultural and aesthetic advancement of society.

The European Commission, in its Green Paper (2010) on "cultural and creative industries", acknowledges a potential role in the switch from factories to creative communities; whose raw material, identified in the imaginative capacity to innovate through creation, is the basis of production and distribution of goods or services in which cultural values are transmitted. Fashion and Design draw from culture as a driving force for innovation, playing a critical and distinctive role within society, translating culture as a pre-eminently functional character within the production process.

The production of goods in Fashion and Design is not only and simply an *affaire de goût incontestable*, a precept, but it represents a value that runs through society and that makes tangible the culture of an era, expanding, interpreting, declining it, in short, making it in "alive" and "usable" material.

2 Aesthetic Experientialisation of Goods

The actuality embodied by physical and virtual, tangible or intangible objects, particularly in the fields of Fashion and Design, arise from very complex processes of design and production thought processes. Fashion and Design feed on the persuasive dynamics of "widespread aestheticization", in some ways widely experimented and codified in contemporary societies; it represents the concrete evidence of the "value of goods" that is transmitted primarily through the perceptual mechanism of aesthetic asset, then through the evaluation of the technical-functional contents, and finally of the economic one and its environmental and social impact.

Today we are experiencing a particular phase of widespread aestheticisation which paradoxically fully interprets the current economic model, and which imposes an unquestionable paradigm shift; from a type of 'robust industrial capitalism' to an 'aesthetic and sensitive one' defined (…) artist". Artist capitalism has the character of creating economic value through the aesthetic and experiential value" (Lipovetsky and Serroy 2013) (Fig. 1).

Therefore, assuming the characteristics of the 'new' as values that cannot longer do without and that must be present in all production, sales and distribution systems,

Fig. 1 The widespread aestheticisation. Vendor of tea capsules in a shopping mall. Jinan, China 2018

ultimately the consumer/user is led to an experience of goods which are increasingly contaminated by aesthetic-emotional models as a tool for personal affirmation that occurs through the purchase.

The main features of the aesthetic-emotional experientialization are to be ascribed to phenomena concerning the aestheticization in a broad sense of the places, the means of communication and distribution, accomplices of a type of seductive representation to induce the continuous purchase of goods.

Some aspects are highlighted that have an increasing weight in the dynamics of the offering and they can be distinguished in this way: the growing weight of "markets of sensitivity" in line with the nature of goods which have a leverage on affective and communicative-symbolic factors; The Design Process, intended in the sense of new value configurations, through the design of tangible and intangible goods that moves from the logic of design-driven (Verganti 2013), result of the connection of economic-managerial, socio-cultural and aesthetic-communicative knowledge; the capillary aestheticisation of the places of commerce, distribution, private life, work and entertainment; the constant "de-differentiation" of the economic and aesthetic spheres, with the consequent hybridization of methods and operating practices.

If then, the economy, hybridizing itself with aesthetics and sensitive experience, discovers in the imagination, in the uncoded knowledge of peripheral experience (Granovetter 1988) and in design-driven innovation, new levers for the value creation, it is essential that a sort of intuitive and aesthetic "artist economy" be configured, confirming the rise of the immaterial and communicative register.

However, far from the occurrence of conditions of widespread beauty that would increase the aesthetic and cultural level of goods and living environments in general, this phenomenon does not guarantee that there is a correspondence between aestheticization and contents; "the artist dimension of capitalism has to do with the objectives and business strategies, not with the results obtained" (Lipovetsky and Serroy 2013).

It is, therefore, an economy that aims at exploiting the new values which emerge from society that they obey at the arising of orientations, even divergent, in the demand for goods that satisfy explicit or latent needs. Fashion and Design fully fulfill the demand for goods strongly characterized by aesthetic-emotional values, enclosed in all possible partial worlds. Through the creation of different "genres", the connections with the consumer/user are multiplied; incorporated in a single stratification, the autochthonous characters are mixed with the stylistic systems and languages of global cultures.

Unexpected openings towards the outside bring back into the production systems the concrete experientialisation of the dialogue established within the design process that collects and organizes the elements useful for the project; it systematizes and deploys knowledge, sets up dialogue between the actors, produces value within the 'transesthetic industrial culture' (Sbordone 2018) of "artist capitalism" (Lipovetsky and Serroy 2013).

3 Productions of Sense

It can, therefore, be said that Fashion and Design derive value from the generation of "pervasive aestheticization", transforming it into a "marketable sense" (Lipovetsky and Serroy 2013), thanks to which a real cognitive supply chain is set up, based on heterodox ways of value creation processes.

The identification of a heroic phase of the production of contents inseparable from their identity as form-goods refers to the amplified economic value by symbolic-aesthetic values; a phase that depends on the invention of economies based on the values of daily life that stand between the economies of luxury and the emerging ones of 'diversified homogeneity' (Sbordone 2018).

The "generalization of aesthetic strategies" (Lipovetsky and Serroy 2013), i.e. the aestheticization extended to all productive sectors, is the theoretical foundation of the supported "artist" economy and which, vice versa, favors the development of the "diversified homogeneity". The latter, by executing the commercial strategies of "creative capitalism", mixes and crosses different fields and genres in order to propose an extremely differentiated offer.

The commercial strategy of 'diversified homogeneity' permeates all sectors, experimenting one sector with another by hybridization the distinctive features: fast fashion and design mass market adopt the stylistic codes of luxury; reception areas and showrooms become places of communication of brand values; restored historical centres become sets, aspiring to provide film contests or set for events; the use of museums and of the digitalised artworks provides an immersive and disneyfied growing experience; advertising often becomes a sequel to episodes like a mini television series; fashion shows leave the catwalks become performances; the domestic environment moves between the spectacularization of hypertechnological solutions and the most sophisticated exhibition of a personal and unique style (Fig. 2).

The category of 'diversified homogeneity', as an oxymoron, gives the idea of the product-goods that combines the double soul; the industrial production of mass consumption and the goods that dialogue with the market, to adhere perfectly to the aesthetic-emotional needs of each consumer/user. Critical capacities are inhibited and the real experientialisation of goods, made up of direct personal knowledge and experimentation, is replaced by a mass of passive spectators, individually (and unconsciously) preys by an acclaimed 'seductive addiction' (Sbordone 2018) (Fig. 3).

The ability to distinguish between the production of goods that still have a meaning, denotes a dynamic of conscious appropriation and direct responsibility that can lead to diametrically opposed mechanisms, i.e.: the ones dedicated to the claim of aesthetic-ethical principles through the achievement of shared projects in a social context, supporting the values of everyday life; with aesthetic-emotional mechanisms, the others, according to the tendency of hedonism *tout-court*, in pursuit or anticipation of self-gratification.

The drifts are known and widely tested, in order to hinder the spread and the indefinitely deepening in all the production systems which act indiscriminately through the improperly recalled value levers, it is necessary a reorientation of the markets.

Fig. 2 The diversified homogeneity. The fashion shows leave the catwalks become performances. Vertigo Sylvio Gardina performance, Roma, 2018

Fig. 3 The diversified homogeneity. Teapots set on shelves for the coffee break. Guandong, China 2016

The risks and opportunities will be interchangeable, in the name of an alleged 'mirror economy' (Sbordone 2018) of the real one whose distinction mechanisms (of the true or false type) for an eventual selection will be definitively abolished because they are no longer so distinguishable.

On the other hand, the claim of aesthetic-ethical principles with the possibility of achieving shared objectives, would limit the parabolic and illusive effects; the affirmation of the commitment in personal terms is disposed to a relational dynamic that arises from the social change and acts on the local level to extend to the network. This represents the critical aspect of the de-regulation of aesthetic and economic categories, to which one responds with the appropriation of one's own sphere of action that promotes the creation of forms of participatory design.

The mix of values between social engagement and new forms of value creation is interchangeable, in relation to contexts of life, to peripheral ties, in short, to the rebirth of everyday economies that build global interfaces (Fig. 4).

Fig. 4 The diversified homogeneity. The designer Nendo materializes air with breeze of light by Daikin. Milano, Italy 2019

4 The Dyschronic Creativity of Fashion and Design

In general, the systems of Fashion and Design, discount and promote the consequence of the continuous switching between real events and "mirror events"; ultimately, Fashion, above all, has anticipated for a long time the occurrence of phenomena of "aesthetic inflation" due to "hyper-production", emblematic aspects of contemporary phenomena (Lipovetsky and Serroy 2013).

Fashion and Design, among the privileged areas of present-day manipulation, are based on local and global, artisan, industrial or digital production methods, all converging in the sole objective of semantic actualization of society.

In particular, the processes of Fashion in their complex phenomenology, are the perfect example of the experience of the contemporaneity, lending themselves to the definition of the sphere of human activities which introduces a substantial discontinuity in the flow of linear time, a phase shift translated into a substantial "de-synchronization".

The statement on actuality or inactuality, the non-being-more-fashionable, calls into question the temporal linearity of events that are no longer so automatic. The actuality or otherwise of the fashion could be enclosed exclusively in the moment in which an outfit is created or when it becomes a prototype to move on to the various stages of processing. Conversely, it refers to the catwalk or to the moment in which it appears in magazines or when it gets to the distribution, and you see the dress worn that crosses the places and therefore the space and time of everyday life.

It shows a "dyschrony" (Sbordone 2012) or a real "de-synchronisation" between living one's own time and being already out, representing an unreachable threshold of which the markets are aware and act as a lever on the mechanisms of personalisation and the belonging to selected groups or elites. The conception of the transition of time in Fashion for Agamben (2008) is related to a de-synchronized time with a 'dyschronic' character, stating that "fashion has its time to manifest itself and is constitutively in advance of itself and, precisely for this reason, even always late, it has incessantly the form of an elusive threshold between a not yet and a not anymore". In this sense, if the actuality of fashion involves a mismatch, being fashionable means at the same time already being out of fashion.

The spasmodic search for actuality is supported by the industry which, by producing at a rapid rhythm and completely upsetting the custom of seasonality, conveys on the markets huge quantities of always updated goods, causing an acceleration and a proliferation of "seductive addiction" (Sbordone 2018).

In general, this condition, known and explored by the fashion system, causes the frustration of never corresponding to one's own time. Instead, who fits perfectly, cannot be contemporary because he/her is not able to live it in full consciousness. This peculiar "displacement" favours the realization of another important condition of non-contemporaneity, that particular dynamic enclosed in the "quoting"; it is emblematic the case of fashion that returns, the '50s, '60s, or '70s, inspiring collections characterized by the phenomenon of "vintage".

The observation and the unconditional adoption of these phenomena, with their reappearance, inevitably involves the continuous revision of the past, that is infinitely interpretable and manipulable. The plurality of visions is the ground from which the "dischronies" (Sbordone 2012) feed; a sum of layers which hardly can be separated and from which elements in chaotic temporal order emerge.

This process of re-invention from the re-emerging past is at the basis of the dynamics of the creation of so-called 'mirror events' (Sbordone 2018). In the actuality of fashion, the process of de-sacralizing of the original is a common practice, the recognition of manipulation is accompanied by a sort of appropriation of the authorship of the original models, authorized by the ability to re-invent them and thus make them recognizable only as "cultured quotations".

The dynamic of re-invention of the original model to innovate, is the engine of "dyschronic creativity" (Sbordone 2018) that anticipates the past, and reflects the ability to push forward into the contents to extrapolate new 'marketable sense'.

5 Advanced Territorial Organizational Model of Widespread Creativity

The current processes of value creation, through the production of goods, are generated in environments where the relational dynamic is established in terms of networks between the different stakeholders of the production, training and distribution system. The social and cultural impact of product-goods, at the centre of a system of relations promoted by ICT, introduces new organisational models in all phases of value creation.

The Design system, with all its implications, acts on the new design oriented processes aiming at the realization of continuously updatable experiences with reference to knowledge-based goods. The model to implement starts from the need to share expectations and desires corresponding to a system of relations that it is translated into a 'decentralised model of production, distribution and consumption'.

Knowledge-based merchandise is a combined expression of the individual's commitment and of 'profitable' local production conditions; capabilities are identified, which are made opportunities to become new forms of commitment and thanks to the creation and availability of new methods in production systems, their value is deployed through the "cognitive chain" in multifunctional networks.

The manufacturing, training and research structure of Design in productive territories is configured as many 'system-areas', where an aggregative logic is established and it promotes dialogue and cooperation between the actors. A sort of local organizational model that focuses on the availability of specific and distinguishable resources, aware of the fact that it facilitates the implementation of innovative activities, if integrated into production systems, would avoid competition on costs with all the implications of the case. The centrality of the interaction between the production, training and research systems is essential to introduce knowledge and creative-based innovation, which translates into a clear process of identification of capabilities; through research, analysis and listening, the objective is to relaunch local production based on cooperative and integrative dynamics.

5.1 Persistence of Tacit Knowledge and Capabilities

The development that comes from below produces resources that are not pre-established and predetermined, but derive from dynamics which are produced locally; the local communities regain strength by recovering their own values to lead the processes of transformation, and become consolidated communities of 'sense' that encourage the rise of 'capabilities'.

These are the result of individual commitment reinforced by knowledge supported by training and continuous experimentation within collective learning in territorial laboratories and digital networks. The capabilities introduced in the productive network guarantee the realization of so-called 'dynamic competitive advantages'; in fact, they are the nucleus around which the local actors experiment innovative solutions for the continuous evolution of the contexts. Through the availability of knowledge and skills, accompanied by a prospective vision, we learn and develop integration processes of ideas and design practices; it is highlighted the need to connect integrated supply chains in a perspective of complementarity, incorporating local skills and capabilities within the production-creative systems.

The development of local integrated supply chains marks the watershed with the previous economic model and it is grafted onto the current model of "hyper-complexity": human needs represent "the agents, both beneficiaries and successful bidders of progress, but they could also be directly or indirectly the main means of any production".

5.2 Systemic Innovation Processes

The processes of innovation in Design have a strong accent of immateriality, in fact, the new developments are not only of a technological nature, but those that concern intangible contents play a fundamental role: the enrichment of the narrative, the meaning and value which takes place through design, the reorganization of processes and work commitment, marketing and branding. The shape of a system of relationships considers the presence of actors of production, training and research, gathered according to an aggregative logic following the rules of networking.

For a long time, the Italian University has been investing in research and training in areas of excellence typical of Made in Italy: degree courses at several levels dedicated to design and fashion contribute to create the conditions for the establishment of a "creative ecosystem" in synergy with research centers and companies (Fig. 5).

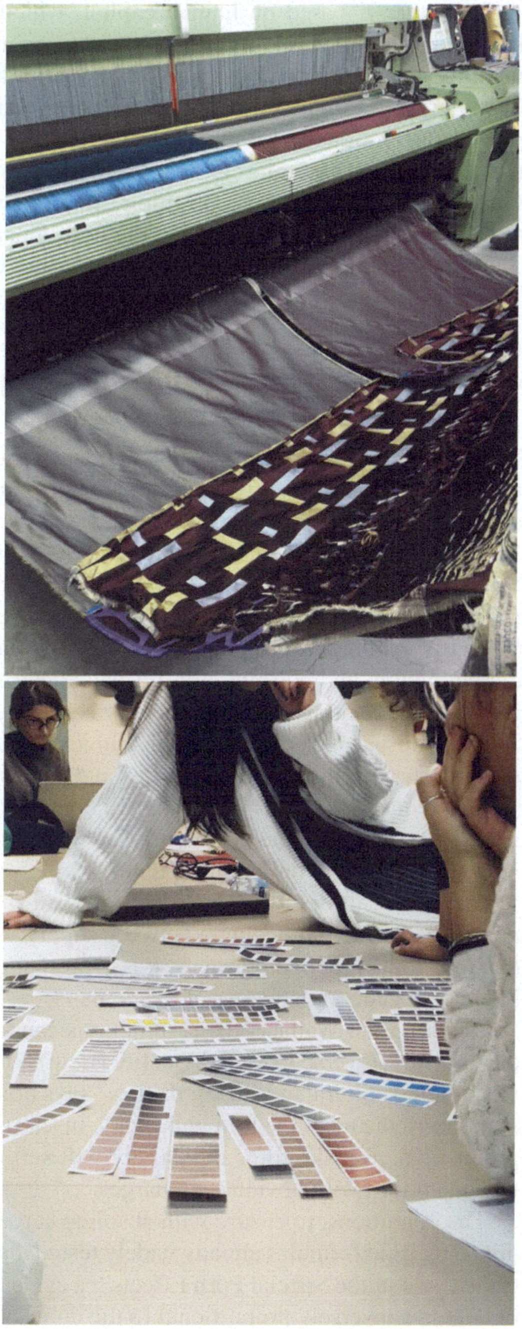

Fig. 5 The creative ecosystem. New Silk Textiles Design made by co-creation processes between University, Research Centers and Factories. Università della Campania, Textile Design Course Master Degree in Fashion Ecodesign (prof. M. A. Sbordone). San Leucio, Italy 2019

The latter is based on human capital, which is the basis of the contemporary knowledge-based economy. The aim is to aggregate universities, businesses and institutions in a sort of "productive ecosystem driven by research and innovation". The aggregative logic sets out in place specific skills for the realization of multidisciplinary interactions and comparisons in terms of operational development strategies that dialogue with specific clusters. From the point of view of the "productive ecosystem driven by research and innovation", it is assumed a type of aggregation according to the rules of the Network, which the training and scientific skills and the network of national and international relations are combined in.

6 Conclusions. Submission to Widespread Aestheticization

The continuous surpassing of phenomena of actualization contents is essentially the expression of a practice that implements small incremental updates; the 'new' advances and it seems to adapt well to the laws that D. F. Wallace identified in an essay from a while ago ("The seminal importance (so to speak) of Terminator 2", in the book "Di carne e di nulla", 2013) and these laws paradoxically seem to pervade the sectors of creative and cultural industries. Production planning is permeated by the pursuit of 'newness' which must be channeled into programmed and consolidated production, dissemination, and consumption formats, determining the offer of a market that is now saturated and which the consumer is "avidly submitted" (Sbordone 2018).

This unusual and controversial dynamic denotes the unconditional submission to a type of offer that the consumer suffers and against which he/her is not the spokesperson for requests in terms of qualitative contents; at the same time characterized by avidity, makes the same consumer, fed with special effects and sequels, an active buyer in pursuit of continuous updates.

If the contents of the productions, based on widely tested formulas and functional now to the profitable encounter between supply and market, are stabilized on methods that lead to well-defined formats, it is clear that the consolidation of the offer is also based on a general flattening of the demand.

According to Wallace, the assumptions can be traced back to a paradoxical law, the "Law of Reverse Cost and Quality (Lcqi)", which fits well to the productions of creative and cultural industries; the law "says quite simply that the greater the budget of a film is, the more the film will suck". In short, the financiers provide megabudget productions to make films that surely will sell out the box office, ensuring in this way the return with obvious surcharges.

The conditions, to ensure with absolute certainty the return of capital, are determined by rigid formulas already widely tested that Wallace encloses "in the insidious triple cycle of the Special Porn Effects": a cycle that confirms the decreasing quality of contents inversely proportional to the contribution of digital special effects, with the consequent progressive abandonment of the narratives in favor of entertainment that causes the definitive discommitment of creative young people.

The general dulling, caused by the incessant spectacularization of everyday life, is one of the searched effects and the sophisticated marketing strategies, based on "distractive valorization", are responsible for an even more lethal effect, a real 'seductive addiction' that concerns the spheres of individual sensibility. The latter, when stimulated, must be continuously fed, of course the goal is to increase the desire for new experiences based on the constant updating of the "special effects" that causes precisely addiction.

Wallace's critical analysis, although focused on the entertainment industry, shows how the decalogue of extractable rules is functional to a well-established practice that instrumentally acts on the 'mature creative industries', extendable to the universe of contemporary productions subject to the same rules even if with different execution methods. By establishing the functioning, the analysis gives us, in short, with impressive evidence, a dimension of contemporary production piled up with events that translate into goods (of the most varied nature) that are based on the assumptions of an extraordinary as well as ordinary and homologated "way to widespread aestheticization".

In other words, Lipovetsky identifies a fourth phase of aestheticization of the world characterized by strict rules of "commercialization and individualization"; if it is true that the dulling caused by the massive production of images with special effects as a 'seductive addiction' as result, this is supported by the production of tangible and intangible goods that determine the phenomenon of "aesthetic overabundance or inflation" (Lipovetsky and Serroy 2013) (Fig. 6).

In general terms, we are in a phase called "transesthetic", a kind of "hyperart" (Lipovetsky and Serroy 2013) in which every form that intrudes into art, it infiltrates into industries and in all the interstices of trade and dailylife. To be implemented in a widespread and homogeneous way, it needs a "generalization of aesthetic strategies" that have an immediate feedback in all parallel commercialization and distribution strategies.

According to Lipovetsky and Serroy (2013), these assumptions follow the "extreme logic of commercialization and individualization" that invade all sectors of the cultural and creative industries; the sectors are those of cinema with films financed by colossal budgets, advertising that become sequels, the overflowing of TV series, shows on television and in large shopping malls, museums, concerts; all these present duplicate realities of hybrid and generalized formats that aim to create temporary but especially memorable 'distracting experiences'.

Fig. 6 The aesthetic overabundance or inflation. Prada collection fall/winter 2018. Milan, Italy

References

Agamben G (2008) Che cos'è il contemporaneo?. Nottetempo collana I Sassi, Assago Milanese

Barthes R (1967) Il sistema della moda. Einaudi, Torino

Barthes R (2006) Il senso della moda. Forme e significati dell'abbigliamento. Einaudi, Torino

du Gay P, Hall S, Janes L, Mackay H, Negus K (1997) Doing cultural studies: the story of the Sony Walkman. Sage Publications (in association with the Open University)

Enzensberger HM (2000) Zig Zag. Saggi sul tempo, il potere e lo stile, Einaudi, Torino

Granovetter M (1988) La forza dei legami deboli. In: La forza dei legami deboli e altri saggi. Liguori (original article: The strength of weak ties, in «Am J Sociol» 78, 1973)

Granovetter MS (1973, May) The strength of weak ties. Am J Sociol 78(6):1360–1380. The University of Chicago Press. Stable URL: http://links.jstor.org/sici?sici=0002-9602%28197305%2978%3A6%3C1360%3ATSOWT%3E2.0.CO%3B2-E

Lipovetsky G (1991) L'empire de l'éphémère. Folio Essais, Paris

Lipovetsky G, Serroy J (2013) L'esthétisation due monde. Vivre à l'âge due capitalisme artiste. Editions Gallimard, Paris

Lipovetsky G, Roux E (2015) Le luxe éternel. De l'âge due sacré au temps de marques. Folio Essais, Paris

Manzini E (2000) Design research: reasons and possibilities. The production of a design knowledge: taking time in the age of real-time. D+R Design Plus Research Conference, Milano

Mc Donagh D, Hekkert P, van Erp J, Gyi D (2004) Design and emotion: the experience of everyday things. Taylor & Francis, New York

Rullani E (2003) The technological evolution of industrial districts (a cura di) (Belussi, Con F, Gottardi, EG), Amsterdam, Kluwer

Rullani E (2004) Economia della conoscenza. Creatività e valore nel capitalismo delle reti. Roma, Carocci

Rullani E (2010) Modernità sostenibile. Idee, filiere e servizi per uscire dalla crisi. Marsilio, Venezia

Sbordone MA (2012) Discronie. Fenomeni del contemporaneo nella Moda e nel Design. Alinea Editore, Firenze

Sbordone MA et al (2018) Listening design. II Design per i processi di innovazione. Collana design experiences design, vol 5. Listlab editore

Simmel G (2001) La moda. Oscar Mondadori, Milano (1895 prima edizione)

Sterling B (2008) La forma del futuro. Apogeo, Milano

Thorstein V (2011) Il consumo vistoso. Editore Centopagine (1902 prima edizione Conspicuous Consumption)

Trabucco F (2015) Design. Bollati Boringhieri, Torino

Verganti R (2013) Design-driven innovation. Cambiare le regole della competizione innovando radicalmente il significato dei prodotti e dei servizi. Rizzoli Etas, Parma

Wallace DF (2013) Di carne e di nulla (Both flesh and not). Einaudi, Torino

Trends Management: The Qualitative Approach as a Methodology

Sandra Regina Rech

Abstract The potential of trends management is originated in the ability to understand the changes in human behavior and to translate this knowledge in guidelines that can bring innovation. The qualitative approach considers the society composed of individuals and groups, who share meanings according to collective expectations and perspectives. Based on these precepts, the researcher in the field of trends investigates processes, facts and situations in the social scene that, interconnected, may explain the analyzed phenomenon. Through literature review and theoretical problematization, the concepts and approaches referenced in qualitative research and trends management are discussed, evidencing the methodology and guiding assumptions of transformations in the patterns of human behavior.

Keywords Trends management · Fashion · Qualitative research

1 Introduction

In the field of fashion design, the trends management, in its various facets—market, consumption, concepts, among others—, provides interesting information for the product development department. It is a tool used by companies to deal with this requirement, since the future that is to be foreseen, stubbornly, is an indeterminate future and may be subject to numerous interpretations. The continued pursuit of product differentiation as a way to avoid standardized types, commodities, which directly generate profitability in companies with high production scale, continues to be the key to the survival of smaller industries. Therefore, the fashion chain must be restructured, initially, through a direct offensive to the fragility and fragmentation of small companies, a characteristic prevailing in the central Brazilian production centers. This particularity inevitably influences the level of productivity and the competitive insertion of these organizations into the different market niches, since sustainable

S. R. Rech (✉)
UDESC—Santa Catarina State University, Florianópolis, Brazil
e-mail: sandra.rech@udesc.br

CIAUD—University of Lisbon, Lisbon, Portugal

D. Raposo et al. (eds.), *Perspective on Design*,
Springer Series in Design and Innovation 1,
https://doi.org/10.1007/978-3-030-32415-5_21

competitive advantage is the foundation of above-average technical performance in the long term.

In this way, constant and oblique monitoring of trends—uninterruptedly associated with the paradigms on which they are based—becomes strategic information packages and an instrument for the development of products and services and the management of innovation and branding processes. Intangible assets are increasingly useful in the competitiveness of the textile-clothing industry. In this type of business, intangible assets in no small extent include prior and post-production assets, such as design, product design, engineering, marketing, brands (preferably global), logistics, chain management, and coordination capacity.

It is mandatory to unveil at the same time what the consumer feels about existing products or create new needs to arrive at the concept of a new product. The literature points out that, in any case, it is in the direct contact with the contemporary sociocultural panorama that one obtains the necessary references for the elaboration of an updated and commercially situated collection, since the research of trends analyzes and decodes information of diverse areas, such as economics, politics, sociology, arts, science, and technology. The expected result of trend management is not to rationalize that only one direction is correct, positively, but, conversely, to present plausible options for the future. With this, we seek to research and analyze research techniques that assist in the monitoring and application of trends to guide organizations in the construction of an innovation process.

It should be noted that trends management is a new area of research within the Academy and a disciplinary field that aggregates concepts and tools from different disciplines. Also, they make it possible to understand the oscillations of consumption patterns and human behavior, within the scenario of why and how they are externalized, as well as providing guidelines for the planning of marketing strategies based on the analysis and interpretation of these manifestations. Thus, not only do they aim to translate these standards into clues of innovation, allowing for better management and policy generation, but also presuppose the practices of identifying and monitoring the trend. Consequently, to analyze trends, transversal knowledge is necessary, which is opposed to a Cartesian line of thought. There is, in this process, an approach with the Social Sciences and Human Sciences in the use of methodologies that allow the identification and analysis of manifestations present in contemporary society.

Qualitative research, likewise, expresses plurality and a variable field encompassing diverse approaches, techniques, procedures, and resources under philosophical and methodological conceptions that aim to research, interpret and explain the social world, at the same time representing possibilities and challenges for the production of knowledge. The qualitative approach of the research considers the society composed of individuals and groups, who share meanings according to collective perspectives. Starting from this methodological design, and from a sequence of hypotheses, the researcher investigates processes, facts, and situations in the social scene that, interconnected, can explain the phenomenon analyzed. The comparison method associates inductive and deductive research aspects, through the collection, codification and simultaneous analysis of the data.

The result of the association between these two areas of research produces stable and reliable information about the behavioral transformations of the actors of society, aiming to generate insights, from the social movements, and the identification of opportunities, through the connection with the needs, wishes, and desires of consumers. Given this context, the general objective of this chapter is to verify the importance of qualitative research at the trends management in the design of fashion products. Through literature review and theoretical problematization, the cardinal premises which guide the qualitative methodology in trends management are evidenced.

2 Qualitative Research

It is profitable to produce texts that address the defense of rigor and quality of qualitative researches, despite the discussions on insufficiently scientific and rigorous principles on the part of qualitative researchers. The truth is that, for the most part, the debate on the quantitative and qualitative methodologies, and also on their paradigms, results in a continuous focus on conflicts, since the nature of the research is defined by methods, and not by research techniques. The interaction between theory and practice is implicit in ways, which, in turn, configures and elucidates the direction of the investigation. So, the methodology is intimately linked to nature and the study object.

2.1 Fundamental Concepts

According to Gerhardt and Silveira (2009), quantitative research has its matrix in positivist thought and tends to highlight the rules of logic, deductive reasoning and the measurable particularities of human experience. Under another perspective, the qualitative research emphasizes the holistic, dynamic and individual dimensions of social practice, to understand the whole context of those who are experiencing the phenomenon. From the late 19th to mid-20th century, tracing a historical evolution of the quantitative and qualitative approaches, it was noted that the first dominated the investigative processes in Social Sciences instead of the second. In general, the implementation of qualitative researches is associated with objective data and numbers, while the base of the qualitative analysis is composed of personal material, words or images. Moreover, the distinction between quantitative and qualitative research is in the use of closed questions (quantitative hypotheses) instead of open questions (qualitative questioning). However, broader ways of visualizing the degree of differentiation of both are based in the basic philosophical inference that the researchers directly to the study, in the various strategic research models used and in data collection and analysis tools.

Vieira (2013), likewise, reports that it should be understood that the qualitative research, instead of being monolithic, expresses plurality and translates a different field that includes several approaches, techniques, procedures and resources under philosophical and methodological conceptions, representing, at the same time, possibilities and challenges to produce knowledge. The author reports that possibilities involve "plurality and diversity in the face of an increasingly postmodern economic and social reality, fragmented, hybrid, hyper-real, with reversion between production and consumption, decentralized subjects and juxtaposition of opposites", therefore they do not admit inflexible or predestined investigative and analytical process. The challenges, in turn, require "clarity of ontological conception, epistemological understanding capacity, wide chronological horizon, theoretical depth, and analytical consistency" (Vieira 2013).

Rodrigues (2014) says that "qualitative research strategies are based on Social Sciences, this being the best way to research in Design since it is research through consumers and consumers". Social researchers seek to probe why changes in variables, not just how variables change, specifying the causal factor as the independent variable and the effect of the variable as a dependent variable (Kawamura 2015). Therefore an understanding of cause and effect is fundamental because it enables researchers to anticipate how a pattern of behavior is produced and reproduced, establishing relationships of time, chance, interpretation, and convergence between concepts, generalizing so-called connections of theories.

Flick (2009) states that qualitative research is relevant to the study of social relations because of the multiplication of the spheres of life. The idea of the author emphasizes that today, the researchers confront new contexts and social perspectives derived from the rapid social oscillation and resulting diversification of the social universe. Therefore, traditional deductive methodologies are not sufficient given the individuation of objects. Thus, the research should be guided by inductive strategies, considering that the theories are optimized from empirical studies.

2.2 The Dichotomy Between Qualitative and Quantitative Research

It is important to emphasize that an essential point in the discussions about methodologies is the dichotomy between qualitative and quantitative research. However, this is not the case for mutual exclusion. Creswell (2014) points out that the two approaches, quantitative and qualitative, should not be considered as opposites, rigid or dichotomous, as they represent different purposes on a continuum. Guerra (2011) ratifies this opinion and affirms that qualitative and quantitative research should not be seen as opposing, since there is a mutual influence, despite the epistemological, theoretical and methodological gap between the two.

One way to understand the gradations between the two types of research lies in the underlying philosophical assumptions that researchers use in the study, the kinds of research chosen, the strategies adopted, and the methods and tools for data collection and analysis. For example, the selection of quantitative data, through questionnaires, can indicate information in the triangulation of the research and add credibility to work. The analysis of the qualitative data resulting from interviews focuses groups or video can determine the meaning of the study.

However, it is known that "qualitative research, in general, privileges the analysis of micro processes, studies of individual and group actions and performs an intensive examination of the data collected by several specific methods" (Gutberlet and Pontuschka 2010). The importance of qualitative research lies in the production of knowledge that, besides being useful, points to creativity, guided by an ethical project. Given this, for the development of qualitative research, a competent theoretical base is indisputable, besides the methodological rigor and creativity of the researcher throughout the research process. That is, the qualitative researcher must conduct the whole process of study supported by theoretical and methodological references, governed by a creative ability and not only by the technique.

2.3 Validation of Qualitative Research

It is necessary to live in a prolonged and intensive way in the study environment and among the participants to validate and legitimize qualitative research, with a consistent and reliable study. Gutberlet and Pontuschka (2010) describe that in these face-to-face moments the ability to see and hear is required, in addition to seeing and understanding, "about what has been the task of the researcher. The triangulation and the checking by the participants to give rigor to the qualitative research, besides winning the confidence of the participants".

Thus, the approach, methodology and research techniques are part of the same whole. Gerhardt and Silveira (2009) discuss the link between the three axes of research:

1. The rupture: in the first constituent axis of the methodological stages, a break is proposed with preconceived ideas and "with false shreds of evidence that give us only the illusion of understanding things";
2. The construction: the "rupture is only effected when we refer to an organized conceptual system, capable of expressing the logic that the researcher supposes to be the basis of the object in a study. This phase is the moment of the configuration of the research path: explanatory proposal of the object, necessary operation, and expected results;
3. The observation: is the stage of verifying the facts, since a "research proposal has the right to scientific status when it is likely to be verified through concrete reality information".

Ullrich et al. (2012) explain that the "the word 'qualitative' in scientific researches refers to the emphasis in the processes and meanings, in which the rigorous or measuring exam in terms of quantification, valorization or sequence does not correspond to the questions proposed in these studies". Based on the differentiation between quantitative and qualitative research, the authors show five premises of qualitative research:

1. Regarding the use of positivism: concerning the "adoption of statistical observation and analysis procedures, instrumentalizing and quantifying the research procedures", either through the use of a quantification software in content analyses of interviews or through the structuring of hypotheses for future testing;
2. Regarding the acceptance of postmodern sensibility: it considers the approaches of critical theory, of constructionism, of post-structuralism or of postmodernism in the use of "alternative work methods such as verisimilitude, emotionality, personal relations, ethics, praxis, multivocal texts, dialogues with subjectivity, in which the criteria of scientific reason and truth are questioned";
3. Regarding the capture of individual and social perspective: "the recognition of the perspective of the other's point of view, or the perspective built by social authors", reflecting the richness of observations and interviews;
4. Regarding the investigation of daily life restrictions: it allows "to discuss social actions immersed in specific contexts, as well as the relation of the emic with the ethical";
5. Regarding the guarantee of detailed descriptions: it allows to understand the minimal details of daily life, marked by social relations through the descriptive detailing of the world.

It is also possible to identify modulations in the construction of qualitative research dynamics in the following levels:

1. Ontological, on the perceived nature of reality—is the locus of ideas and structures of analyses;
2. Epistemological, based on the relationship between researcher and researched—configurations or games of the research questions;
3. Axiological, focused on the role of values in research—the methodology establishing the specific ways of examining problems.

In summary, it can be stated that, in addition to the primordial rigor with data collection techniques, the analysis of the process-interactivity perspective must also consider the participants' movement in the research environment. Both the researcher and the researched, plan the activity of data collection, highlighting the research process with the purpose of exhausting the network of meanings pertinent to the analyzed phenomenon. This commitment produces solid, reliable information that results in an in-depth analysis.

3 Qualitative Research and Trends Management

Concerning the qualitative research, generally associated with the studies of Human or Social Sciences, the principal axis focuses on the analysis and interpretation of patterns or paradoxes related to social processes. According to Kawamura (2015), researches focused on human behavior; in general terms, cover two fields: 1. Macro, of deductive approach; and 2. Micro, which reflects an inductive approach. In other words, ideological conflicts, of macro perspective, are investigated through quantitative methods, while questions regarding symbolic interactionism, of micro perspective, tend to use qualitative methods.

The author also highlights that lay people in trends management, fashion or clothing may believe that this area has no relation with Social Sciences. However, the author sanctions that, to earn respect in the academy, scientific investigation in these study areas is necessary, since, daily, it is possible to find information and subjective opinions on trends or fashion in any vehicle of information, in most cases, without reference to any reliable source as a base. Therefore, human behavior cannot be investigated in precisely the same way that Natural Sciences, Chemistry or Physics are studied, nor "can no problem or behavior be understood in isolation, it is necessary to consider the dynamics in operation in the global system" (Vieira et al. 2014).

Denzin and Lincoln (2006) ensure that the problem is related with some factors, such as the researchers' ability to observe the social field; the establishment of the world concepts by social subjects; the several facets of the world obtained through research processes, epistemologies and forms of representation. Trends management is a complex set of elements, with nonlinear patterns and independent subjects, which interact in several ways. The interest lies in investigating social behavior, by the type and form of feedback that this social system issues, in addition to particularities and duration of the feedback cycle of the information in the order.

According to Morin (Pimenta 2013), "when we are researching, thinking, analyzing the complex, we will never achieve completeness, and doubts, more interactions, researches and questions to be processed will always be present". "Totality is the non-truth", highlights Morin (2015), who lays out three principles to elucidate complexity:

1. Dialogical, which "allows to keep duality within the unity. It associates two terms [order and disorder] at the same time complementary and antagonistic";
2. Organizational recursion, based on the recursive processing, "a process in which the products and effects are, at the same time, causes and producers of what produces them", that is, the social sphere results from the interactions between individuals;
3. Hologrammatic, connected to the recursive, which, in turn, is linked to the dialogical principle. "The idea […] goes beyond the reductionism that only sees parts and beyond the holism that only sees the whole" and that "immobilizes the linear spirit" (Morin 2015).

3.1 Fashion and Trends Management

Lipovetsky (2010) sets fashion and, therefore, trends management, as a mirror of society, since it is possible to research it in several places and moments: in the streets, on the internet, at the supermarket, in an art exhibition or music festival, in releases of the cinematographic industry, in various reports, in people's customs and habits. To paraphrase Pastoreau (2014), it is the society that architect trends, that defines its senses, that structures its codes and values, that plans its praxis and designates its inferences.

We know that observing human behavior in its natural environment is essential. "Understanding cultural and social elements becomes necessary to provide the empirical information needed to discern the nature and weight of the various manifestations of trends, such as new products, services, behaviors, styles, and representations," given that there is a conceptual difference between the manifestation of tendencies and object derived from the tendencies (Gomes 2015). The analysis of trends enables the discovery of new consumption behaviors and new directions, highlighting the acute social and cultural reality that, through an individual mentality, detects the collective unconscious. Tovar (2014) asserts that there are two ways to investigate trends:

1. From the bottom to up: highlighting the relationships between small manifestations in various aspects of human life and consumption categories, encompassing global trends;
2. From top to bottom: examining major historical events or international brand strategies at a massive level, permeating the lives of consumers in their everyday dimension.

In other hands, it is possible to assert that all thinking, and action is intrinsic to the ways of thinking, apprehending, interpreting and absorbing of social subjects through a process of interaction with their context. By drawing an analogy with Trends Studies, the five dimensions of the ecosystem perspective allow the trend analyst to consider the facets of behaviors, mentalities, and symbols associated with trends in their social environment, in a relationship of the individual with the collective, because the "all is in part, which is in the whole" (Morin 2015).

We agree with Morin when he affirms that the framework [of the Studies of Trends] is being constructed at the same time as the improvement, without a predetermined model. As the poet from Seville, Antônio Machado, said: "Walker, there is no road, there is road when walking."

4 Conclusions

Trends management is not constituted as a stable and unified area, so, they do not always present uniformity among its fundamentals. The scope is to stitch clues together, in a search for deciphering and unraveling social mentalities and behaviors, focusing on the investigation of past and present moments to understand the future. In short, they represent an essential part of the process of designing new business and communicational strategies, as well as new policies, social projects and, consequently, innovations.

The qualitative research, likewise, expresses plurality and a complex field that includes several approaches, techniques, procedures, and resources under philosophical and methodological conceptions, representing, at the same time, possibilities and challenges to produce knowledge. The result of the association between these two research areas produces robust and reliable pieces of information regarding the behavioral transformations of the social actors, aiming at the generation of insights, from social movements and at the identification of opportunities, through the connection with the needs, desires, and yearnings of consumers.

References

Creswell JW (2014) Investigação Qualitativa e Projeto de Pesquisa: escolhendo entre cinco abordagens. Penso, Porto Alegre

Denzin NK, Lincoln YS (2006) O Planejamento da Pesquisa Qualitativa: teorias e abordagens. Artmed, Porto Alegre

Flick U (2009) Desenho da Pesquisa Qualitativa. Artmed, Porto Alegre

Gerhardt TE, Silveira DT (Orgs) (2009) Métodos de Pesquisa. Editora da URFGS, Porto Alegre, p 46

Gomes NP (2015) Comportamento de Consumo, Elites Sociais e Moda: contributo para a consolidação disciplinar dos Estudos de tendências. Tese (Doutorado). Programa de Cultura e Comunicação, Faculdade de Letras. Universidade de Lisboa, Lisboa, p 66

Guerra IC (2011) Pesquisa Qualitativa e Análise de Conteúdo – sentidos e formas de uso. Principia Editora, Parede, Cascais (e-book)

Gutberlet J, Pontuschka NN (2010) Pesquisa Qualitativa sobre Consumo: experiências interdisciplinares. Olhar de Professor. Ponta Grossa 13(2):217–224

Kawamura Y (2015) Doing research in fashion and dress – an introduction to qualitative methods. Bloomsbury, London

Lipovetsky G (2010) O Império do Efêmero: a moda e seu destino nas sociedades modernas. Dom Quixote, Alfragide

Morin E (2015) Introdução ao Pensamento Complexo. Sulina, Porto Alegre, pp 74–75

Pastoreau M (2014) Preto – história de uma cor. Lisboa, Orfeu Negro

Pimenta AC (2013) Resenha: Introdução ao Pensamento Complexo De Edgar Morin. Revista Científica da FHO|UNIARARAS 1(2). http://www.uniararas.br/revistacientifica/_documentos/art.4-001-2013.pdf. Accessed 17 May 2016

Rodrigues SVG (2014) A Importância da Participação dos Consumidores no Processo de Design. Dissertação (Mestrado). Programa de Pós-Graduação em Design de Equipamento, Faculdade de Belas-Artes. Universidade de Lisboa, Lisboa, p 75

Tovar PR (2014) Coolhunting y Tendencias: a la caza de grandes ideas. Starbook Editorial, Madrid

Ullrich DR et al (2012) Reflexões teóricas sobre confiabilidade e validade m pesquisas qualitativas: em direção à reflexividade analítica. An Rev Admin PUC-RS 23(1). http://revistaseletronicas.pucrs.br/ojs/index.php/face/article/viewFile/11329/9676. Accessed 02 May 2016
Vieira FGD (2013) Perspectivas e Limites da Pesquisa Qualitativa na Produção de Conhecimento em Marketing. Revista de Negócios. Blumenau 18(1):10–24
Vieira EJ, Martins HC, Gonçalves CA (2014) Teoria da complexidade: um estudo em organizações em diversas perspectivas. Rev. Econ. Gestão 14(36). http://periodicos.pucminas.br/index.php/economiaegestao/article/view/P.1984-6606.2014v14n36p85/7255. Accessed 02 May 2016

Method to Create Fashion Collections—From Practice to Teaching

Gabriela Teixeira

Abstract One of the challenges for fashion designers is the development of successful collections. They must be consistent with needs like brand values and innovation. The experience of working in the garment industry, allied to the practice of teaching in design institutions, allowed the creation of a Method to Create Fashion Collections. For its development, we studied different methodologies from design and other areas with the aim of analyzing and organizing them in specific stages. The method presented in this study could be used as a tool for professionals, teachers, and students to encourage a global vision in the planning of collections. Divided into 6 Steps this Method comprises: Collecting Information—Brand and Market; Choosing Innovation Guidelines; Creating the Consumers Profiles; Inspire to Create; Developing the Collection—Results and Processes; Analyzing and Improving—Short, Medium and Long Term. Therefore, it works as an open tool for better application in view of the needs perceived by the professional of the area, so it can be adapted according to each case study. Its purpose is to allow a global view of a diversity of parameters so that they are applied into goods or services.

Keywords Fashion design · Brand values · Market · Consumers profiles · Fashion collections

1 Introduction

Throughout our lives we are faced with challenges, some of these in the course of our professional activity. These challenges allow us to analyze experiences to develop proactive attitudes in the search for solutions. One of these moments occurred when it was necessary to reflect on how to teach students in an undergraduate degree in design to plan collections effectively, considering various parameters and current complexity.

G. Teixeira (✉)
Faculdade de Belas-Artes, Universidade de Lisboa, Lisbon, Portugal
e-mail: gabiltdesigner@gmail.com

Centro de Investigação e de Estudos em Belas-Artes, CIEBA, Lisbon, Portugal

D. Raposo et al. (eds.), *Perspective on Design*,
Springer Series in Design and Innovation 1,
https://doi.org/10.1007/978-3-030-32415-5_22

There were in mind three different realities to consider. First, a girl or boy with the dream of wanting to be a famous fashion designer. Second, a young person working at a fashion industry and living day by day the challenges of making decisions, search trends, dealing with suppliers, monitoring the production of products, and remaining tasks. And third, a professional thinking how to show future fashion designers the complexity of this profession and the daily life of a fashion designer working in the industry. From these realities emerged two starting questions: How to assist in the preparation of students of fashion design courses for the professional activity of designing projects? How to effectively organize collections planning steps and pass them on to students? The search for answers to these questions led to the elaboration of the method presented in this chapter.

The method proposed is a tool for professionals, teachers and students that encourages a comprehensive vision in the planning of fashion design collections. This means that it not going to be a simple or quick method to develop projects. On the order hand, it tries to stimulate a more complex way of thinking for designers or students—which is going to require reading, observation, analysis, debate, creation and verification. Because the design should be consistent with social, economic, cultural and environmental issues, brand values, innovation, ethics, quality, durability and others (Bonsiepe 2011; Kazazian 2005). In today's complexity it is a challenge to teach and transmit to students these values that are the basis of the design planning activity, taking into account that designers need to understand theirs place in society and acknowledge it as part of the design process.

As a framework for the presentation of the method, a didactic step-by-step was chosen that aims at an easy application for both students and professionals. Each of the 6 steps of the method contains the expected objectives, the indication of texts or theories to read, the suggestion of a set of activities to be carried out and an evaluation of each step. After the explanation of the method, the conclusions reached with the elaboration of the study are indicated. It is important to say that this method is constantly improving.

2 Method to Create Fashion Collections

The Method to Create Fashion Collections was developed based on a practical experience in clothing industry and also in the teaching for students of undergraduate courses in fashion design, allied to a theoretical knowledge based on relevant authors within the area of design and other related.

Aiming to respond to the challenges and issues formulated in the introduction of this chapter, an analysis of the different approaches proposed by the studied theorists was carried out. This analysis, together with the practical experience, made it possible to organize, select and integrate theories and methodologies into a method suitable to fashion design, which is composed of 6 steps, as described in the following list:

Method to Create Fashion Collections

Step 1: *Collecting Information—Brand and Market*
Step 2: *Choosing Innovation Guidelines*
Step 3: *Creating the Consumers Profiles*
Step 4: *Inspire to Create*
Step 5: *Developing the Collection—Results and Processes*
Step 6: *Analysing and Improving—Short, Medium and Long Term.*

The steps of the method are intended to effective execution of the steps considered necessary for the development of a collection for the fashion industry by raising the awareness of the designer about important issues related to the development of products and service. It also aims to encourage the designer to form a theoretical background that contributes to the reflection of their choices by promoting coherence with the brand identity and the new scenarios that emerge (De Moraes 2010), encompassing so many aspects in a collection is challenging.

But conscientious choices would need to be on the agenda of a professional who projects goods that will be acquired, used, and discarded by people (Bhamra 2007), since there is a prior, during and after a product. It is worth mentioning that the method allows for open adaptations. The steps can be enlarged or reduced in their objectivity, application and outcome. The professional should carefully consider this change and the result that they will bring. This reflection should be carried out with the team or group—the importance of collaborative work is emphasized. For the implementation of the 6 steps, a minimum of three months are estimated for an effective execution, for practice with students and for the designers that already work in a company.

To make the steps of method more dynamic, they have the following components: 'Objectives for Designer | Student'—explain what the stage will provide; 'Reading Suggestions'—to quote methodologies and theories that should be deepened; 'Step Activities'—suggestion of collection planning exercises; and 'Step Assessment'—to draw conclusions about the activities developed. With recurrent use of this method, certain activities or steps can be skipped, reevaluated or improved. This means that it is in continuous construction, that it is not linear and that its dynamic will be conditioned by the designer according to the needs of the brand, the market, consumers profiles and other factors observed.

It should be emphasized that the method is suitable for brands that intend to develop projects with some innovation or differentiation for the market. Thus, for fast-fashion brands, adaptations or an effective coordination of the research and creation teams will be necessary. This is a decision to be made by students or designer(s) of the brand. We will then explain the steps in the method.

2.1 Step 1—Collecting Information—Brand and Market

Goals for Designer | Student: The objectives of this stage involve understanding the brand identity, mission and strategic positioning; research the new scenarios and analyze which will bring impacts to the brand; work in a team and network to evaluate or reevaluate creation criteria; understand the region for which the collection is intended; and realize which brands or competing products.
Reading Suggestions: The design process has similarities with the management activity. This means that the designer has (or can develop) the ability to manage design processes and production management. This type of professional has the option for management responsibility in 3 phases. The first as the 'Designer as Differentiator' promoting improvements in the product or service, in the production process and in the packaging—which can bring a financial increase of the company, the sales value and the value perceived by the customer. The second is the 'Designer as a Coordinator' who seeks to innovate the products or services, coordinates management processes and interactions and works the possible conflicts in the team. The "Designer as Transformer" focuses on innovation for the client or user by anticipating future scenarios and trends (Mozota 2011).

The author De Moraes (2010) designed a tool for gathering information and elaboration of concepts that can help the designer to develop better projects. The tool is titled *Metaprojeto*. With its tool it is possible to decode the existing scenarios in order to stimulate the differentiation through the design. The aspects that make up the *Metaprojeto* are: (a) Marketing; (b) Productive and Technological; (c) Typological, Ergonomic and Formal; (d) Socio-environmental; (e) Sociocultural; and (f) Product-Design System. For this stage of the method, the determining factors are (a) Marketing—defining the scenario, understanding of the identity, mission and strategic positioning of the company; and (f) Product-Design System—understand if there is the participation of the designer in the process of defining the communication channels (catalog, website, social media, packaging) and distribution channels (markets to be reached, points of sale, participation in fairs and showrooms).

Sant'Anna (2005) presents in his book in the area of marketing a series of questions that aim to bring clarification on important points in the development of a project—this type of questionnaire is called a briefing. It has been formulated for advertising however, it can also be applied in the conception of design projects. The questions asked in the briefing allow collecting information about the product, the market, the competitors and the consumer public. So, it should be done early in the project, be clear and concise.

Seivewright (2009) and Renfrew and Renfrew (2016) address another important factor in the development of collections—macro trend research. This type of research is used by the fashion designer and also in projects in other areas, since there is a need to understand the new scenarios that will arise and translate them into material goods (aspect also defended by Mozota and De Moraes). The macro trends are future scenarios projected as a reflection of current social, cultural, economic and political conditions. They serve as guidelines to prepare companies for the changes that are to come, as well as to provide new business opportunities.

It is advised at this moment to deepen the aforementioned theories, by reading and discussing the texts with the group or team. The designer needs to develop his sense of criticism in order to understand the old and new behaviors and events, and thus to reflect and make coherent decisions in the creation of projects. For this it is important to read about diverse subjects; people watching; attending museums, theaters, cinemas, cafes, restaurants, squares, fairs; and stimulate creativity with varied activities and techniques.

Step Activities:

- Choose the brand to develop the collection (student). Select three brands in the market and present them to the coordinator with the intention of defining which of these will be more appropriate for the gathering of information. Possible question: Why not create your own brand and use it as a case study? Because this will involve more study time and other knowledge relevant from other disciplines.
- Answer questions that aim to perceive the relationship between Company-Product-Designer-Competition. They were elaborated based on the *Metaprojeto* (De Moraes 2010) and the Sant'Anna briefing (2005) and are grouped in Company and Product; Company and Designer; and Company and Competition.

Questions about Company-Product-Design-Competition ***Company and Product***

- Describe the types of products developed by the company.
- Do these products meet a special need or do they respond to a consumer problem? If yes, which one or which ones?
- Identify the identity of the company in the following aspects:
  - Be Conservative, Innovative or Vanguard;
  - Be Passive or Active;
  - Create trend or Follow trend.
- Does the Company have a strategic positioning? If so, what is it?
- What are the points of selling of the product? Own store, multi-company, website, other?
- Has there been any improvement in the product, process or packaging recently?
- Considering the life-cycle of the company in the market, it is in which phase would you position it: introduction, consolidation or reactivation?
- Does the company have mission and established values? If so, what are these?

Company and Designer

- Does the designer participate in the decision-making process regarding the communication and distribution of the product or the collection?
- Is there consistency between the collection that is developed and the way it is advertised on the points of sale?

Brand and Competition

- What brands and products are the competition?
- Point out the advantages and disadvantages of competitors.
- What improvements could be made to the product, packaging or brand to differentiate from competition?

- Search Macro trends | Future Scenarios

Understand the scenarios that will arise and which will impact the company or product. This research should be started at this stage and completed in step 4, since it requires some reflection about information already accessed and other data should be collected in the next steps. Some agencies gather trends research, such as WGSN (www.wgsn.com), Future Concept Lab (www.futureconceptlab.com), The Future Laboratory (www.thefuturelaboratory.com), among others. The company can also hire an agency to conduct the research of the scenarios and the scenario can tailor the information according to the needs of the company.

Step Assessment: Observe and analyse the answers found together with the group or team of the company. The aim is to understand the points that stand out the most and if any aspect about the company or product needs better definition or coherence. The information you collect must be stored in a folder, notebook, or online file. Does the company have well defined values, mission and positioning or do they need to be better elaborated? Does the product that the company develops bring some kind of innovation? The answer to this second question will be the basis for the next step, which aims to define the innovations for the company.

2.2 Step 2—Choosing Innovation Guidelines

Goals for Designer | Student: As objectives for stage 2, we intend to generate information about the possible paths for innovation in design; research examples of companies that innovate; expand the criteria that should guide projects in the area of design; and select the driving force for innovation in creating the collection.

Reading Suggestions: In the various social media are observed subjects that are usually on the agenda, among which are innovation and sustainability. So how can the designer apply innovation and environmental sustainability to product creation? The designer and theorist Bonsiepe (2011) argues that project management should be based on criteria that involve human, technological, environmental, social, cultural and economic aspects—depending on the type of project to be developed. It criticizes the fact that today the designer does not respond effectively to solve the problems observed in society, remaining mostly in purely aesthetic issues. In the book "Design, Cultura e Sociedade", among the different themes presented, the author proposes the use of driving forces for innovation—which serve as possible ways for the designer to design innovative projects. These driving forces can be based on Technology, User, Form, Invention, Symbolic Value, Tradition, Mechanical Engineering, Ecology, Branding, Trends, Art, or Criticism.

In the book "Design for Environmental Sustainability" Manzini and Vezzoli (2008) argue that the designer can guide the creation of products by environmental criteria. Thus, they present three possible scenarios for sustainability: the sufficiency scenario—with cultural innovation; the efficiency scenario—with technical innovation; and the effectiveness scenario—by combining technical and cultural innovations. They also explain the Life Cycle Design tool, which is composed of the phases of pre-production, production, distribution, use and disposal. This tool aims to verify and minimize the environmental impacts in the development of projects. Its application can be done in whole or in only one or more phases, so that the designer can efficiently manage all the necessary specifications for each phase.

The *Metaprojeto* of De Moraes (2010), mentioned in step 1, presents as one of the factors that compose this tool the Socio-environmental aspect. This aspect aims to reflect on the raw materials used in the production of a product—that is, to perceive its degree of toxicity, whether they are biodegradable or compatible with each other, besides analysing the inputs and outputs of resources and energy in the productive process and its environmental impacts. The author also suggests that the designer reflect on the creation of new lifestyles that are more sustainable when developing a project.

Step Activities:

- Look for other companies that use driving forces for innovation in the development of products or services. These examples can serve as inspiration for what kind of innovation the group or designer wants to bring to the company. When selecting the driving force, it is important to do so considering the information about the market in which the company is inserted.
- Describe how this innovation can be implemented in the development of the collection. One suggestion is to use the brainstorming technique or mental map to generate ideas.
- Look for environmentally sustainable fashion companies and see what measures they take to develop their products. What raw materials do they use in production? Are they toxic or biodegradable? Do they encourage the reuse or recycling of the products they sell?
- Reflect on which scenarios for sustainability and which phases of the Life Cycle Design tool can be applied to the development of the collection.

Step Assessment: Analyse what ideas have come up for company innovation. What was the biggest challenge of the step: researching companies that innovate, proposing a path to innovation, or defining issues related to environmental sustainability? Discuss with group or team a way to address these challenges. In the next steps of the method it will be possible to deepen the chosen innovation, since the consumer's profile, the theme of the collection and the production processes to be used will be studied in more detail.

2.3 Step 3—Creating the Consumers Profiles

Goals for Designer | Student: The objectives of this stage are to purpose reflections on the challenges of researching consumer groups; raise awareness of the need to establish consumers profiles; research cultural, economic, demographic data of the target region of the company; and develop a semantic panel for the selected consumers profiles.
Reading Suggestions: This is a stage that can generate controversy due to the changes and extensions that have been occurring in the understanding of the classifications of target audience and consumer groups, such as the definitions presented by the theorists Frings (2012), Morace (2012, 2018), among others. It is considered valid in this method to approach the question of creating consumers profiles for the company, because design projects are aimed at people. However, it is reinforced that this profile will vary according to the country, the region and even the city to which the brand directs its production. Thus, the designer needs to understand the specific characteristics according to the cultural, geographic, economic and demographic aspects of each target market.

Seivewright (2009) in the book "Pesquisa e Design" points out that brands can establish a muse or customer profile for the development of a product (or collection). Defining aspects such as age range, habits, experiences, profession, lifestyle and financial income. It also suggests that when defending consumers profiles it is important to conduct market research.

Morace (2012) defends in the book "Consumo Autoral" that currently exists the consumer author. This is more and more demanding and seeks innovation in a design that combines creativity, art, spiritual elements and technological aspects. In this way, the designer must design products that respond to varied needs—comfort, humanistic and cultural load, advances in science—so that these goods are stimulating and durable. At a recent research, the author argued that a company should choose not only one, but many different consumers profiles because today there are a segmentation dynamism at market (Morace 2018).

Sources that present the most recent studies on consumer groups are the bureaus of trend research already mentioned in step 1.
Step Activities:

- Define the consumers profiles for the company. Describe elements such as age group, habits, lifestyle, values, professions, financial income, places they frequent and other relevant points.
- Create a semantic panel with inspirational images of the people who make up the group as well as the places and objects that relate to the idealized profiles. This panel will serve as a reference for product development.
- Perform a cultural, economic and demographic research of the main regions that the brand sells its products. For the student, one can do the research of a region (city or capital of the country) that the brand has as a market, so that the information is more easily collected.
- Conduct interviews with consumers that are part of the consumers profiles.

Step Assessment: The information collected on the consumers profiles needs to be analysed in conjunction with the driving force of innovation selected in the previous step. At this moment it is possible to deepen some elements on the innovation to be implemented in the collection. Observe the coherence between consumer needs and expectations according to the cultural, geographic and demographic data collected. This information will serve as elements for the development of the collection's products. But first we must define the theme of inspiration for collection, subject of the next stage.

2.4 Step 4—Inspire to Create

Goals for Designer | Student: The objectives of step 4 are define the theme of inspiration for the creation of the collection; elaborate a semantic panel; assemble the color chart; define raw materials to be used in the collection.
Reading Suggestions: The designer and author De Moraes (2010), in the *Metaprojeto* tool, identifies the Sociocultural factor as a means of achieving a differential within the homogeneous and globalized production, by the conception of products that value the characteristics inherent to a culture or region. Examples are local handicrafts and raw materials, popular traditions, landscape, people, shapes, colors, among many others. The purpose is to know the richness of the local repertoire so that it serves as reference in the conception of creative and stimulating goods. The author also presents the Technological-Productive aspect. In this, one observes the material characteristics of the product, its composition and the productive processes applied. What reinforces the designer's need for reflection in terms of materials and processes composing the product, besides the aesthetic and functional issues and *estésico* factors (sensibility, emotional, sentimental).

For Seivewright (2009) the selection of colors is one of the primary points in the creation of the collection, since it also delimits the spirit and the season. Color is often the first element perceived by a person and is associated with his/her affective memory, in addition to having different meanings according to culture. Thus, colors have a great importance for translating personality, taste, feelings and social status (Fraser and Banks 2004).

We know the importance of color to create fashion products. But let's now reflect on the placement of Seivewright (2009)—the delimitation of color by season, recalling the concepts already presented as the author of Morace (2012) and the criteria of sustainability of Manzini and Vezzoli (2008). It is suggested that color is more than a trend element of a season. That can be used in a way to value the product, to be a reference to the theme of the collection and have a relationship with the consumers profiles.

Step Activities:

- Point out possible inspirational themes to create the collection. Select the one that can bring a differentiation to the profiles of the selected consumers.
- Study the chosen inspiration theme. Collect information, search for images and videos, photograph various elements, visit places that have a relationship with the theme, interview people, visit museums, buy objects or related materials, observe the colors and shapes that stood out.
- Assemble a semantic panel on the theme, with the best elements collected.
- Define which colors will be used in the collection and assemble a board with a number between four and eight colors—dividing them into major and minor ones. The panel should be organized in white background and contain the name of each selected color.
- Define the raw materials that will be used in the collection based on the theme and the colors chosen. For the students: search for fabrics, supplies and other elements that can be used to make the products. It is suggested to visit specific stores (and if possible, factories) to perceive the raw materials by touch and visually. For the designer of a brand: research and contact suppliers of fabrics and fittings to know materials and launches, attend trade shows and other events in the industry. It is possible to develop unique fabrics or fittings for the brand, but this will require more investment.

Step Assessment: Notice if there is a coherence between theme, colors and materials chosen. Correlate the type of innovation selected to the theme, the raw materials and the productive technology. Thus, the question is: Does the brand have the technology to process the selected fabrics and supplies? Remember that these elements will imply in the final cost of the product. The consumer profile is also associated with the quality of the product and its cost. For designers working in companies, pay attention to the time required to request fabrics and fittings for assembly of a prototype, approval of the material and request to the supplier (considering the time of delivery of the same for production). By achieving the objectives of this step is possible to proceed to the creation of the products of the collection.

2.5 Step 5—Developing the Collection—Results and Processes

Goals for Designer | Student: The objectives expected in this step are define how many pieces will make up the collection; generate alternatives to products; select the best alternatives for production of the collection; prepare the technical file; produce the prototypes; approve the parts for production; generate cut order for production.
Reading Suggestions: In the research that included creating a tool to analyze the environmental impacts of the clothing industry (Teixeira 2012), are present some of the phases that make up the production processes of a company in the area. Among these phases are creation, production of prototype, cutting, sewing, stamping, among many others. Thus, the designer needs beyond the theoretical and critical background to understand the complexity of the day to day operation of the industry to make choices in a coherent and conscious way. Especially because many companies outsource their production to other cities, countries or continents. The activities of this stage are intended to address some of the procedures required for the creation and production process.

The Design Thinking method (Ambrose and Harris 2009) was elaborated so that the designer learns from each stage, develops creativity and realizes the feasibility of projects. It consists of the phases: Define, Research, Ideate, Prototype, Select, Implement and Learn. In the 'Ideate' phase, the research carried out (scenarios/macro trends, consumer profile), reticulation, budget and originality (intended innovation) should be considered. In the 'Prototype' phase, the objective is to select the possible solutions for design, verify its technical viability, understand operation of physical object and visualize the handling of the design concept.
Step Activities:

- For the accomplishment of the activities of this stage it is important to reflect on several aspects together: chosen innovation, requirements of the profile of the consumer, differentiation for the market, relation with the selected theme and technological capacity of the brand.
- Define which shapes and types of products will be created (trousers, skirts, blouses, dresses, shirts, shorts, bikinis, handbags, swimwear, shoes, bags, among others) and the quantity that will compose the collection.
- Generate alternatives for the types of parts to be produced. It is suggested to generate two to three alternatives per part model to be produced.
- Select the alternatives that best meet the defined criteria.
- Elaborate the technical file of the product that will be prototyped. The model data sheet may vary from company to company. It should contain information on raw materials, technical drawing with product measures, production processes, material costs, process costs, and other data relative to the prototype.
- Produce the prototype of each approved alternative. The purpose is to check the finish, the suitability of the fabric(s) and/or the material(s), and the modeling. Ideally, the brand has one model to prove prototyped parts. If modifications are

required in the prototype, the ideal is to carry out a maximum of one or two more attempts.

- After the prototype is approved for production, you must create the document for the cut order—which will describe the quantity and colors that a part will be made. Example: 300 model X women's pants will be produced, in sizes 34–44, in blue, black and gray.
- Compose the catalogue of the developed collection. Select the products and compose the looks, choose the photographic models and define the layout. The chosen layout and location for the photos should be related to the theme collection.

Step Assessment: The student or designer must follow the production process of the prototype. All changes made to the prototype must be informed to the development team and documented in a technical sheet. Should be noted that a well-made prototype and a detailed technical sheet will ensure an efficient execution in the production line of the approved product, avoiding wasted material, time and unnecessary financial expenses.

2.6 Step 6—Analysing and Improving—Short, Medium and Long Term

Goals for Designer | Students: Among the objectives of step 6, we intend to analyse the whole process of development of the collection by applying the proposed method to perceive the positive and negative aspects; seek feedback from consumers; analyse the information and define improvements for the next projects in short, medium e long therm.

Reading Suggestions: In the Design Thinking method (Ambrose and Harris 2009) the last phase of a project is Learn. The intent is to identify what worked well or what needs to improve according to the information collected with the consumer about the proposed design solution and also according what was learned during the design process. This feedback provides critical analysis and allows for enhancement of future projects.

Step Activities:

- Analyse with each group or team each of the steps of the method—in terms of activities performed, difficulties and learning, what could not be done and why, and positive points that stand out;
- Conduct a consumer survey to verify the degree of satisfaction with the products created and the company. To make the collection of information more effective, online surveys can be developed, interviews conducted and discounts or promotions offered to clients who are willing to contribute their opinion;
- Discuss the results of the process analysis and feedback from the consumer to establish the guidelines of the next projects to be developed;

- With the new application of the method, it is suggested to deepen theories, to maintain a continuous research on the scenarios, to look for new ways to innovate and to broaden the study of the consumer profiles that the brand intends to achieve.

Step Assessment: At this stage it is desired that the designer I student has achieved a degree of criticism for the project developed and a broader view over new possibilities for creation in the area of design. What goals are set for improvement for the brand I company that can be established in the short, medium and long term?

3 Conclusions

The development of this method aims to contribute for the student I designer to understand the steps needed to develop collections for fashion industry. Design methods must be in constant mutation, as challenges arise everyday and will have impact in the necessary procedures. With constant improvement it is possible for the designer to search for proper means so projects are more coherent with the requirements of the several consumer profiles, meeting untapped needs. In order to propose solutions with different innovations it is decisive to have a analytical, critical, curious, creative and collaborative attitude.

The topics discussed were considered fundamental to the design project for fashion industry, although they don't exhaust the full range of possibilities. There is opportunity for continuous research in the area. The exchange and dialog on this and other matters involving design projects are important for the consolidation of our activity.

References

Ambrose G, Harris P (2009) Design thinking. AVA Book, Lausanne
Bhamra T (2007) Design for sustainability: a practical approach. G. P. Limited, Hampshire
Bonsiepe G (2011) Design, Cultura e Sociedade. Blucher, São Paulo
De Moraes D (2010) Metaprojeto: o design do design. Blucher, São Paulo
Fraser T, Banks A (2004) The complete guide of color. Ilex Press, London
Frings G (2012) Moda: do conceito ao consumidor. Bookman, Porto Alegre
Kazazian T (2005) In: Kazazian T (ed) Haverá a Idade das Coisas Leves: design e desenvolvimento sustentável. Senac São Paulo, São Paulo
Manzini E, Vezzoli C (2008) Design for environmental sustainability. Springer, London
Morace F (2012) Consumo Autoral: as gerações como empresas criativas. Estação das Letras e Cores, São Paulo
Morace F (2018) Consumo Autoral: os novos núcleos geracionais. Estação das Letras e Cores, São Paulo
Mozota B (2011) Gestão do design: usando o design para construir valor de marca e inovação corporativa. Estação das Letras e Cores, São Paulo
Renfrew E, Renfrew C (2016) Developing a fashion collection. Bloomsbury, London
Sant'Anna A (2005) Propaganda: teoria, técnica e prática. Pioneira Thompson Learning, São Paulo

Seivewright S (2009) Pesquisa e Design. Bookman, Porto Alegre

Teixeira G (2012) Desenvolvimento de uma ferramenta para a análise do impacto ambiental dos processos de produção de uma indústria de vestuário de médio porte. Universidade Federal de Pernambuco. Retrieved from: https://repositorio.ufpe.br/bitstream/123456789/11461/1/Mestrado%20Design%20-%20Gabriela%20Lyra%20Teixeira.pdf

Nisa Quartz Inlaid Pottery Workshops: An Action Model in Design Teaching

Ana Helena Grácio, Miguel Aboim and Cátia Rijo

Abstract Design teaching, as design thinking involves a heuristic methodology of creative imagination and functionalism that seeks to address unique circumstances with factual and appropriate formulations. In the learning process, design thinking cultivates the necessary cultivation of memory through experiences and instructional encounters that yield behavioral and functional impact. In a workshop model, effective design involves the simulation of the assemblage of useful and advanced media content and other instructional technology that leads to the optimal attainment of learning and noticeable behavioral impact. With Nisa quartz inlaid pottery research project in mind, the design approach can be implemented in the precise understanding of the historical and socio-economic circumstances prevailing in the project locality, and the impetus for change that solicits project appraisal. The grand approach has to be adapted to the prevailing local conditions while also recognizing that the way forward for the community has to be a difficult path that leads to better outcomes.

Keywords Nisa quartz inlaid pottery · Workshop · Design · Collaborative process · Design teaching

1 Introduction

From an ethnographic and anthropological perspective, the preservation of cultural traditions can be achieved when the cultures make economic sense to the community and when the community sustains an element of pride in its traditional culture, which is cherished as part of the identity of the people. Design can, therefore, be invoked

A. H. Grácio (✉) · C. Rijo
ESELx—Lisbon School of Education, Lisbon Polytechnic, Lisbon, Portugal
e-mail: hgracio@eselx.ipl.pt

C. Rijo
e-mail: crijo@eselx.ipl.pt

A. H. Grácio · M. Aboim · C. Rijo
CIAUD—Faculty of Architecture, University of Lisbon, Lisbon, Portugal
e-mail: mab@fa.ulisboa.pt

D. Raposo et al. (eds.), *Perspective on Design*,
Springer Series in Design and Innovation 1,
https://doi.org/10.1007/978-3-030-32415-5_23

to yield to the task of repackaging the crafts form to correspond with contemporary trends and futuristic trends while preserving elements attuned to the traditional production.

The risk of extinction of Nisa quartz inlaid pottery culture is witnessed because fewer people are willing to venture into the art, and only three potters and their wives remain in the trade. The deduction from these basic facts is that cultures becomes extinct when its popularity wanes due to changing sociological circumstances. Therefore, an appropriate stimulus package in a workshop model can help bring in new enthusiasm and cultural prestige by crowing and rewarding the participants.

Pottery integrates man daily life from immemorial times, since the capacity of clay, being firstly moldable and easily able to assume a final shape for which it was imagined, offers an extraordinary utilitarian capacity, as represents a dazzle for those who take actions in this process, during its creation, use and appropriation.

The technique of quartz inlaid pottery consists in the incrustation of quartz in the clay in order to make it more resilient and increase its porosity, providing greater resistance to different temperatures and higher cooling capacity, particularly in water storage. In Portugal this technique was frankly rooted in national territory and throughout the Portuguese Diaspora (as in Macau and Cape Verde), particularly in the sixteenth and seventeenth centuries, mainly due to the fame and consumption of Estremoz[1] water vases throughout Europe, particularly in the major royal houses and the papal court.

Today, quartz inlaid pottery is closely linked to Nisa village, where the technique was developed in a very particular mode, especially since the second half of the twentieth century, assuming motifs treated with a detail similar to the filigree. However, the emerging of industrialization as disturbed several craft expressions, since crafts as historically been accommodated to a constant and repetitive manufacturing, saw its foundations shaken by the contraction in demand, devaluation of their products or their transformation into collection object, e.g. devoid of the use value (Branzi 1999).

Quartz inlaid pottery remains nowadays in Portugal, Nisa, a small village in the countryside, near the Spanish border. Its main distinguishing feature is the task performed by "*pedradeiras*", women involved, that when inlaying small white quartz stones in clay, draw motives based on flora, fauna and local mystical symbology.

Today only three potters and their respective women, with ages from 65 to 73 years, remain active. The lack of new persons interested in learning or develop this technique, confirms the extinction risk of this ancient material culture, what reinforces the need for an urgent effort of research and knowledge preservation.

The starting point of this study is based on the knowledge of history and its various appropriations that over time have traced the path of what is now known as quartz inlaid pottery. Considering the turning point now recognized, in its social, and the derived UNESCO Convention recommendation for its urgent inventory for

[1]Estremoz is a Portuguese city in Évora District in Alto Alentejo region, in Portugal, is the seat of a municipality with 14,318 inhabitants. Its white marble deposits are internationally known as Estremoz Marble.

safeguarding this cultural heritage, there is a vital need to develop not only a careful inventory of the parts in production, but also understand and provide knowledge that could project the quartz inlaid pottery and its specific features and techniques to new contexts in contemporaneity.

Assuming that this form of craft could be a real contribution to the discipline of design, it seems that this legacy still suffers from a careful construction and theoretical reflection to enrich the knowledge about the heritage itself and for its social and cultural reality able to revert for a more sustainable design reflection and practice. Through interaction and cooperation in areas, such as design and craft, that sometimes seems so close, but at the same time so distant, it was intended to achieve and implement in design disciplinary field, new models for understanding the practice of the craft know-how, assuming the challenge of evading to extreme distances—where there is no room for dialogue—or unequivocal—where no differences are assumed—between both areas. The quartz inlaid pieces have some technical gains but also some weaknesses. Meaning, it as a great capacity to resist to different temperatures, and an exceptional ability to cool water, even in very hot environments (i.e. temperatures reaching 40 °C), as are those who can be found in the interior of Alentejo region during the summer season. However, quartz inlaid pottery is not traditionally fired at very high temperatures (maximum 700 °C) which results in low resistant pieces to heavier touches or crashes. New techniques have been studied and tested in order to overcome these limitations, and thus give rise to new opportunities for this cultural heritage.

Although knowledge of this heritage is at risk of extinction, we believe it is possible—by the use of participatory methodologies within the community, craftsmen and design students—to find new opportunities for the future of quartz inlaid pottery, by using collaborative and cooperative tools.

In design, the basis of a critical design theory or hypothesis is that whatever ephemeral or grounded artifacts that could be elicited in the program, there have to be an agreement that no grounds for fundamentalism would suffice. Under the consideration of total quality management approach, the collaboration between the consenting entities has to make an explicit commitment to the project objective. In this manner, a singularity of objective that is comprehensive and framed to the operational environment of the enterprise allows collaborating parties to take roles. Events, inputs, mechanism and outcomes of a design process should assume a compact enterprise model (Frow et al. 2015). Depending on the levels of complexity within the project, and the degree of uncertainties and indeterminate variables, a complete theory should be coined that spells in great detail pertinent concerns and variable for the parties and for the intended product. The program should be deployed on the paradigm of an evolutionary agile mechanism that advances while also learning new directions and possible frontiers of adaptation and survival in deriving the outcome.

Nisa workshops concept involves the collaborative engagement between local artisans, designers, and students to facilitate an ethnographic engagement so as to bring about greater participation in the rich artistic cultural heritage of the community in designing several kinds of art forms with a sustainable mark on the environment while also creating source of employment to the artists. Each section of the teams would have to yield some forms of experience and knowledge that are shared with the rest of the teams and eventually promoting an extended engagement whose aim is the continuity of the culture of communal art. While such art may serve aesthetic purposes, the fact that it has got ideas useful for the future of the community regarding green development, it is worth supporting and promoting. Moreover, it should be developed so that scientific tools and concepts are incorporated to improve the quality of the art forms for a market bursting with innovative tastes and preferences.

The design of Nisa quartz inlaid pottery workshops on the field, should enable the analysis of the traditional methods so that it can be infused with more trendy and futuristic elements found in the contemporary best practices in pottery, aiming to reach a viable and suitable methodology in design teaching capable to preserve and promote Nisa quartz inlaid pottery.

2 Perspectives and Interactions Between Crafts and Design

From a design outlook, The Nisa Quartz Inlaid Pottery Project has traditionally involved the participation of women, since "*pedradeiras*" use small white quartz stones into the clay to draw images found in the local flora and fauna to yield political mystic symbolism in their pottery. The risk of extinction of this culture is witnessed because fewer people are willing to venture into the art and only three potters and their wives remain in the trade. The deduction from these basic facts is that cultures become extinct when its popularity wanes due to changing sociological circumstances. Therefore, an appropriate stimulus package in the workshop model can help bring in new enthusiasm and cultural prestige by crowing and rewarding the participants.

Furthermore, creating greater stimulus in the art can be achieved by adding new trendy methods so that there is a higher value in the products and hence making the product win in the market. Inferior quality products are doomed to stagnation and failure in the market. The design of the workshop should enable the analysis of the traditional methods so that it can be infused with more trendy and futuristic elements found in the contemporary best practices in pottery.

To adequately safeguard the memory of the cultural heritage in The Nisa Quartz Inlaid Pottery Project, much more than mere preservation of the production needs to be done. The team needs to be facilitated on marketing, elements of social entrepreneurship that corroborate popular participation of the community and means of livelihoods. The curriculum design, therefore, has to incorporate elements of the social and cultural history of the community and the change impetus that has led to fewer and fewer people taking up the art. If the dwindling numbers are attributed to social or cultural changes, then useful proposals can be advanced to invigorate change in a new direction. However, if the difference is attributed to economic circumstances, suitable recommendations should also account for better production and better marketing to revamp the value in the market and hence newly improved attention from the community in the art. Participatory and collaborative methodology unleashes the full potential and synergy of creative possibilities among the different teams brought to the seminar because, in the interaction, new learning and new methods can be cultivated among the community stewards.

Due to the triumph of the scientific revolution in design, it was thus an obvious step for designers to try to integrate scientific methods into the design process so that they could be accepted as severe associates in the sphere of the productive industry. It is definitely a creative process, even in as much as scientific and technological tools are deployed in such a fruitful venture. Nevertheless, the design does not take place in space, with an entirely free selection all kinds of material resources, and designers are limited in their scope of the creative process (Chamberlain et al. 2012). Each design object is the outcome of a unique development process influenced by extensive arrays of artistic conditions and procedures. Socio-economic, technological, and cultural developments, in particular, along with the historical milieu and the environment of production expertise, yields just as important a role as ergonomic and the environmental demands, economic and artistic-experimental aspirations just like the underlying political interests. Dealing with design, therefore, always entails reflecting on the circumstances under which it originated and theorizing the effects on the productivity of the artistic process.

As the world grapples with developmental challenges and an ecologically constrained process of production and consumption, design thinking has to be drawn into the spheres of planning and simulation of societal development. From a sociological perspective, development projects that seek to yield progress in people's lives have to embrace sustainability and the necessity to scale up traditional processes in light of scientific and technological innovations that lead to higher productivity while not damaging the sociological and ecological dynamics. A possible convergence can be attained in the simulation of resource planning and the human capital endowment of the community such that optimality is given priority over other competing priorities. The workshop model presents a plenary session where stakeholders share ideas on the possible project outfits and components attuned to the higher value to the society. The model also allows for the facilitation of expert knowledge through facilitators who make a presentation to the entire group in the forms a lecture session. All such instructional encounters should be furnished with the right kind of media and content attuned to the curriculum objectives.

While the discipline of design may only conjure the notion of material production, it invariably also has a great deal to do with critical communicative functions. The technical performance of instructional mediation in the workshop model also has related social functions to produce operational outcomes. The conciliatory influence of the instructor when facilitating the workshop is to cultivate new states of cognition while delineating those, which are deemed repugnant, and urging participants to recognize and eliminate them in their private discourses. The design methodology relies entirely on abstraction and to some extent on empirical intuition to yield solutions in critical milestone expectations. In this regard, design methodology starts with the description of the initial state of the learning and behavior threshold and simulating the instructional encounter through a suitable curriculum design to predictable outcomes of the process. Moreover, since curriculum contents may conform to different performance conditions, the initial entry and inventory conditions have to be adequately determined to facilitate sound learning encounters attuned to the cultural precepts of the learners.

2.1 Fundamental Concepts

Nisa Quartz Inlaid Pottery Workshop Project involved the collaborative engagement between local artisans, designers, and students to facilitate an ethnographic engagement so as to bring about greater participation in the rich artistic cultural heritage of the community in designing several kinds of art forms with a sustainable mark on the environment while also creating source of employment to the artists. Each section of the teams would have to yield some forms of experience and knowledge that are shared with the rest of the teams and eventually promoting an extended engagement whose aim is the continuity of the culture of communal art. While such art may serve aesthetic purposes, the fact that it has got ideas useful for the future of the community regarding green development, it is worth supporting and promoting. Moreover, it should be developed so that scientific tools and concepts are incorporated to improve the quality of the art forms for a market bursting with innovative tastes and preferences. Moreover, the project has to make clear its objectives and principles so that participants have a clear understanding of their roles and expectations in the encounters.

Learning art can be modeled in the form of participation, in which all the three teams work in turns to produce material art while also promoting a new sensibility of aesthetic value for the broader community and therefore soliciting popularity and massive participation. Alternatively, formal learning can be interchangeably initiated with practical sessions in which the local artisans showcase their methodology and designers can bring in expertise to either improve the local production mechanism or overhaul processes that are redundant and possibly unproductive. A progressive engagement thus would yield optimal quality and optimal value in the final products thus leading to better market performance and profits to the community. Moreover, designers can have a far better chance to analyze the process in the artwork and incubate production procedures that would be far better and far marketable from their wealth of technical expertise. Blending all these optional approaches into the workshop model would produce a far more sustainable and productive ethnographical value to the community regarding the esteem of their culture and traditions.

Craftsmen and designers should produce prototypes of the expected outcomes and test them on various considerations depending on the eventual use and form of the product or artifact. Multi-stakeholder approach suffices on this basis because each section has their areas of competencies and experience which is relevant to a completely versatile and monumental product. A multi-component and multi-implement approach would suffice because of the intricate connections that occur between variable stakeholders. In the model, teamwork involves attention, satisfaction, confidence, and relevance. At the psychological level, the team should invent approaches to deal with anxiety or complexity that hampers motivation and innovation (Cui and Jiang 2016). Attaining preliminary wins in the project liberates the members to get bold with the project. It also boosts the intrinsic motivation of the team. Visual stimulation using models can be deployed frequently because they foster visual intelligence and enrich brainstorming to tackle any challenges.

3 Conclusions

In conclusion, needs matching from the onset of the project throughout should abridge the necessity to satisfy the essential attributes of the artifact being produced. The agile project paradigm approach combined with a mechanistic and scientific erudition to yield authentic materials and design components that eventually meet the needs is an essential consideration. A holistic theory should be simplistic and utilitarian but based on the engineering elements and requirements of the project.

The workshop model as an instrument able to contribute for knowledge safeguarding and as an innovative approach, seems a viable and suitable methodology to preserve and promote Nisa Quartz Inlaid Pottery. The suitability of the method lies in its capacity to solicit the widespread participation of various sections of each team involved, where experts and students gather for promotion of an art form. Furthermore, as a cultural artifact, the art form needs to be preserved, and design innovation can only be integrated to the extent that sustainability and cultural integrity of the art is not entirely compromised. Culture represents a modality of interaction and interrelationships. Therefore, the essential aspect of the workshop model is that it allows for the extension of the interrelationships between local artisans with designers and students so that far greater productive potentialities can be promoted in the long term. The producers should have a global consumer outlook in mind because, under the new globalization, aided by e-commerce, the channels of commodity supply chains traverse the entire planet.

From a post-industrial era perspective, the concept of design with a strong social consciousness seems to be an idea whose time has come, and ignoring it or doing nothing is likely to be perilous for the contemporary designer, especially given the knowledge economy and information society we live in today. In this regard, designers have an obligation to identify and understand the various existing collaborative process associated with social design. In particular, designers need to embrace the platforms of action research and organizational transformation. In addition, it is imperative for the designers embrace the methodic approaches of co-creation, participatory design, and social innovation in their design processes.

References

Branzi A (1999) Introduzione al design italiano: una modernità incompleta. Baldini & Castoldi, Milano

Chamberlain P, Bonsiepe G, Cross N, Keller I, Frens J, Buchanan R, Meroni A, Krippendorff K, Stappers PJ, Jonas W, Schneider B (2012) Design research now: essays and selected projects. Walter de Gruyter

Cui Z, Jiang H (2016) Design, analysis, and simulation of a planar serial–parallel mechanism for a compliant robotic fish with variable stiffness. Adv Mech Eng 8(8):1687814016660927

Frow P, Nenonen S, Payne A, Storbacka K (2015) Managing co-creation design: a strategic approach to innovation. Br J Manag 26(3):463–483

Computational Design: From Algorithms to Digital Production for Traditional Manufacture Sectors

Annalisa Di Roma and Alessandra Scarcelli

Abstract The paper aims at demonstrate the adaptive value of the digital crafts, regarding design and realization methods. In order to achieve some outcomes from a practical approach, some experiments are described. The discussion focuses on the relationship between the parametric tools supporting the design process (devoted to textile design and colour theory) and the aesthetical and technical outcomes. The typical way in which the craftsman "makes"—based on the informal transfer of models to production—shares the adaptive and repeatable approach of the parametric design. Therefore, through the specifications of the materials, the techniques and the realization tools, the seamless process is achieved.

Keywords Parametric tools · Design process · Textile design · Colour theory · Parametric design

1 Introduction

Parallel to the history of the industrial product, developed through the definition of principles of useful form, serial production, ergonomics, etc., the industrialization of the production processes and the evolution of the tools for the realization, in the aftermath of the second industrial revolution, have played a crucial role also in the fields of the artistic industry. Evidence of this is the cultural debate of the late nineteenth century, centred on the relationship that art and technique establish in relation to the theories that pertain to the form, colour, methods and techniques of production. The development of the technical-creative tools and of the prototyping and digital production systems creates in the computerized logic a new relation

A. Di Roma (✉) · A. Scarcelli
Politecnico Di Bari, Bari, Italy
e-mail: annalisa.diroma@poliba.it

A. Scarcelli
e-mail: alessandra.scarcelli@poliba.it

D. Raposo et al. (eds.), *Perspective on Design*,
Springer Series in Design and Innovation 1,
https://doi.org/10.1007/978-3-030-32415-5_24

between project theory and production and reproduction practices (Di Roma 2008): the technique no longer resides exclusively in the finished object, but in its conception pre-formal. In fact, the management of form and colour in the ancient art industry did not assume mediation between the design phase and the realization phase; the author of the conformation of the product often identified himself with those who technically carried out its production. The division into phases, already within the development of the serial processes of realization of the artistic industry, has made it necessary to elaborate geometric models to support the transmissibility of the idea to the different material performers.

Digital modellers, who in the last twenty years have made possible the so-called digital revolution as a question related to the representation and communication of the project, through its virtual restitution, today assume a central role in the prefiguration process of the artifact, as demonstrate technical-formal control tools able to synthetically manage all the main phases of conception and prototyping, providing key decision-making for the optimization of the form and its realization process. The geometric definition of shape and colour values, together with the implementations of parametric software and the generative logic of some cad and raster applications, favour the development of new formal expressions; the ideation process thus becomes «un vero e proprio sistema di informazioni: informazione culturale del prodotto, informazione sul suo uso, informazione linguistica e informazione visiva» (Branzi, 1984, p. 117).

2 Problem

The translation of the ideal form into a technical material form within the artistic industry has always been expressed through the use of tools able to realize the conceptual model and make the craftsman's gesture repeatable in the series of objects to be reproduced, adapting from time to time the ideal model to the contextual needs of specific sizing or functionality (Di Roma 2016).

At the base of the informal process of interaction between ideation and production was the mathematical model that managed the proportion between the parts, the variation of the rays and the continuity of tangency of the curves, the symmetry groups of the plane. It is no coincidence that in the field of ornamental art (Carboni 2001, p. 79) the different historical civilizations have expressed the language on the basis of their own mathematical code.

On the instrumental level the question of technical reproducibility is characterized by the use of shapes, models, seals (of repertoire or designed ad hoc by the same architect of the work) necessary for the transfer of the "form"; or from those mechanical aids (such as the lathe, the drill and the wheels of the potters) useful for the mass production process (Benjamin 1966, p. 20). This mingling between the possession of the mathematical model and the knowledge of how to produce is at the basis of the freedom of artistic will (Riegl 1893; Focillon 1939; Panofsky 1961) and expresses itself through the extreme flexibility of the production methods of the artistic industry.

In reference to the aesthetic intention of the artisan-artist, Hauser states that "the technical solution is itself part or variant of the visual aesthetic solution" (Hauser 1955, p. 95). Therefore, the instrumental technical evolution of contemporaneity, in the digital field, favours a new relationship between art and technique, between formal ideation and realization, re-enacting those processes of limited craft series, interrupted in their millennial tradition by industry (Branzi 2008, pp. 10–11), within the so-called 2.0 craft.

3 Methodology

3.1 Parametric Design Digital Craft

The generative modelling, thanks to the contribution of coding, has given great development to the theme of the new design concept and to the same digital craftsmanship. This new modelling process has the characteristic of approaching computer programming, but with an easy approach thanks to visual components. The system allows to implement the normal operations of generation, transformation and even substantial evolution of the models through reversible processes, preserving the memory of the original geometries.

Algorithms allow designers to overcome the limitations of 'traditional' software and achieve a much higher level of complexity and control (Fig. 1).

In the algorithmic modelling the user has the possibility to create three-dimensional objects through the description of the system of relations at the base of any complex geometry. This description takes place through the development of a node diagram (visual algorithm) according to an associative logic, within specific editors that operate in parallel with the modelling software. Therefore what is manipulated is not the object, but its construction process and its data.

The algorithmic modelling tools allow the generation and control of complex shapes at any scale: from architecture to design. The parametric design systems are dynamic modifiable in real time, through the variation of the parameters defined during the construction of the diagram: this leads to immediate advantages in relation to formal exploration, control and rationalization of the shape (Fig. 2).

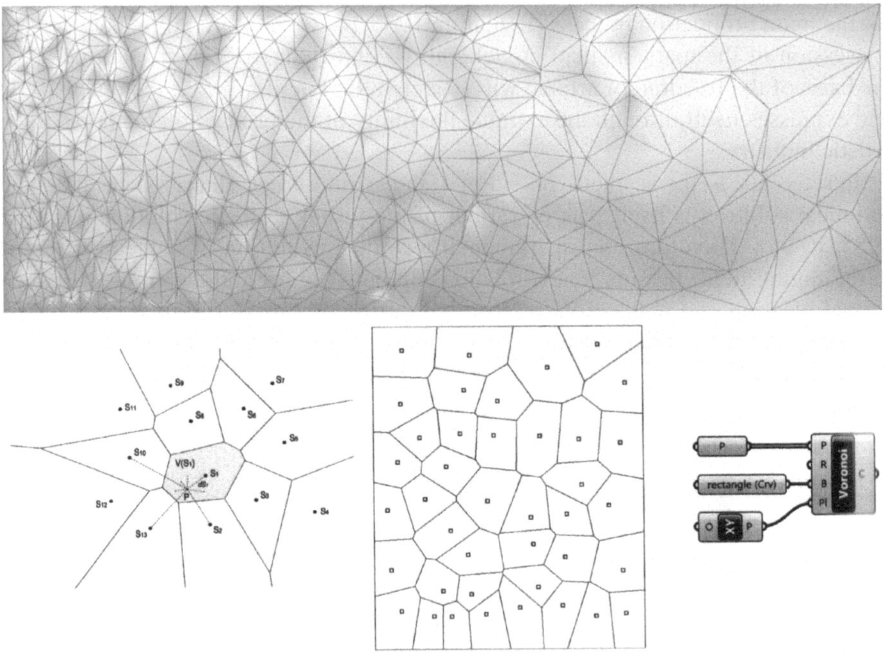

Fig. 1 Parametric design: Delaunay triangulation, generated by the Voronoi algorithm

3.2 *Color Parametric Design*

In the design process of an artefact, the role of color has always been ambiguous (Batchelor 2001), both for the complexity of a very elaborate theory,[1] and for the instrumental difficulty in identifying and accurately reporting the selected colors. In traditional production processes (Scarcelli 2016), following the development of the dyes chemical industry, the artisan chooses the color variations of his product from a range of dyes (or already colored semi-finished products) within a commercial catalog; even the designer has the same restrictions: he must choose based on a predefined range, influenced by the laws of the market, which condition those of taste (color matching), or vice versa.[2]

---

[1]For a deeper understanding of the various theories of color, applied to the world of design, a comparison with the pioneering texts of Itten (1982) and Albers (1971) is fundamental, in which the critical apparatus is always accompanied by graphic experiments.

[2]«Il controllo qualitativo del colore è uno dei problemi centrali della cultura ambientale moderna. Prima ancora che della forma o della funzione di un oggetto, noi ne percepiamo l'identità cromatica, a tal punto che l'insieme dei colori che ci circondano costituisce uno specifico livello d'uso dell'ambiente stesso. All'interno di questa problematica, il colore deve essere oggi inteso come oggetto di una ricerca progettuale autonoma; fino ad ora il colore è stato l'ultimo attributo del prodotto industriale: il designer, dopo averne progettato la forma e la funzione, sceglie il colore

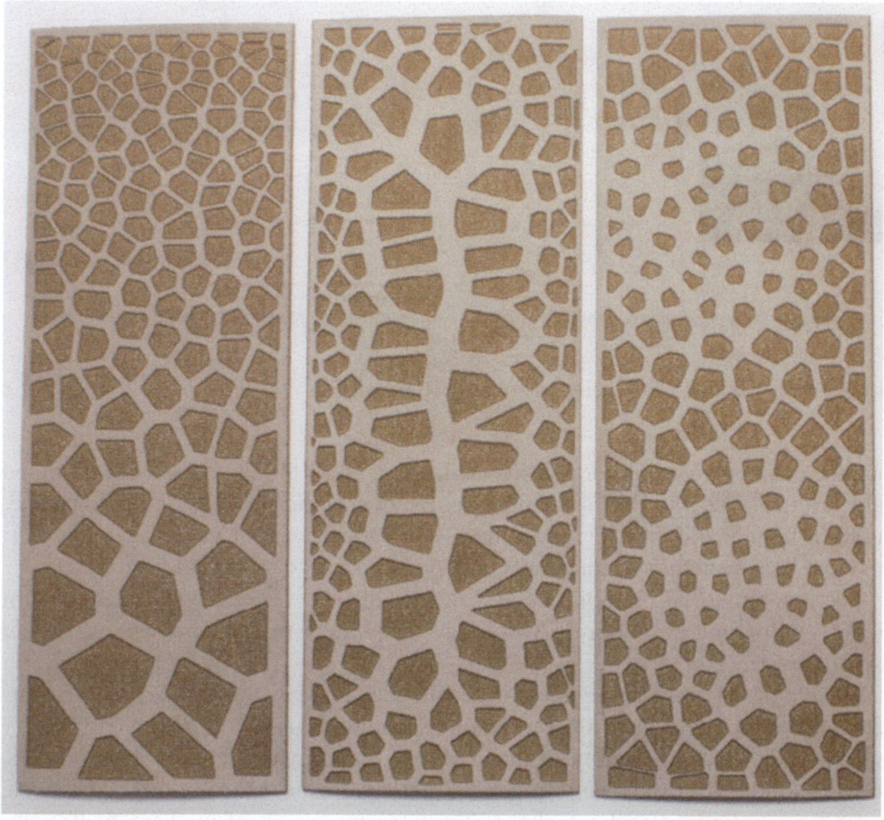

Fig. 2 Some outcomes obtained using computational design applied to textile laser cutting production

It is not possible to identify and quantify existing colors. From the dawn of time, in order to classify colors, artists, physicists or chemists have introduced very heterogeneous systems: in the principles and in the shapes, plane or three-dimensional, more different, for example the Itten hue circle or the Munsell tree.

Nowadays, the choice of a color system over another depends essentially on the scope of application and transformation processes that attribute perceptual qualities to the material: dyeing of the material through dyes (plastic fibers and textiles) or surface deposition of pigments by paints or inks (printing and painting processes).

The adoption of computerized drawing systems has multiplied the management and control tools of the project, also from the chromatic point of view: each dedicated software has a color selector.

This tool associates each color with a numerical code, which identifies it and makes it unique and universally distinguishable, within its own coding system. In

---

come ultimo segno da collocare su di una struttura che possiede già tutte le qualità fondamentali.» (Andrea Branzi (1984), La Casa Calda, pp. 102–103).

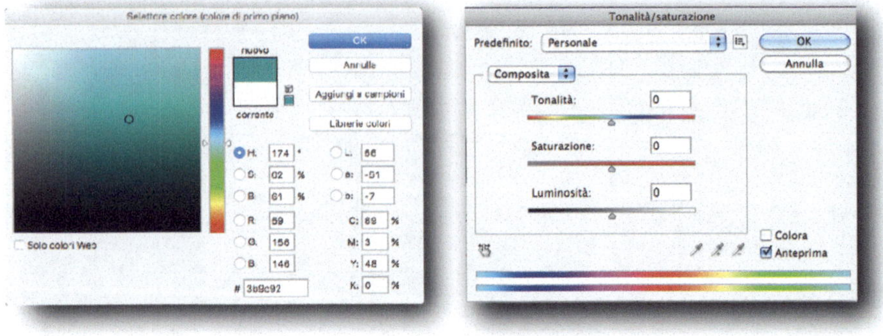

Fig. 3 Parametrization for the color according to RGB and HLs system

fact, the different systems adopt different languages, so the same color is associated with several codes: the HLS system has a triad of specific numbers, in the same way the RGB system and the Lab space; the CMYK attributes four values to each color; the Html hexadecimal system adopts an alphanumeric code instead (Fig. 3).

The combinatorial possibilities associated with the various systems can return 16,000 colors (and even more). Absolutely many more than the human eye is able to distinguish.

The coding of a tint through a numerical code distinguishes the parametric character of the chromatic variation: by changing a single number it is possible to modify the superficial spectral quality of the artifact in its virtual definition. This process, now consolidated in the design practice of design, especially in the graphic area, undergoes continuous updates, due to the implementation of mathematical codes within the digital development environments.

Similar to the definition systems of shapes and patterns, susceptible to changes based on algorithms that control their geometries, color can also be managed digitally, by writing parametric codes that link each individual color value to a specific point/pixel of the physical/graphic space. In particular, using the parametric patterning tool, included among the cad application tools such as Grasshopper, it is possible to interpolate a predefined image with a geometric grid, so as to associate each RGB color value of the map with the corresponding point-position in the grid. The image, ultimately, is discretized in points with numerical values, and then translated into a vectorial geometry.

Another tool allows this association between the polygons of a reticular mesh—organic or geometric—with selected colors, with the possibility of parametrically varying the tonal, luminosity or saturation values. In this way both continuous and discrete results are possible, resulting in a raster graphic, no longer vector. This approach presents interesting developments concerning the transition operations between the different colors, which can occur by gradient, giving rise to organic matrices, or by contrast, with respect to discontinuous geometries. The parametric application adopted for the transitions is not only linear, but presents articulated solutions linked to the complexity of the mathematical codes used.

The size of the capability of color parametric systems in digital design processes can be found in the results of Digital Art: if the artistic culture abandons its work tools to test the opportunities of new IT tools, it is the sign of a change not only instrumental, but specially linguistic.

4 Results

This section shows the results of some experiments aimed at the implementation of parametric information systems for the development of the decorative pattern and for the development of the colour gradient. In particular, the path has been focused on the understanding and innovation of artistic craft processes, mediating the scope of unique artifacts (or limited series) with the scope of large-scale industrial production.

In particular, with reference to the theme of the material and the conformation processes of the manufactured product, the interest in traditional production has been focused, reaching the integration of enabling technologies towards the artisanal productivity of type 2.0, through the use of 3D printing aimed at making moulds (excavation, embossing, graphic impression, etc.).

The experimentation theme was addressed to the aesthetic characterization of the surfaces made in ceramics, textile and leather; the methodology adopted has marked the process in five phases of planning and realization.

The approach adopted aimed at defining the expressive and communicative potentiality of a three-dimensional decoration through graphic-visual, tactile and chromatic considerations, in relation to the triad consisting of material, tools and production process. The graphic theme of "gradient" has been elaborated on the base of a parametric 3d model.

In the first phase a visual graphic composition is developed, responding to precise principles of perceptive nature such as "rarefaction and thickening", "direction", "order" (Fig. 4).

The original compositional principle is thus implemented on a running surface, through module iteration operations, by reflection, rotation or transition (Figs. 5 and 6).

The third phase concerns the association of colour values of tonal scale and contrast to the graphic-visual composition, and the subsequent transposition onto the material support (Figs. 7 and 8).

The procedural implementation of tools takes place in the last two phases: the graphic composition is a 3d model for the production of files useful for the realization of negative—positive moulds, using three—dimensional printing processes

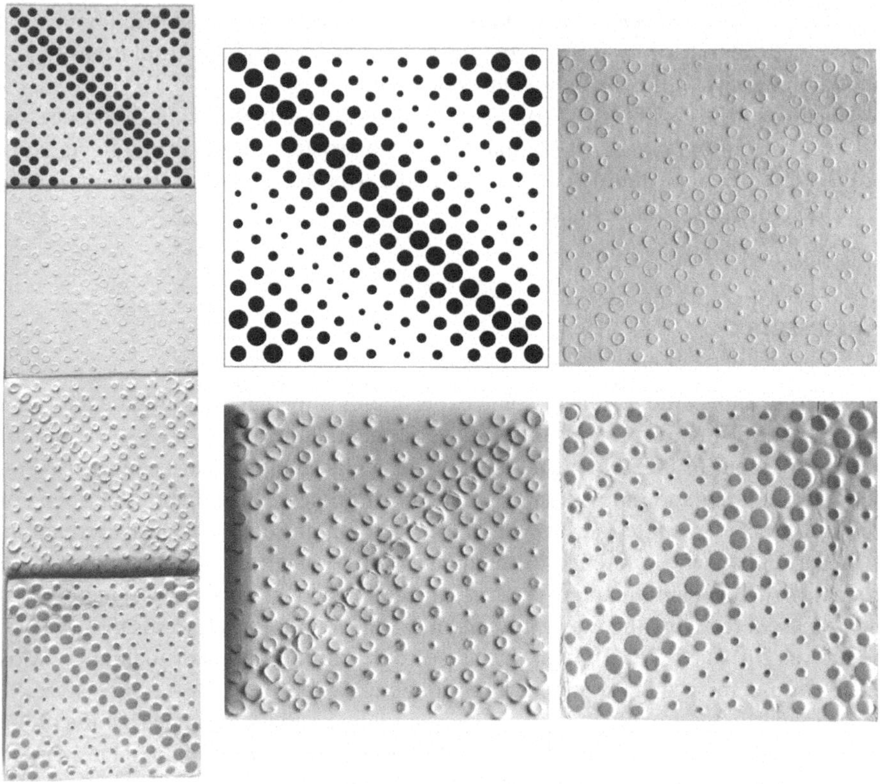

Fig. 4 The original compositional principle is thus implemented on a running surface, through module iteration operations, by reflection, rotation or transition

with filament deposition in PLA. The moulds thus obtained will be used to define the surface quality of the physical panel, experimenting "mould" procedures on rigid supports (Fig. 9).

5 Conclusions

The parametric design combined with the production of models through print output and electronically controlled processing represents a phase of evolution of the contemporary art industry linked to digital outcomes (Oxman 2007). The vector and bitmap software applications through the generative logic offer new operative tools and possibility to elaborate innovative formal languages, coherent with the mathematical models that support them through the algorithms. In accordance with the

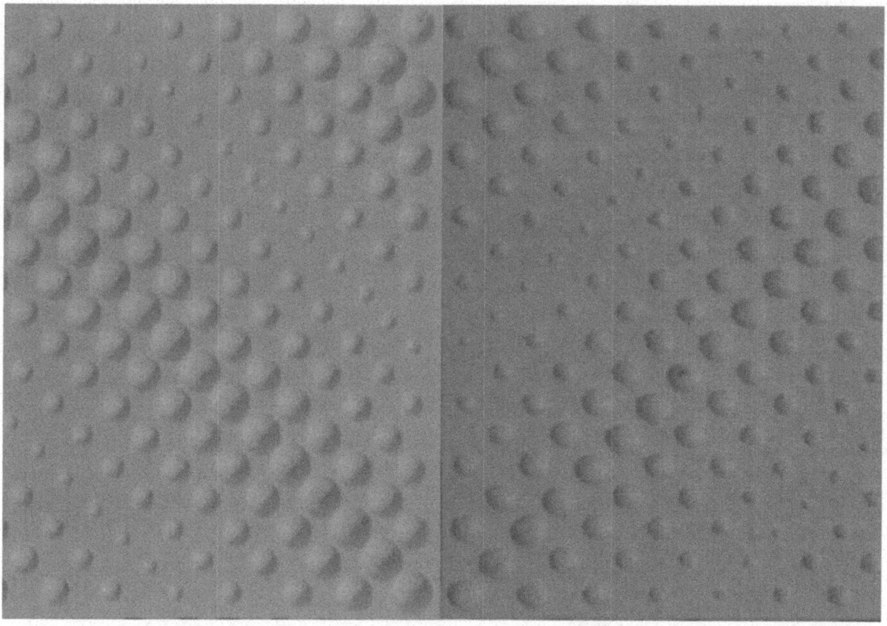

Fig. 5 Implementation of the graphic design on a tridimensional surface using a parametric model. The tiles are produced using FD 3d print system

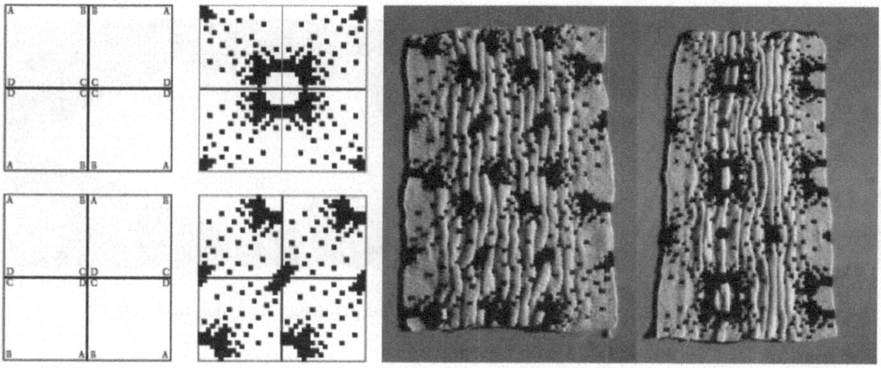

Fig. 6 The original compositional principle is thus implemented on a running surface, through module iteration operations, by reflection, rotation or transition starting from a graphic design and translating it in a textile weaved surfaces

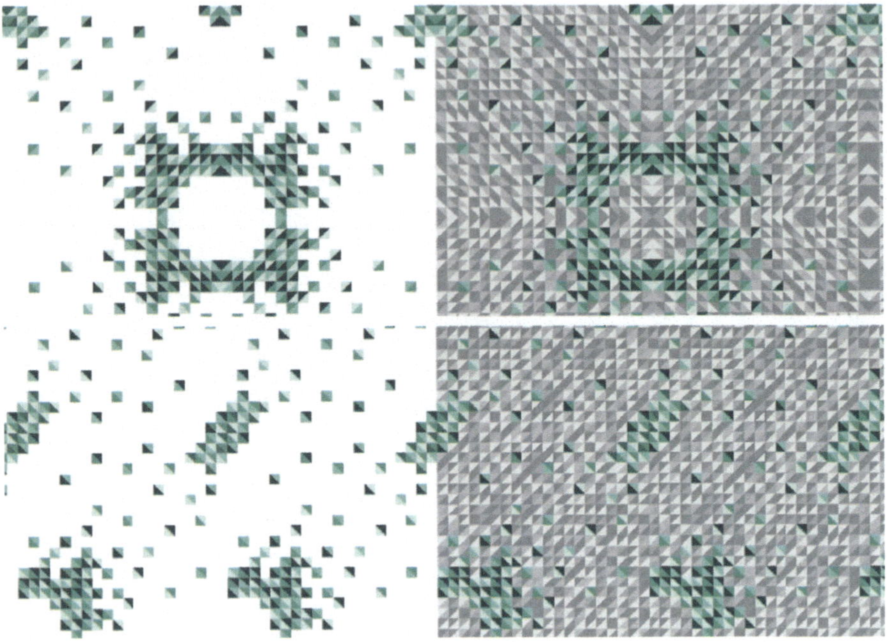

Fig. 7 The original compositional is overlapped with a color graphic schema

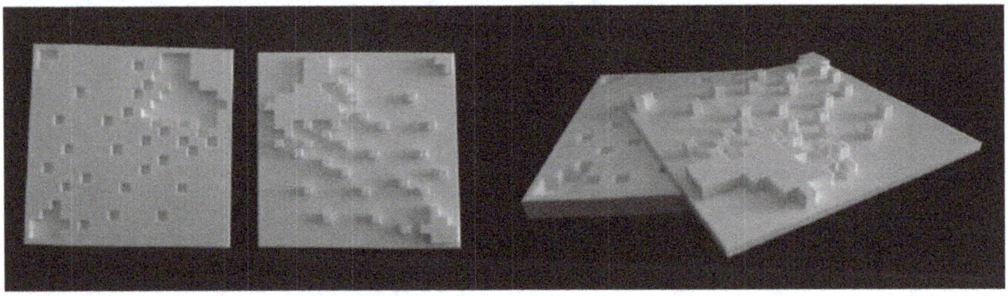

Fig. 8 Cast tool for the embossy realized using a fdm 3d printing machine and PLA

Kunstwollen by Riegl, the instrumental technical equipment supports the will of art and configures new operating methods in which the designer, the computer technician and the machine operator share the artistic process of material transformation. This essential process of evolution starts from the adoption of 3d modelling software for design, digitizing, plotting and cutting components. The considerable development had also adopted new materials, among which the polymer taken over since the 60s as innovative materials for all the components for accessories, and a whole range of technical fabrics and synthetic specially developed. Until you get to the present, where the introduction of 3D printing is marking a real "new industrial revolution", including in the Fashion Sector (Fig. 10).

Fig. 9 From the graphic composition to the 3D modelling—to the 3D printed mould for the embossy of the tissue

Among the many techniques used today in the industry of fashion design, laser cutting techniques are having a grate development and application. Laser processing and 3D printing are, in fact, contributing not only to redefine the aesthetic canons of the formal production in that area, but also to bring innovation to a new level of complexity.

Generative modelling, thanks to the contribution of coding, has given great development to the theme of the new concept of design and the same digital craftsmanship. This new modelling process has the character of computer programming, but with an easy-to-use approach through visual components.

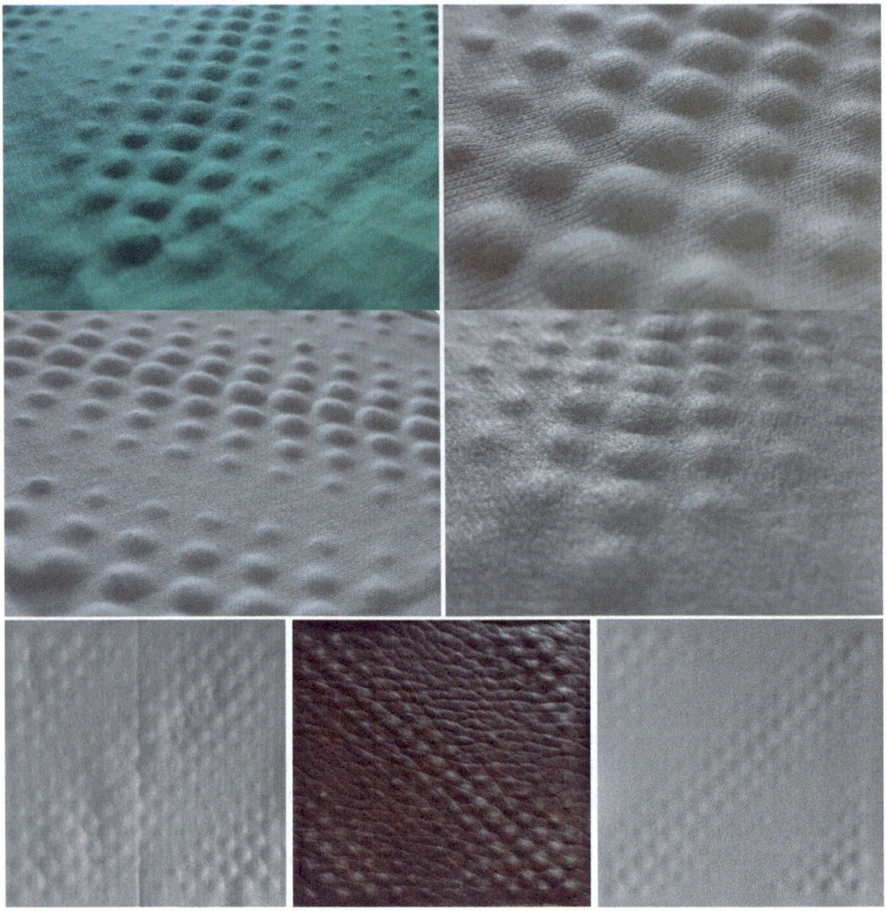

Fig. 10 The present picture shows the different effects of the embossy produced on different surfaces: cotton, felt, polyester, leather

References

Albers J (1971) Interaction of color. Yale University Press, New Haven

Batchelor D (2001) Cromofobia. Storia della paura del colore, Bruno Mondadori, Milano

Benjamin W (1966) L'opera d'arte nell'epoca della sua riproduzione tecnica. Einaudi, Torino

Branzi A (1984) La casa calda. Esperienze del Nuovo Design Italiano. Idea Books, Milano

Branzi A (2008) Introduzione al Design Italiano, una modernità incompleta. Baldini e Castoldi, Milano

Carboni M (2001) L'ornamentale: tra arte e decorazione. Jaca Book, Milano

Di Roma A (2008) La produzione in serie dell'ornato architettonico: dall'industria artistica ellenistica alla prototipazione con processi cad–cam. Aracne, Roma

Di Roma A (2016) La cultura materiale tra tradizione e innovazione. In: Di Roma A, Scarcelli A, Minenna V (eds) Cultura materiale e Design. Favia, Bari, pp 11–30

Focillon H (1939) Vie de forms. Presses Universitaires de France, Paris

Hauser A (1955) Storia sociale dell'arte, vol III. Einaudi, Torino

Itten J (1982) Arte del colore. Il Saggiatore, Milano

Oxman N (2007) Digital craft: fabrication based design in the age of digital production. In: Workshop proceedings for international conference on ubiquitous computing, Innsbruck, Austria, pp 534–538

Panofsky E (1961) La prospettiva come forma simbolica. Feltrinelli, Milano

Riegl A (1893) Stilfragen: Grundlegungen zu einer Geschichte der Ornamentik. G. Siemens, Berlin

Scarcelli A (2016) Il colore materiale del progetto. In: Di Roma A, Scarcelli A, Minenna V (eds) Cultura materiale e Design. Favia, Bari, pp 31–44

From Interior Design Imaginary to Jewellery Concept

Mónica Romãozinho

Abstract In this project, we had started with the experimentation around housing typologies inspired by nature, by its asymmetric but at the same time modular shapes and its capacity to metamorphose. The purpose was not the construction itself but the transposition of this process to jewellery practice would imply the materialization of these spatial concepts. We had explored the movement of undulating surfaces, the relation between shadow and light, scale variations, the opposition between empty and full spaces, chromatic or texture contrast. A jewel can convey different appropriations and emotions such as a habitat, poetic and intangible space regardless of the economic value of its constructive solutions. Jewellery can be an extension of our selves just like interior design and that's the reason why we produce unique pieces or limited series. This article focuses on the methodology and results of our research underlying the creation of a contemporary jewellery line, focusing its main principles: spatiality, mutability and versatility. Sketches and cardboard models were intersected with 3D software and 3D printing without forgetting traditional jewellery skills. Another complementary approach was the assemblage technique applied to the most recent pieces.

Keywords Interior design · Jewellery · Versatility · Mutability · Reuse

1 The Meaning and Functions of Jewellery in Everyday Life

Jewellery is much more than an industrial product and the same happens with interior design. Jewel is a complex object. It is obviously permeable to fashion and trends along history, but it can establish a true connection with our character helping us to communicate our own individuality, working as an asymmetric detail in the way

M. Romãozinho (✉)
CIAUD, Lisbon School of Architecture, Universidade de Lisboa, Rua Sá Nogueira, 1349-063 Lisbon, Portugal

IPCB, Castelo Branco, Portugal

D. Raposo et al. (eds.), *Perspective on Design*, Springer Series in Design and Innovation 1,
https://doi.org/10.1007/978-3-030-32415-5_25

we dress that denies the standardization of fashion. On the other hand, jewellery doesn't have to answer to a utilitarian purpose, although it should be ergonomic in our point of view: "Through the centuries, jewellery has developed a variety of different functions, being able to adorn and also bear symbolic meaning." (Passos 2019a). Chanel soon understood that jewellery was much more than a communication of a social status: "The purpose of the jewel is not to make a woman seem rich but to adorn her. It is not the same thing." (Morini 2000).

Nowadays, researchers such as Ana Passos are mostly interested about the anthropological meanings of jewellery especially those related to affection and memory. The sentimental jewellery is the main field of this researcher who takes a customer's sentimental jewellery and reworks it into something new that can be worn every day (Passos 2019b). The theme is pertinent, and she refers the global success of Pandora, the Danish brand created in 1982 because of its marketing strategy based on the sharing of life stories through the use of jewellery that mark significant life events (Passos 2019b). This area probably results from the feeling of being lost because of the diversity of products that market offers nowadays. Sentimental jewellery is born from an eclectic attitude not from a pastiche creation. The same happened along history of decorative arts if we look to the 19th century. I think looking back to the past and rework that same past with a critical attitude can be an answer if, looking to our pieces, we recognise the differences between old and new; at the same time, we must question that same past just like avant-garde artists did it in the beginning of the twenty century not in an aleatory way. We can feel disoriented with the excess of jewellery and accessories trends but that doesn't mean that bringing the past is the answer, something which is happening nowadays with some furniture design, for example. Maybe bringing some part of it is an efficient way to connect with collective imaginary and make people feel there is a continuity between past and present but in an eclectic process. We don't do sentimental jewellery, but we must understand that History knowledge is a part of everything we make and it's impossible for us not to recall important references that belong to architecture, design and art fields. We always avoid any explicit transposition of this universe or even miniaturisation. It can consist in small details or even in a methodology. We try to communicate a specific message and awake emotions just like in architecture or interior design projects.

2 Genesis of the Project

A jewellery piece can constitute itself as a platform for our individuality, a function that approaches itself precisely to a house, a room, an intimate space. The 1930s, for example, were marked by "the CIAM (Congrès internationals d'Architecture Moderne) for Architecture, UAM also aspired to a "truly social art suitable for progress and able to integrate present-day industrial forms and technologies struggling against classicism and tradition" (Cappellieri 2010a). Raymond Templier's, for instance, followed the principles of the International style, reflecting a functional purity, mechanical geometry, rigorous volumetry and also an alternation of volumes and voids, mass and lightness (Cappellieri 2010a). Cappellieri tell us precisely that first Templier's and then Jean Després' jewelers recall the rarefied abstraction of Le Corbusier's Ville Savoye or Mies Van Der Rohe's Barcelona pavilion."* (Cappellieri 2010a). In this project, it is clear the important role of architecture and interior design for the development of a concept. Sabine Pagan is a contemporary jeweler who studied cross-disciplinary concepts between contemporary architectural and jewellery practice, considering Peter Zumthor's work Therme Vals as a case study for understanding the significance of sensorial experience on human beings and its relevance to memories, so important in jewellery field. She refers the obvious contrast between scales and connection with human body: "On one hand, the architectural space is designed to shelter the human body and enables specific actions such as working and sleeping to take place within it. On the other hand, in the context of jewellery, the human body is used as a platform to showcase attributes in accordance to social, cultural and historical contexts." (Pagan 2011a). For Pagan "whilst the human body differentiates one practice from the other, it also appears to bring them closer together, for the body does not only reference a physical entity but also relates to the frame of mind. Together architecture and jewellery contribute to the 'Self' and its human characteristics such as emotions and our capacity for memory" (Pagan 2011b). The aim of her research is to examine the value of a phenomenological approach in architecture and understand how such methodology might be applied within the context of contemporary jewellery (Pagan 2011c).

Pagan's jewellery project is related to a concrete study case while the present research project consists of the design of the jewellery collection "Possible but improbable spaces" that is strongly related to our own architecture and interior design drawings envisioning new concepts for a more human and sustainable society, which is part of a post-doctorate in Design developed within the CIAUD (Lisbon School of Architecture, Universidade de Lisboa).

It started with the experimentation around space, its principles, its geometry and crosses with the Architecture and the Interior Design at the conceptual and methodological level. Our initial incursion focused on the design of organic architectures inspired by nature, with asymmetrical and irregular contours resulting from tensions between colours, materials and transparencies. The main compositional principles that approximate these designs of so-called organic architecture, because the essential organisation arises from the internal needs and external conditions, in which each architect tries to connect their own volumetric and geometry of space projected to the topography of the land itself. What intrigues us is to draw inhabited spaces as opposed to cells inhabiting anonymously, seeking to explore a strong sense of identity that we have tried to transpose to the scale of jewellery. We rethought the territory through the design of fluid, experimental, irregular spaces, wrapped in organic membranes that adapt to the morphology of the terrain (Fig. 1). There is space and time to ramble and contemplate nature. We standardised some components, namely interior volumes, coverings or furniture that also arises from the geometry of these spaces yet evoking the sense of intimacy. Carpets model spaces, in the image of "tatamis".

The translation from architecture and interior design drawings or cardboard models to jewellery hand-scale pieces was taken over considering fundamentals such as

Fig. 1 Example of experimental drawings that were in the genesis of the jewellery research project: Utopia #VI-11. 2015. Mónica Romãozinho

irregular and oblique contour shapes, repetition and modularity, opposition between transparencies and opacities that relate to exterior and interior, the unfolding of the interior into volumes, the border between public and private, sense of movement or rhythm, the contrast between full and empty, the contrast of colours or textures, determinant factors as shadow and light. Essentially, we intend to create an open project, which means to explore a transversal speech to all the pieces of distinct lines within the same collection.

3 Architecture and Art as Main Influences: Between Eclecticism and Avangarde

This possible connection can be feed by our knowledges, references and practice in architecture and interior design. One of the master inspiring architecture masters is Carlo Scarpa (1906–1978), because of his relation with the history of architecture and arts, because he could find beauty in particular details invisible to other eyes and mostly because when he started working on a new project he never repeated himself, searching at the same time for an "uncompromising argument for utopian vision, for a kind of architecture that is committed to art and culture only – even if this should go against reality or against those in power" (Noever 2003). This is the way we want to work when we are creating a piece of jewellery. The public can consider this jewellery collection as a contemporary work, but its essence is completely eclectic. On the other hand, architecture and art are references for everything we do because they are not so ephemeral as fashion and even my fashion references are much closer of avant-garde designers like Rei Kawakubo, Yoshi Yamamoto and specially Issey Miyake who first opened doors to post-modernist interpretation, playing with barriers between Orient and Ocident, fashion and anti-fashion, modern and anti-modern (Kayamura 2003a). They don't produce simply clothes but poetry just like Japanese architecture which was also a great reference for Scarpa.

Crane considered that any artistic movement can be regarded as avant-garde when "its approach to aesthetic content of its artworks follows the following criteria: redefines the artistic conventions, uses new tools and new artistic techniques, redefines the nature of art object and belongs to the palette of objects that we can consider as an art work." (Kayamura 2003b). This definition can be crossed with Cappellieri's perspectives for the future defined in her book which is related to the "Avant-Craft" tendency.

Nowadays, it is indeed possible to distinguish various trends inside jewellery field as a result of different researches and methodologies. In this perspective of paradigm change, Alba Cappellieri exposes five possible scenarios for the future: Marvellous manufacture, Everyday Beauty, Avant Craft previously referred, Precious Technology and Collective Creativity. Marvellous manufacture highlights jewellery strictly made by hand with precious materials, what we recognize as high jewellery (Cappellieri 2016a). The second scenario represents the ability of jewellery to communicate beauty, harmony, emotion and all areas can converge here such as craftsmanship, architecture, fashion, design. It's a ready-to-wear jewellery, more accessible, that we can use according to our mood or clothes. This attitude results in multiple and expressive languages and in different meaning of preciousness as well (Cappellieri 2016b). Avant-Craft is close to the so-called author's jewellery, resulting from independent and experimental research, disconnected from the logic of production, communication and distribution, by crossing heterogeneous languages, materials and techniques (Cappellieri 2016b). Precious technology involves new production processes and materials usually distant to this field as titanium or carbon for instance. Several researchers explore the incorporation of new technologies especially useful as answers for people with disabilities, difficulties felt by older people manipulating pieces, and we have got the example of biotechnology as an answer to a more sustainable jewellery that can be made by the manipulation of fibers adaptable to different environments. The last scenario is centered in creative, production, distribution and communication processes.

Our approach is mostly born from the intersection between Avant-Craft and Technology. We carry an artistic work intersecting technology and traditional skills, considering its ergonomic connection with the body and pursuing its sense of identity as a platform of communication of ourselves. We recognize the role of art in the historical evolution of jewellery design. In the 1930s and 1940s, jewellery oscillated between experimental jewels of bijoutiers-artistes, and "haute" traditional jewels particulary in French context but, it is evident the increasing connection between art and jewellery exemplified by the 1946 MOMA exhibition in New York (Cappellieri 2010b) and we must remember Elsa Schiaparelli, for instance, who involved artists in jewellery field such as Alexandre Calder, Christian Bérard, Salvador Dali, Jean Cocteau mainly related to Dadaism and surrealism currents. Their innovation was determinant but the problem was that some of their works resulted from the miniaturization of details or compositions retrieved from their art experiments. Mario Pinton, founder of the School of Padua and an expert about gold would assume precisely that "he was not interested unlike many artists such as Dalí, Fontana, Man Ray and Max Ernst, in the miniaturization of an artistic expression already encoded in painting or in sculpture" (Cappellieri 2010c). Our work is inscribed in the avant craft perspective, much closer to art, architecture and interior design, less permeable to changes than fashion design and we produce limited editions although our design process allows mass production.

4 Creative Process: About Changeability, Versatility and Colour

However, we felt these first experiments were incomplete. If we wanted to rethink the connection between jewellery and users, its positioning and relation with the body, maybe we had to start looking at jewellery as a set of dynamic objects, even mutable, just as a spatial solution does not necessarily have to be imposed on the public. We were not only worried about answering to the traditional functions but about responding to different occasions, states of humour or even events through minimal but versatile pieces.

The same assumptions were explored along design history. For instance, we must remember the originality of the "Pass-partout" necklace from Van Cleef that consisted in a "serpent chain" made of yellow gold, whose clips could be autonomous and applied to hats or clothing" (Cappellieri 2010d). In the same decade, we can recall the Gabriele Chanel's only collection, presented at the 1932 *Bijoux de Diamants exhibition* (London, 1932) composed by transformable pieces inspired by stars, planets and comets before her collaboration with Fulco Santo Stefano della Cerda, Duke of Verdura. Chanel worked in partnership with Iribe, this time producing strictly precious, articulated and transformable pieces (Cappellieri 2010e). Pieces such as the open necklace in the shape of a comet are not only transformable allowing several functions (necklace, brooch, a hairdresser) but also asymmetric, consisting simultaneously in an open structure in order to create a new relation with the body without losing simplicity. As Chanel declared: "My (jewels) are flexible and detachable [...] You can take apart the jewelry and use it to match a hat or fur. In this way, the set of jewelry is no longer an immutable object. Life transforms it and bends it to its needs" (Doulton 2019). The same philosophy was a reference and inspiration for the Chanel's 1.5 collection of camellia-themed high jewels presented at Paris Haute Couture 2019 (Doulton 2019).

We explored the possibility of mutation by designing a necklace that results from reusing a PVC cylindrical object that symbolises the idea of an organic habitat always changing like the organisms of nature (Fig. 2) without any submission to an orthographic system. The pre-existing object was the starting point, our pre-existence, just as the understanding of the terrain, of the "genius loci" is the lever for an architectural project. At the same time, these projects are not simply wearable artefacts but spatial structures as well, that we can contemplate, rotate and (re)discover according to each new observation angle.

The necklace focused in the present article unfolds in two main solutions, through the insertion of a sliding horizontal plank (the floor of this structure) inside the cylinder or through the movement of the cylinder to a new position (behind our neck) and the simultaneous combination between a ramp and the copper module present in both solutions. We enjoyed playing with these objects, manipulating minimal elements resulting from varied production processes and crossing them in the same geometrical composition as if it were an *assemblage*. At last, our approach is connected to contemporary authors such as Giampaolo Babetto (1947) contrary to a product massification, however interested about abstraction, modularity and economy in his whole design concept that is born from a "cartesian precision" and a "passionate research on the material": "All his creations come from the idea that the jewel is a work of art, based on the study of geometric forms, modular construction and texture" (Cappellieri 2016c). He introduces movement in the geometry, putting the accent on the relationship between inside and outside, full and empty (Cappellieri 2016c).

We chose to explore some of these strategies, combining them with changeability in the design of a brooch (Fig. 3), a piece that can play a more interactive relation with shirts, coats or capes, assuming a functional role. With this brooch that can assume a more neutral or contrasting character, we wanted to recall de idea of a house viewed

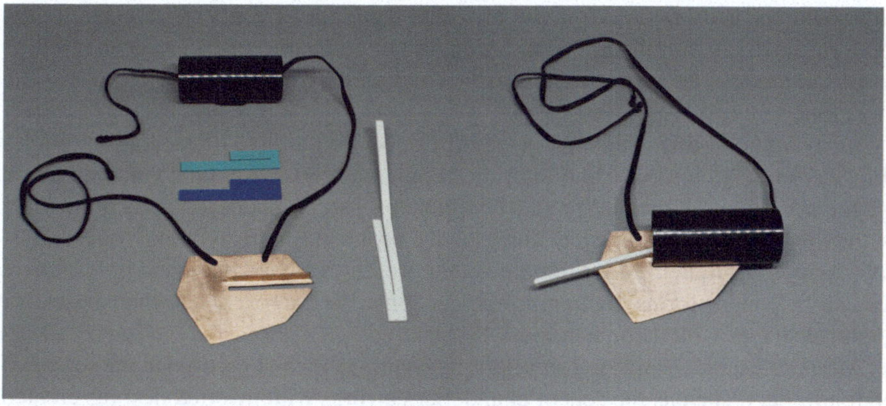

Fig. 2 Collection "Possible but improbable spaces". Line "Abstract spaces". Mutable necklace (single piece). PVC (reuse), PLA and copper. Mónica Romãozinho

Fig. 3 Collection "Possible but improbable spaces". Line "Abstract spaces". Mutable brooch (limited edition). PLA and copper. 2018. Mónica Romãozinho

in parallel perspective defined by successive plans and spaces that are not complete revealed. This translucent shelter receives not only a coloured "wall" (Blanc Gris or turquoise), but also a horizontal line sequence (a "brise soleil") separating the interior from the exterior and giving rhythm to the whole composition and, at last, a shape in copper. Line is considered as a matter of fact as a fundamental component in the same way that Aleksander Rodchenko (1891–1958), a constructivist designer, stated in his work: "Both in painting as in any construction in general, line is the first and the last thing" (Rodchenko 1979).

The ring was developed from a detail of other piece of this line, a brooch. We've tested several solutions around the idea of articulating a shell with an interior volume that could be easily replaced: in the first one, the blade of copper is attached to the PLA volume by a rail, in another solution, the thin metallic surface was also folded in its end engages in a cut made in the second piece. The present solution was the most functional and intuitive (Fig. 4).

An example of this is the prototypes developed in the field of earring typologies (Fig. 5). As in Architecture and Interior Design, we started from the volumetric and the relationship with the place and a complete project is supposed to reach the scale of the door handle, answering to constructive problem. In this typology of earrings, we explored asymmetric PLA modules that present recessed parts and holes where we can fit copper forms containing the ear pins. We can exchange their bases for identical shapes with distinct colours.

Although market is offering solutions of asymmetric earrings even in the so-called high jewellery, such as the sets of chandelier earrings with inverted designs created by De Grisogono (Liu 2017), for The *Tubetto* collection we can conclude changeable solutions an unexplored point of view. At last, we wanted to play not only with organic

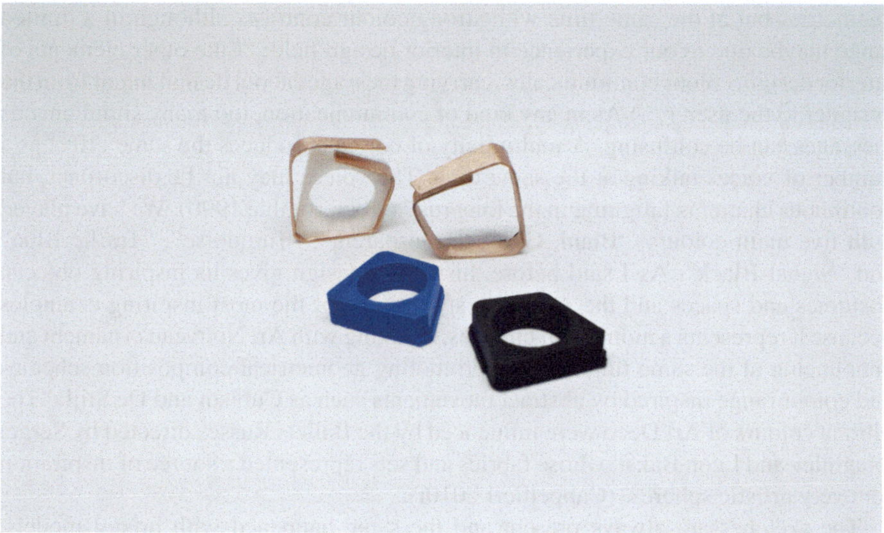

Fig. 4 Collection "Possible but improbable spaces". Line "Abstract spaces". Mutable ring. 2019. Mónica Romãozinho

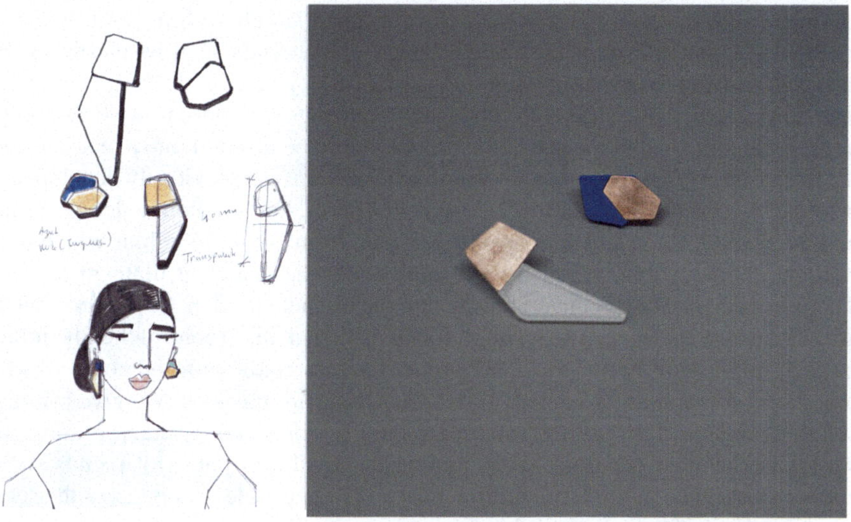

Fig. 5 Collection "Possible but improbable spaces". Line "Abstract spaces". Mutable earrings. 2019. Mónica Romãozinho

geometries but at the same time with strong colour contrasts although in a limited range maybe due to our experience in interior design field: "Like other elements of interior design, colour communicates, carrying messages about design intent from the designer to the user. (…) As in any kind of communication, too many simultaneous messages can be confusing. A multiplicity of colours produces the same effect as a number of voices talking at the same time. The voices may not be discordant, but continuous chatter is fatiguing in the long run" (Abercrombie 1990). We have played with five main colours: "Blank Gris", "Transparent", "Turquoise", "Traffic Blue" and "Signal Black". As I said before, history of design gives us inspiring objects, costumes and spaces and the Art Deco style is one of the most inspiring examples because it represents a moment of changes, breaking with Art Nouveau ornament and announcing at the same time its end, promoting geometrical composition schemes and colour range inspired by abstract movements such as Cubism and De Stijl: "The vibrant colours of Art Deco were influenced by the Ballets Russes directed by Sergei Diaghilev and Leon Bakst, whose fabrics and sets represented a source of inspiration for every artistic sphere." (Cappellieri 2010f).

The sketches are always present and the same happened with bristol models, three-dimensional modelling and 3D printing, in order to find solutions and test fittings, density and quality levels, minimum section bearable in the perforated zones, minimizing the expense of PLA (biodegradable polymer termed polylactic acid), a biodegradable material that is obtained from plant starch such as corn, which has a low environmental impact. The advantage of materials such as plastic or resin also derives from their extreme lightness as it allows you to rotate the pieces causing different effects. In the case of the changeable earrings, we chose to weld a silver post to the copper blade, but the pin of the brooch would be inserted into a recess made in the back of the same piece in a manner like other earrings previously developed. 3D printing has opened the door to all these possibilities, but its colour palette is limited, and the quality of the filaments is also variable from colour to colour which makes printing unpredictable. We also must think about the matter of filament recycling. There are machines that convert plastic into the filament used in 3D printers, but they are still much expensive. The solution found is, first at all, to control the first minutes of each impression because the 3D printer process starts with the contours of the object and sometimes, we detect initial problems with recesses. We give priority to solutions that work by fitting, minimizing the use of bonding, incorporating some wastes in our own creations safeguarding all possible residues of metals such as copper that can be melted, re-entering the production cycle. Copper was the second material chosen for its high melting point but mostly because of its reddish colour, due to the mutations that it undergoes over time and oxidation, gaining unpredictable tones. The cut of copper forms is manual using the goldsmith's saw, followed by a brushed finish.

5 Conclusions

The exploration of new concepts, materials and creative processes allowed us to achieve a collection that is reaching a different public with a taste for contemporary jewellery but without economical possibilities of "consuming" "haute" jewellery. We have explored principles such as the versatility, the mutability that provide a manual experience of transformation, exploring a geometrical language, irregular forms, opting for subtraction instead of excess, banning any use of ornament, searching for simplicity and elegance. But above all we must underline the idea of conveying a critical narrative of how our cities' architecture and interiors evolve. They are limited series pieces so that whoever uses them feels that they communicate their individuality such as a house, although they don't play a functional role. In this sense, we cannot fail to evoke the words of Gijs Bakker, one of the major exponents in the context of contemporary jewellery: "I like jewellery because it is absolutely superfluous. I like jewellery because it is never a priori functional. I like jewellery because like clothes, it is closest to our body and says something about the wearer. A painting is hung on the wall and can be ignored. A piece of jewellery is worn and creates an impression" (Cappellieri 2010g).

References

Abercrombie S (1990) A philosophy of interior design. Westview Press, Boulder, p 113

Cappellieri A (2010a) Twentieth-century jewellery: from Art Nouveau to contemporary design in Europe and the United States. Skira Editore S.p.A, Italy, p 34

Cappellieri A (2010b) Twentieth-century jewellery: from Art Nouveau to contemporary design in Europe and the United States. Skira Editore S.p.A, Italy, p 36

Cappellieri A (2010c) Twentieth-century jewellery: from Art Nouveau to contemporary design in Europe and the United States. Skira Editore S.p.A, Italy, p 43

Cappellieri A (2010d) Twentieth-century jewellery: from Art Nouveau to contemporary design in Europe and the United States. Skira Editore S.p.A, Italy, p 35

Cappellieri A (2010e) Twentieth-century jewellery: from Art Nouveau to contemporary design in Europe and the United States. Skira Editore S.p.A, Italy, p 24

Cappellieri A (2010f) Twentieth-century jewellery: from Art Nouveau to contemporary design in Europe and the United States. Skira Editore S.p.A, Italy, p 20

Cappellieri A (2010g) Twentieth-century jewellery: from Art Nouveau to contemporary design in Europe and the United States. Skira Editore S.p.A, Italy, p 147

Cappellieri A (2016a) Brilliant! I futuri del gioiello italiano. In: XXI Esposizione Internazionale della Triennale di Milano. Corraini Edizioni, Mantova, p 33

Cappellieri A (2016b) Brilliant! I futuri del gioiello italiano. In: XXI Esposizione Internazionale della Triennale di Milano. Corraini Edizioni, Mantova, p 35

Cappellieri A (2016c) Brilliant! I futuri del gioiello italiano. In: XXI Esposizione Internazionale della Triennale di Milano. Corraini Edizioni, Mantova, p 121

Doulton M (2019) Chanel 1.5: one camellia five ways to wear it. The Jewellery Editor. Retrieved from: http://www.thejewelleryeditor.com/jewellery/article/chanel-15-one-camellia-many-ways-wear-it-diamond-jewels-2019/. Accessed 30 Jan 2019

Kayamura Y (2003a) La révolution japonaise dans la mode parisienne. In: XXIème CIELE: mode in Japan. 5 continents editions, Milan, p 47

Kayamura Y (2003b) La révolution japonaise dans la mode parisienne. In: XXIème CIELE: mode in Japan. 5 continents editions, Milan, p 48
Liu M (2017) Jewelry, matching just isn't cool any more. The New York Times, pp 1–5. Retrieved from: https://www.nytimes.com/2017/11/22/fashion/jewelry-assymetrical-earrings.html
Morini E (2000) Storia della moda XVIII-XX secolo. Skira, Milan, p 182
Noever P (2003) The other architect. On the sublime of architect. In: Carlo Scarpa: the craft of architecture. Mak Studies, Vienna, p 10
Pagan S (2011a) Contemporary jewellery: a phenomenological approach to making informed by architecture. In: ACUADS conference 2011: creativity: brain–mind–body: a view into the future of Australian Art and Design Schools. Retrieved from: http://acuads.com.au/conference/article/contemporary-jewellery-a-phenomenological-approach-to-making-informed-by-architecture/. Accessed 21–23 Sept 2011
Pagan S (2011b) Contemporary jewellery: a phenomenological approach to making informed by architecture. In: ACUADS conference 2011: creativity: brain–mind–body: a view into the future of Australian Art and Design Schools, p 1. Retrieved from: http://acuads.com.au/conference/article/contemporary-jewellery-a-phenomenological-approach-to-making-informed-by-architecture/. Accessed 21–23 Sept 2011
Pagan S (2011c) Contemporary jewellery: a phenomenological approach to making informed by architecture. In: ACUADS conference 2011: creativity: brain–mind–body: a view into the future of Australian Art and Design Schools, p 3. Retrieved from: http://acuads.com.au/conference/article/contemporary-jewellery-a-phenomenological-approach-to-making-informed-by-architecture/. Accessed 21–23 Sept 2011
Passos A (2019a) Jewellery: from material to affection. J Jewel Res 2:2
Passos A (2019b) Jewellery: from material to affection. J Jewel Res 2:4
Rodchenko A (1979) Line. Alexander Rodchenko. Museum of Modern Art, Oxford, p 12

Furniture as a Piece of Design and as a Piece of Decorative Arts

Graça Pedroso

Abstract Most of the time, when it comes to furniture developed until the early 20th century, furniture is referred to as a piece of Decorative Arts. However, when furniture is produced industrially it is already considered an object of Design. This division between furniture as a Decorative Art or as a work of Design is little discussed, although some manuals on history of design already cover periods prior to industrial production, putting that furniture and the one that is considered Design together in the same definition. A piece of furniture is constituted by several components that are organized in a system. That system can be analyzed on several domains, such as constructive, formal, functional, socio cultural, economic, semantic and symbolic. In one or more of these domains we can find a decorative art, but not in the piece of furniture as a whole. This study results from the observation of different pieces of furniture from different periods. Each piece of furniture is studied in its different domains allowing an intrinsic synthesis. Using this method of studying the piece of furniture we intend to question if furniture should be seen as a Decorative Art object or as a Design object.

Keywords Furniture · Decorative art · Design · Domain · System

1 Portuguese Case Studies

Written works about furniture are usually directed to a historical perspective directed to artistic styles, which are mainly ways for a creative expression that, at each historical moment, reflects the collective liking and the way of living of a society. This kind of study is very useful in order to establish a correct knowledge of the different eras and consequently to the understanding of the relationship between furniture and other arts. But when it comes to the understanding of the singularity of a piece

G. Pedroso (✉)
Instituto Politécnico de Castelo Branco - Escola Superior de Artes Aplicadas, Campus da Talagueira, Avenida do Empresário, 6000 Castelo Branco, Portugal
e-mail: graca.m.pedroso@ipcb.pt

D. Raposo et al. (eds.), *Perspective on Design*,
Springer Series in Design and Innovation 1,
https://doi.org/10.1007/978-3-030-32415-5_26

of furniture, this kind of study is inadequate. Valuable in order to find tendencies and common characteristics between pieces of furniture, it is not effective for the understanding of that which remains exclusive of each piece of furniture.

Present chapter is focused on the Victorian period and on the following ones, though a few pieces of furniture from other periods are presented in order to clarify some contents.

Subsequent to the exhibitions of Paris, in 1819 and 1834, the Exhibition of Art and Industry of all Nations at the Crystal Palace of London, in 1851, presented pieces of furniture made exclusively for the event. They used good materials and they were perfectly made but, as the main idea was to attract the public, they were over-decorated and its big dimensions stayed away from the essential notions of scale and proportion. Luxury was achieved through the use of a great diversity of profusely ornamented materials.

For a long time, this kind of furniture became the Victorian Style negative image. At the time Industrial Age had reached its apogee, less creative furniture had emerged. Revivalisms of different historic styles and poorly achieved imitations were usual. Even so, some little innovations appeared concerning to the seat furniture, such as the creation of new typologies. This furniture corresponds to an ever changing society, whose objects acquired an excessive ornamentation never seen before. It was the answer for a new increasing consumption medium class.

According to David Raizman "Never before were the decorative and industrial arts made the focus of such widespread attention, crossing boundaries of class and presenting a kind of visual excitement that stirred the popular imagination" (Raizman 2003). Within this revivalist liking, the Neo-gothic Style became more and more relevant. In 1835, Augustus Welby Northmore Pugin published "Gothic Furniture in the style of the Fifteenth Centrury", in which he created a new interpretation of this style, quite different from that of his contemporaries. He was interested in the integrity of form, structure and construction, ignoring the exterior aesthetics, as the excess of ornament (Pugin 1835).

In furniture, ornament represents a certain liking and the culture of an era, but other domains are also relevant for the understanding of a piece of furniture in its whole. For such an approach a different methodology is needed. A methodology that must point to the piece of furniture, analyzing its whole as an organized system composed of several domains and allowing an interpretation of furniture through different points of view. Not only the functional, the formal, the decorative and the constructive points of view, but also those related to symbolic, semantic, social, cultural and economic contents. As Lucy Smith said (2004), a piece of furniture is the answer to a certain need, obtained from an intention, built by an assertive choice of materials and construction processes, identifiable but that identifies itself, and that is suitable for a certain society, or for anyone of its sectors.

As an approach upon this method, let's observe a chair. As many authors' opinion, chairs are the most representative kind of furniture of an era. Back in 1953, George Nelson said: "every truly idea – every innovation in design, every new application of materials, every technical invention of furniture – seems to find its most expression in a chair" (Fiell and Fiell 2002). History is plenty of examples in which the power and the social relevance of the user were reflected in furniture. The chair is probably the kind of furniture that shows this the best, because in chairs this is obtained not only by the use of rich materials but also, and most of all, by the height of the backrest (Fig. 1).

The furniture collection of the Museu de Artes Decorativas da Fundação Ricardo do Espírito Santo Silva, in Lisbon, includes a chair called "cadeira de vestir", which means a chair to be dressed. It is representative of the mid eighteen century, the *Rocaille* period, which, in Portugal, took place during D. José kingdom. This name, "cadeira de vestir", became from the fact that, in solemn occasions, an unattached textile cap, made in velvet or brocade (gold-cloth), bordered with gallons and fringes, was placed over the seat and the backrest.

Its structure is made of carved walnut, partially gilded with gold sheet. Both cushion and backrest are removable and upholstered with damask embroidered with gold thread. The backrest is fixed by wooden latches pinched to the stiles. These

Fig. 1 Chair to be dressed. Museu Escola de Artes Decorativas da Fundação Ricardo do Espírito Santo Silva (Freire et al. 2001)

are surmounted by pinnacles made in a molded bronze or brass league. Joints are mortise and tenon and in many cases they are reinforced with pegs. It has a shellac and beeswax finishing. The carving of the armrests prolongs over the seat rail, in order to hide the joints, which reveals the constructive refinement. Legs are cabriole legs and have ball-and-claw feet, quite common in Portuguese furniture from the 18th century. Stiles, front legs and stretchers are structural components. Backrest and seat are functional components. The armrests are both structural and functional components.

Ornament is carved in curves and volutes drawing asymmetrical and sinuous vegetal motifs. At the middle of the front seat rail, it has the *Rocaille* style characteristic reversed seashell. As usual in Portuguese furniture till the 19th century, the author and the workshop are unknown (Fig. 2).

As shown by the existence of ball-and-claw feet in the front and in the rear, this chair is made to be placed in a central space; this means it is a chair of great refinement, to be used by someone with high social status. Although made at a time

Fig. 2 Armchair (set of four seats). Museu Nacional da Ajuda Inv. 2400

the woodworks were highly appreciated, this chair still denotes the given appreciate to the rich textiles, used in order to improve and promote the piece of furniture.

This chair was handmade and the one who create it must have been the one who build it. During many centuries there was a tradition concerning to the legacy of knowledge: from father to son, based on an almost familiar based work, and from master to apprentice, based on the small workshop.

Far from its ancient functionality, in our days, this chair is seen like a fine example of Portuguese Decorative Arts within the *Rocaille* style. Anyhow, it remains an answer to a certain need, obtained from an intention, built by an assertive choice of materials and construction processes, identifiable but that identifies itself, and that is suitable for a certain society or for anyone of its sectors; just like in the day it has been produced. This means it remains an object of design, despite the fact that it is seen as a decorative arts object, in nowadays.

Observing another chair we might see that any change in any of the domains that, in a whole, constitute the totality of furniture, might result in changes within all the others. As Lucie-Smith said: "L'un des attraits de l'histoire du mobilier, réside dans le fait que ses perspectives sont en continuelle mutation" (2004).[1]

This chair is part of a set of four throne chairs from the second half of the 19th century, made by Elysio Santos & Cª Lda, from Lisbon (Silva 1912). In nowadays this chair is part of Palácio Nacional da Ajuda collection, as it belonged the Portuguese Crown who lived there. Just like last one, it was used in protocol occasions, but this one is a revival one, made in the D. José style *Rocaille*.

The chair is made of gilded carved walnut. Cushion and backrest are removable and are made of red silk velvet. The backrest upholstery is connected to the stiles by unmovable metallic latches. Joints are mortise and tenon and in many cases they are reinforced with pegs. Just like we've seen in last chair, armrests' carving prolongs over the seat rail, hiding the joints, revealing the same constructive refinement. The structural and functional performances are also the same. Legs are cabriole with feet winding inside, resting on flat bottoms.

Ornamentation is composed by flowers, volutes, symmetric flowers and the reversed seashell at the middle of the front seat rail. The crest rail is topped with the Portuguese royal family blazon. Side seat rails are undulated and have flowers and leaves at the middle.

This chair might be considered as a revivalism of those that belonged the ensemble of majestic chairs used during the festivities of the inauguration of the equestrian statue of king D. José at Lisboa in Praça do Comércio, in 1775 (Pinto 1952, Fig. 209); although those have been made in Brazilian tulipwood and with other measures, principally because the backrest is higher. In the case of these chairs, the nobleness of the wood and height of the backrest accorded to the status of those who did make use of them. In the second half of the 19th century Brazilian tulipwood became to be scarce, so in order to achieve an identical nobleness, the nut wood has been entirely gilded. In Portugal, the fact of being gilded means we are in presence of a religious

---

[1] "One of the attractions of the history of furniture, lies in the fact that its perspectives are constantly changing".

Fig. 3 Elysio Santos & C^{a} Lda. Great Modern Cabinet-maker—Electricity-Powered Furniture Factory (Silva 1912)

cult object or a luxury object, and the presence of the royal blazon assures they were used by the royal family or the body of courtiers (Fig. 3).

The set of throne chairs from the second half of the 19th century described above was made in the beginning of the use of industrial processes in Portuguese furniture. At that time the recourse to machine work is punctual and is simultaneous to the one of the artisan. Being revivalist pieces of furniture, they have inherited the ornamental vocabulary of Portuguese *Rocaille* and therefore no creative process has been needed. After the introduction of the machine in the furniture production process, the one who created the piece is not necessarily the same who produced it. These chairs were made at Elysio Santos & C^{a} Lda woodwork shop, where the use of electricity was already a reality for one hundred workers. They usually produced replicas of pieces of furniture from the past. In the middle of the 19th century, these chairs have had a determinate functionality, hence they are design objects but, in our days, they are not used and so, they are considered decorative arts objects representative of that time.

Concerning this matter, it will be interesting to pay attention to an illustration for the cover page for the catalogue of the Great Exhibition, 1851, London, published in The Journal of Design. According to Raizman (2003), in this illustration we may see an allegorical figure of Peace, in front of a globe, flanked by a designer and a craftsman. The first one is a man with a long hair, dressed in a tunic, holding the drawing of a calyx, surrounded by books or folders with drawings. The second appears holding the calyx on hand. Surrounding him, we may see tools and a few

Fig. 4 Drum topped table. José Aniceto Raposo. Museu Escola de Artes Decorativas da Fundação Ricardo do Espírito Santo Silva (Freire et al. 2002)

made products. This illustration hints the necessity of cooperation between designer and artisan, showing that between them there was a difference that had never been noticed before. This way, the designer is the one who creates but does not produce and the craftsman is the one who produces but does not create (Fig. 4).

The Museu Escola de Artes Decorativas da Fundação Ricardo do Espírito Santo Silva, already mentioned above, within its collection, has a Neoclassical Portuguese drum topped table from the end of 18th century made in D. Maria I style. In this case, its creator was the same who produced it. It is a multi-purpose table. It is used, not only to place and store objects, but also to write, to read and to play cards. It is made in hawthorn wood and it has turnery work. Marquetry and inlay are made in Brazilian rosewood and natural European boxwood, tinged green. Nowadays, the table top is lined with green velvet. Its drawers have circular handles made in brass league. Its four legs are curved and join to the central leg by rabbet joint assemblies. At the top of this central leg there is a mechanism that allows the rotation of the table top. All other joints are dovetail assemblies.

The legs are structural components. The drawers and the table top are functional components.

On the lower face of the drawers' box we may see, for twice, the author initials: JAR, and in its interior, after removing one of the drawers, there is a label: "Joze Aniceeto Raposo a fez ao Loureto em Lisboa".[2] Next to it, it is also possible to see the inscription RPC 48, which means that the table's provenience is from the Collection of The House of the Crown – Royal Palace of Cintra (Fig. 5).

[2]"Joze Aniceeto Raposo has made this table, Loureto, Lisbon". (José Aniceto Raposo, a well-known Portuguese master-carver was one of the few Portuguese masters that signed his work).

Fig. 5 Inside detail of the drum topped table. The stamp and the label of José Aniceto Raposo might be seen. Museu Escola de Artes Decorativas da Fundação Ricar-do do Espírito Santo Silva (Freire et al. 2002)

Its ornamentation with flowers and pearled borders, and the crosslines made with a contrasting wood are characteristics of the neoclassical style.

Although with his apprentices' cooperation, José Aniceto Raposo is simultaneously the creator and the producer of this piece of furniture (Fig. 6).

Even though this table is now part of a museum collection and as so considered a decorative art object, in 18th century it was an answer to certain functional needs that remain in nowadays. It is a timeless piece of furniture because since its production it would be in use. So it is an object of design, although seen as an art decorative object due to its ornamentation.

The articulated chairs, whose structure is quite simple, usually composed by few components and simple but efficient joints and mechanisms, are used since ancient

Fig. 6 On the left: Articulated chair. Museu Escola de Artes Decorativas da Fundação Ricardo do Espírito Santo Silva (Freire et al. 2001). On the right: Articulated chairs. Author: Álvaro Siza Vieira. Producer: SPS Sucrs. 1956 (Santos 2003)

times by Egyptians, Etruscans, Romans, or others, and they remained till today. This kind of chairs remembers an itinerant life: they are easy for transport; they are suitable for small space storage; they are equipments used for short time, being used only when it is needed. But they are not able to confer status. They are pieces of timeless design; its efficacy had remained for centuries, everywhere, independently from the materials and tools used, the workmanship, the workshop or the factory. If ornamented they are considered decorative arts objects, if not, they are design objects.

Created in 1903, the *Société dés Artistes/Décorateurs* had promoted annually an exposition, the *Salon D'Automne*. At the time, this initiative intended to frame the Decorative Arts within Art in general. This society was embodied by many artists whose ambition was to retrieve the dignity of decorative arts, lost in revivalisms and excess of ornament. In their opinion, decorative arts would be renovated, and furniture shouldn't be an assortment of luxury objects but a useful set of equipment instead.

In 1971, Tony Selmersheim concludes that in the beginning of the 20th century "*L'art décoratif etait dans le marasme*" (2005).[3]

Since 1911, the presidents of the three main French cultural associations bound to decorative arts: *Union centrale des Arts décoratifs*, *Société des artistes décorateurs* and *Société d'encouragement à l'art et à l'industrie*, wished the realization of an exposition dedicated to modern decorative arts. This exposition would take place in 1925. The *Exposition Internationale des Arts Décoratifs et Industriels Moderne* would present two alternatives to the exhibitors: Luxury furniture and simple furniture industrially made using new materials and without any ornamentation. The last one would be presented by a recently created organization, l'Esprit Nouveau, inside a pavilion with the same name. The interest of Le Corbusier in this organization and his contribution for this event has been relevant. In his opinion, modern decorative arts should present unornamented pieces for which the use of the machine was of great value: "*La machine, phénomène moderne, opère dans le monde une réformation de l'esprit*".[4] "*Pourtant, intact, le facteur humain demeure, la machine étant conçu par l'homme, pour de besoins humains*" (Le Corbusier 1996).[5]

Pieces of furniture are included in the class of objects Le Corbusier classified as "*objects d'usage*", "*destinés à faciliter les taches ingrates de la vie quotidienne*" (1996).[6] They concretize the answer to the necessity that motivates the creative act, which considers all intentions involved and uses pre acquired knowledge, in order to obtain the object. By its turn, the object institutes itself as a prolongation of man and influences him. Sena da Silva explains this relation: "*… os objetos começam por ser 'resposta para uma necessidade' e acabam por determinar um 'modo de viver'*" (Gulbenkian 2009),[7] and as so we may add that there is a way of use for them.

---

[3]"Decorative art was in the doldrums".

[4]"The machine, a modern phenomenon, operates in the world a reformation of the mind".

[5]"Yet, intact, the human factor remains, the machine being designed by man, for human needs".

[6]"Objects of use intended to facilitate the unpleasant tasks of everyday life".

[7]"At first objects are answers to a necessity, thereafter they define a way of living".

In 1978, another Portuguese designer of the second half of 20th century, Cruz de Carvalho, said: "nas peças que têm como finalidade contribuir para a realização de funções fisiológicas como a cama para dormir, os assentos para repousar, as mesas para comer, ou para o desempenho de certas tarefas como é o caso das secretárias de escritório e das carteiras de escola, o móvel torna-se como que no prolongamento do corpo ao ponto de constituir com ele um todo único..." (Carvalho 1978).[8]

Merely utilitarian furniture, whose functionality concerns to equipments for civilian use, efficient and economic, made in new materials and with new technology, had determinated "*une grande incertitude au niveau des styles*" (Lucie-Smith 2004) (Fig. 7).[9]

This is the context in which Cruz de Carvalho e José Espinho have conceived unornamented pieces of furniture designed for part of the Portuguese middle 20th century societies' liking.

One of the best-known chairs by Cruz de Carvalho is called "Simples" and has been produced by Interforma during the end of the sixties.[10] This chair has a single functional component comprising seat and backrest, made by a veneer laminating process that uses a split mould with adhesive cured at room temperature, upholstered with foam coated with napa leather. The structure is made in solid wood and joints are made by mortise and tenon joints and dowel joints. These are carried out by a

Fig. 7 "Simples" chair and secretary. Author: Cruz de Carvalho. Producer Interforma. End of sixties (Anúncio publicitário Interforma 1971)

[8]"In the case of pieces of furniture that have as purpose to contribute to the accomplishment of physiological functions like the bed to sleep, the seats to rest, the tables to eat, or to the performance of certain tasks like the office desks and the school desks, the furniture becomes an extension of the human body to the point of constituting with it a single whole...".

[9]"A great incertitude about styles".

[10]Interforma was a Portuguese furniture factory.

driller with multiple heads with a 32 mm distance between them. Dowels are inserted into the joint with glue and hammered into place.

As seen above, this chair has a set of structural components and a single functional component.

An specimen of this chair was present at the 1.ª Exposição de Design Português and, at the time, was profusely divulged in magazines such as Casa & Decoração, both in advertisements and integrated in different interior spaces. There were two options for wood: beech or mahogany, and other two for upholstery: napa leather or textile. There is a variant for this chair that includes armrests; simply obtained prolonging the legs above the seat and applying the armrests on its tops. All other components remain the same in order to keep costs low. Besides these choices in appearance, the "Simples" chair has constituted itself in the origin of many others that used similar components.

In advertisements this chair appeared next to a writing table convertible into a secretary by the addition of boxes of drawers. Its structure could include one or two boxes simply placed over a transversal inferior stretcher and attached to the legs only by screws. There were two sizes for these boxes and within them there were various boxes with different compositions of drawers. This way, this writing table fitted itself to user needs; using few different components and so keeping the costs low (Fig. 8).

During the first half of the sixties, José Espinho has designed for Olaio various versions for resting chairs, each of them with many variants.[11] The "Maple modelo Sol" is one of them. It distinguishes itself from the rest of these models mainly because it has armrests and because it has a high backrest. The height of seat plan,

Fig. 8 "Modelo Sol" armchair. Author: José Espinho. Midle sixties. Producer: Móveis Olaio. Photo courtesy of Maria José Espinho

[11]Olaio was a Portuguese furniture factory.

its deepness and the height of backrest, as well as the angles of inclination between both, and regarding the soil, are suitable for comfortable rest. This chair was clearly thought for resting.

Its lineament did consist of two structural flanks composed by rear leg, front leg, side seat rail and armrest. Both flanks there were connected to seat and backrest, which were functional and structural components. Special hardware was used in these joints.

Flanks' components were made of wood. Mortise and tenon joints were used. Seat and backrest incorporated slabs of wood products and foam enclosed by napa leather.

The target market was the middle class, in great part newcomer from villages to main Portuguese towns in the beginning of the second half of 20th century. It was a growing social sector which had new necessities and different claims and, so, needed innovative functional solutions. These pieces of furniture answered these demands, allowing a significant reduction in productive costs, mainly achieved not only by the rationalization of their frame lineament and of components, but also by a formal simplification. In addition, the use of new materials and constructive techniques was a great contribute. As told before most part of these pieces of furniture components were used in the production of many other models and that also contributed to keep low costs.

Seat and backrest with bigger thickness, compared with all the other models, transmit the user the feeling of a more sophisticated seat. In this chair, the backrest height, rather than a symbolic reason associated with power or social relevance, has not only an ergonomic reason, because it allows the rest of the cervical region, but also a semantic reason, because it appeals to rest. In these cases, the opinion of Clement Meadmore in his book "The Modern Chair" can be evoked. Writing about the "Poltrona Mole", a Sérgio Rodrigues design he told about the existence of communication between the piece of furniture and the user: "Surprisingly rare in modern language is the way in which the chair promises comfort. [...] Despite many modern chairs being, in fact, quite comfortable, few of them give a visual value to this characteristic" (Borges 2007) (Fig. 9).

Concerning to visual value, Indo-Portuguese furniture from the 17th century, might help us see the relevance of semantics in furniture. Indo-Portuguese decorative grammar presents visually not only the production region's culture, but also the culture of the one who ordered the piece of furniture. Only by themselves, these objects represent the Portuguese Diaspora. Without this ornamental grammar the communication between the object and humans would be impossible.

The 17th century Indo-Portuguese portable fall-front cabinet from Mongol India we may see in Lisbon at the Museu Escola de Artes Decorativas da Fundação Ricardo do Espírito Santo Silva, is a small piece of furniture for writing and containing (Ferrão 1990). As small that might be transported easily, it was useful because it stored jewels, documents, relics and other precious goods the user intended to keep save. This kind of cabinet is one of many pieces of furniture for writing that existed in 17th century. They were the most luxurious pieces of furniture at that time and were essential for those who travelled. In nowadays they lost this functional utility and they are

Fig. 9 Indo-Portuguese cabinet. Museu Escola de Artes Decorativas da Fundação Ricardo do Espírito Santo Silva (Freire et al. 2002)

admired principally for their decorative wealth and their historical information. This one, coming from Mongol India, reflects the intellectual curiosity of Grand Mongols, especially those of Ackbar (1556-1605) and their descendents; who, by their cultural policy, have assured a space for hosting and harmonization of cultures and even religions (Freire et al. 2002).

Its dense and colored ornamentation expounds Persian Art naturalism. Unlike in other areas of India, where human figure representation was not allowed, in this cabinet's ornamentation it is possible to see human figures dressed with Mongol vestments; drawn with front view bodies and side view heads. The main symbolism is exposed in its top face, where elephants, symbols of wealth and force, are ridden in the Indian way by two men. At the center there is a tree of life, symbol of fertility. Surrounding we may see small birds and two peacocks, symbols of immortality. Marquetry and inlay works are made in East India satinwood, rosewood, natural and tinged green ivory and brass wires. The small metallic pins that pin these materials to the teak structure also have an ornamental effect.

Far from its ancient functionality, in nowadays this decorative object reveals the History of an era at a certain place of the globe and evidences the cultural encounter between Occidental and Oriental cultures and, this way, it remains a design object with a current functionality (Fig. 10).

During the decade eighty of 20th century, following the post-modern movement created by Etorre Sottsass, the Portuguese architect Tomás Taveira conceived several pieces of furniture that might be integrated in it. The 1989 "Sandman chair" is an example. It is made in wood and wood products, gloss polychrome painted, where brilliant colors stand out. In this chair, ornament is a way of expression and meaning that overpasses its function. Rui Afonso Santos describes it as a "spectacular" chair "in which the multidiscipline emphasis resorts to graphic design (with the appropriation and transfiguration of the famous Sandman Port logo, a graphic image belonging to our collective memory)" (Santos 2003). The formal intention overpasses the needed rationality that functionality and construction impose, standing

Fig. 10 "Sandeman" chair. Author: Tomás Taveira. Produced in 1989 (Santos 2003)

out the sculptural and communicational value of the piece, diluting "the boundaries traditionally established between equipment, painting, sculpture and advertising". As Fusco had said, chairs deny its own nature, offering themselves as bright demonstrative plastic valued aesthetic objects (Fusco 2005). In these cases furniture doesn't use ornament as decorative element, but as an aesthetical element that, just like in any other media, intends to communicate cultural values of society.

This chapter intends to demonstrate the relevance of historical design stud-ies in which might be detected causes evoked by society, whose importance has resulted in specificity in furniture production, thought as preservation of collective memory and, or, as an answer to society and market's needs.

This work's fundamental intention is to demonstrate the importance of an understanding of the piece of furniture as a system, which means an organized ensemble of physical and immaterial elements identifiable by different modes of observation. Each one of these domains have different analyze perspectives and different own objectives, consummated in the specific domains by means of which the piece of furniture might be considered. Their synthesis consists on the totality of Furniture as a discipline.

Thought this way, ornament is part of certain domains of furniture, and it is fundamental for its understanding; however, it does not characterize Furniture in its totality. This means that furniture, especially ancient furniture, does not might be considered as a decorative art, forgetting all other domains than form. All other domains are essential for the understanding of furniture as design objects.

2 Conclusions

What was said above takes us to the conclusion that a systematic classification for furniture is fundamental, whatever its production period is. A piece of furniture must be seen through different domains. One of these domains includes ornament, which is the result of a determinate aesthetical preference from a certain moment in History, and which offers us essential information in order to date the piece of furniture. Nevertheless, other domains, such as the functional domain, the constructive domain, the form domain, the semantic domain, the symbolic domain and the cultural domain, are necessary for the understanding of the piece of furniture as a whole. Moreover, many of these domains are related with the creative act and the needs of the consumers and they consist in a referential for those concerned with furniture, such as those who create and produce it, or researchers and historians. Furniture understood in its whole, should not be seen as a decorative art object from a certain period. Decorative contents consist on one only domain of the object, in other words, of the piece of furniture.

References

Anúncio publicitário Interforma (1971) Arquitectura – Planeamento – Design – Artes Plásticas 120

Borges A (2007) Sérgio Rodrigues. Viana & Mosley, Rio de Janeiro

Carvalho JC (1978) O design e a indústria de mobiliário: intervenção do designer Cruz de Carvalho na Movelnor Braga 1977. Casa & Decoração 40:44–47

Corbusier L (1996) L'Art Décoratif d'aujourd'hui. Flammarion, Paris

Ferrão B (1990) Mobiliário português. Lello & Irmão, Porto

Fiell C, Fiell P (2002) Chairs. Tashen, Köln

Freire FC, Pedroso G, Henriques RP (2001) Mobiliário: Móveis de Assento e de Repouso. Fundação Ricardo do Espírito Santo Silva, Lisboa

Freire FC, Pedroso G, Henriques RP (2002) Mobiliário: Móveis de conter, pousar e de aparato. Fundação Ricardo do Espírito Santo Silva, Lisboa
Gulbenkian FC (ed) (2009) Sena da Silva. Fundação Calouste Gulbenkian, Lisboa
Fusco R (2005) Historia del diseño. Santa & Cole Publicaciones, Barcelona
Silva CA (1912) Grande Marcenaria Moderna – Fábrica de móveis movida a electricidade. Occidente: Revista Ilustrada de Portugal e do Estrangeiro 1211:181–182
Lucie-Smith E (2004) Histoire du mobilier. Thames & Hudson, Paris
Pinto AC (1952) Cadeiras Portuguesas. Augusto Cardoso Pinto & F. F. da Silva Nascimento. Fig 209, Lisboa
Pugin AW (1835) Gothic Furniture in the style of the Fifteenth Century. Ackermann & Co., London
Raizman D (2003) History of modern design. Laurence King Publishing, Londres, London
Santos RA (2003) Cadeiras Portuguesas Contemporâneas. Asa editores, Porto
Direcção Geral do Património Cultural (2019) Ficha de inventário Palácio Nacional da Ajuda: 2392. Available via MatrizNet Cited 14 May 2019 http://www.matriznet.dgpc.pt/MatrizNet/Objectos/ObjectosConsultar.aspx?IdReg=992131&EntSep=3#gotoPosition. Cited 14 May 2019
Selmersheim T (2005) La dissolution de L'Art dans Tout et le « marasme » des arts décoratifs. In: Rossella Froissart Pezone. *L'art dans tout: Les arts* décoratifs en France et l'utopie d'un Art nouveau. CNRS Éditions, Paris. Accessed via Open editions books 24 Apr 2019. http://books.openedition.org/editionscnrs/8431

Further Reading

1ª Exposição de design português 1971, Lisboa
Canti T (1999) O móvel no Brasil: origens, evolução e características. Editora Agir & Fundação Ricardo do Espírito Santo Silva, Lisboa
Nunes G (2012) O mobiliário português de produção em série do terceiro quartel do século XX (1951-1974). Faculdade de Arquitectura de Lisboa, Lisboa
Pedroso G (2013) José Espinho: a diversidade no fazer. Uzina books, Lisboa
Pedroso G (2017a) José Espinho: o homem na sua época. In: José Espinho: Vida e obra. Caleidoscópio, Lisboa
Pedroso G (2017b) Neorrústico, moderno e estética industrial. In: José Espinho: Vida e obra. Caleidoscópio, Lisboa
Pedroso G (2017c) José Espinho, o percurso e a visão nos móveis Olaio. In: Móveis Olaio 1886–1998. Câmara Municipal de Loures, Loures

Design(ers) Beyond Industrial. From Dissociation to Reassociation Through Craft

Paulo Colaço and Demétrio Matos

Abstract Towards the end of last millennium, cultural and technological development allowed humanity to severely alter the way we live, the objects of everyday life and how they were made and used. The advent of industrial production made it possible, for it contributed the rise of the Design professional, responsible for the ideation and project of products, filling the recently created gap between intent and machine production. For decades the seeds of post-industrial design have been identified by researchers, their bloom is now shown evident by growing number of people that try to avoid industrial goods. This has made the non-industrial market to become multimillionaire, attracting an increasing number of people to make products. Craftsmanship holds the roots of cultural variation and carries a long tradition of using sustainable materials and processes, values with increasing acknowledgement by the market. Recent events will be reviewed, which show that the present state of affairs combines both increasing market opportunities and communicational abilities for designers to develop products for non-industrial means of production. This study concludes that within this perspective, designers and craftspeople collaborations have an option to successfully develop sustainable products, while contributing for a more thoughtful and sustainable world.

Keywords Design · Post-industrial design · Craft · Human development

P. Colaço (✉)
ESD-IPCA, Barcelos, Portugal
e-mail: colaco.paulo@gmail.com

D. Matos
ID + ESD-IPCA, Barcelos, Portugal
e-mail: dmatos@ipca.pt

D. Raposo et al. (eds.), *Perspective on Design*,
Springer Series in Design and Innovation 1,
https://doi.org/10.1007/978-3-030-32415-5_27

1 The Rise and Promotion of Industrial Design

Modernity is characterized by a shift in the perspective about the relation between man and the world, which allowed a great scientific and technologic development (Casini 1979), originating a radical change in the way objects were made by the middle of the eighteenth century. It was a systemic radical change happening at a fast pace and induced radical changes in social and economic structures of society. Although previous technology wasn't completely eliminated, by the middle of the nineteenth century machines technology was developed to such degree that had become the main element of most of the technological systems of materials transformation, as well as energy and transportation that it required and produced (Paulinyi 1986). Also creative work within the production of goods was transformed, industrial production divided the ideation of products from its fabrication (Giedion 1948).

It was in this context that design as an activity emerged, filling the gap between commercial intent and fabrication, developing and applying an industrial style, imposed by the adoption of machines technology. This industrial style is based on standardization of materials and processes, the resulting novelties were presented in World's Fairs (Bürdek 2005). A significative change industrial production brought to the creative process was abstract notation mediation, design—the finished work of the designer—is a set of specifications translated into a language, which will then be interpreted by those responsible for materializing them into an object. So the responsibility for the quality of the finished object lays, in great degree, in this process of interpreting specifications and the ability to fulfill the requirements (Pye 1968). This is one of the reasons why there was such effort in developing new materials and automated production techniques. Today the use of standard homogeneous materials, technologies like injection molding and digital modelling and fabrication, reduces the need and eliminates almost entirely the influence of the worker in the quality of the finished product.

In opposition to the industrial perspective, the *Arts & Crafts* movement arose in the middle of nineteenth century, aiming to reunite design with production, which would lead to social renewal of the arts and their aesthetics. Later in the nineteenth century another movement emerged, *Art Nouveau*, led by Henry van de Velde, trying to imprint in the objects the artistic sense of live. Workmanship was valued in an elitist and individualist way (Bürdek 2005).

In the beginning of twentieth century Hermann Muthesius was in charge of bringing Arts & Crafts' values into techniques of proto-industrial design in German arts, crafts and trade schools (*Kunstgewerbe- und Handwerkerschulen*). The goal was to train the new generation of craftsmen and other workmen to German SME (*Mittelstand*), so they could serve the modern economy, aiming to develop design and production of goods and improve German's competitive position in international markets. In 1907 *Deutscher Werkbud* was created, uniting a diverse group of artists, craftsmen, architects, manufacturers, entrepreneurs and government officials, using artistic intervention to improve German production. The tension between artistic individualism supported by Henry van de Velde and industrial types supported by Muthesius was evident in 1914, with the latter imposing his view. For most of the twentieth century, this opposition "was at the core of a modernist teleology in which, supposedly, standardization and an accompanying tendency toward machined abstraction were the inevitable results of the progress of industrial culture" (Maciuika 2011, pp. 99).

After WWI the influent *Staatliche Bauhaus Weimar* was created, direct by Walter Gropius, where art and technique should come together, focusing on creating accessible products with high functionality, which meant combining work and social well-being. The two decades it lasted were not enough to allow its cultural influence to go beyond the elites, but the pedagogical influence was great, many teachers and students had important roles in American and European institutions, spreading the principles of Bauhaus. After WWII, also in Germany, *Hoschule fur Gestaltung Ulm* was created, a school in which the direction and part of the teachers had studied in Bauhaus. Ulm was a technical humanist school where they tried to conciliate the practical, aesthetical and pedagogical aspirations of the time. Lasted fifteen years but also had great influence, creating a significant part of repertoire of the design (Bürdek 2005).

The evolution of industrial design, based on industrial technology and society, matured to the Modern Movement, to which design should be rationalistic, reductionistic and mechanistic, free from subjective speculation (Cross 1981).

2 The Tendency Towards Dissociation

In the first decades of the twentieth century the ability to create value within commercial organizations by adopting an holistic perspective, a strategic mindset and a project approach—what we know now as design—became evident (Bürdek 2005). This ability was also perceived outside the industrialized western world and, in 1957, Indian government requested Eames couple recommendations for a training program that could contribute to the development of their small and medium enterprises (SME), mostly working with low or very low technology. The result was a report—the India Report—which encourages the analysis of what Indian people felt was important to them and recommended an institute of design, research and service to be founded, which could provide advanced training. This report contributed to the establishment of National Institute of Design (NID) in Ahmedabad, in 1961 (NID 2015).

After WWII the industrialized western countries ensured a development model for non-industrialized countries focusing in infrastructure and industrial enterprises, financed by International Monetary Fund (IMF) and the World Bank (Margolin 2015). In the early 1970s the International Centre for Settlement of Investment Disputes (ICSID), now World Design Organization (WDO), asked United Nations

Industrial Development Organization (UNIDO) a plan to implement national design policies in peripheral countries, which was developed by Gui Bonsiepe (Bonsiepe 1973), former student and teacher at Ulm. In this document industrial design was clearly defined as a mechanism for the development of underdeveloped countries, nevertheless ignoring the capitalist nature of the activity (Alpay Er 2015). In an interview in 2015, Gui Bonsiepe states the historical context in which the document was created was very different, there were many significant changes since then, but the basic content—reduce the dependency and increase the autonomy—still maintains the same relevance that it had before (Patrocínio and Nunes 2015).

Design as a development tool has had many interpretations, some advocated solutions based on low technology and intensive labor, as Victor Papanek in his book *Design for the real world* (1972), which was very influential (Margolin 2015). Later Papanek recognized it was a paternalist perspective, common at the time (Alpay Er 2015). From a 1979 conference organized by NID resulted *Ahmedabad Declaration on Industrial Design for Development*, which emphasized the role of industry in the development process, the pertinence of design methodology for this process, and traditional abilities and native materials in alliance with science and technology (Margolin 2015).

In the following decade the United Nations Development Programme (UNDP) increased the concept of development, considering not only economic development, but also human development. In 1987 the World Commission of Environment and Development (WCED), known as the Brundtland Commission, release a report—*Our common future* (WCED 1987)—in which the concept of sustainable development was introduced. The idea was to shift from large scale industrial projects to mitigating the condition of the more disadvantaged, with emphasis on environmental issues. In 1995 the World Commission on Culture and Development (WCCD) prepared a report—*Our creative diversity* (WCCD 1996)—in which is stated that the ultimate goal of development is the physical, mental and social well-being of every human, also addressing cultural heritage, the maintenance of cultural plurality and global ethics. In 2000 the United Nations (UN) presented eight *Millennium Development Goals* (MDG), aiming at reducing poverty and increasing quality of life by 2015, by then only three of them were met (Margolin 2015). They were later increased to the actual seventeen *Sustainable Development Goals* (SDG), addressing economic growth, social needs and environmental protection (UN, n.d.).

Comparing the humanist evolution of these declarations with international trade policy, the contradictions at macro level are numerous (Margolin 2015).

3 The Tendency Towards Reassociation

At the consumer, or micro level, changes are already in place, as the number of people influenced by sustainability is increasing, also rising is the number of people willing to pay more for brands committed to positive social and environmental impact (Nielsen 2015).

A still pertinent answer to sustainability was the proposal of post-industrial design (Cross 1981), which presented a new vision in three domains:

- Products: should be long-lived, repairable, customized;
- Process: should be externalized, inclusive, extensive;
- Designer: should be collaborative, anonymous, participatory.

Post-industrial design resulted from the adaptation of the sane, humane, ecological (SHE) perspective, presented by James Robertson in 1978: *The sane alternative*. In the 2008 preface of 1983 revised version, the author states that, in the thirty years since the original version, humanity moved towards collapse. Also alerts to the fact that the attention given to global warming and carbon emissions, although deserved, distracts us from other problems humanity faces, as food and agriculture; travel, transport and trade; prevailing structures of work and employment, and patterns of production, consumption, and distribution; waste production and disposal among others (Robertson 2008).

These are concerns shared by increasing number of people interested in reinventing their lives, which is one of the factors contributing for society to improve, along with the development of information and communication technology (Manzini 2015). Ezio Manzini tells us this is possible because everyone has the ability to design—diffuse design, but not everyone dedicates the effort needed to become a trained professional—expert design. The author goes on presenting a definition of design's role in this process: "Design for social innovation is everything that expert design can do to activate, sustain, and orient processes of social change toward sustainability." (Manzini 2015, p. 62).

Distributed systems are those in which sociotechnical context promote production closer to consumption, which is now possible because of the convergence between networked systems and creative communities. Distributed systems, along with social innovation, are the main pillars of Manzini's small, local, open, connected scenario—"SLOC scenario proposes a possible future, though it is a future that requires many converging efforts if it is to become real" (Manzini 2015, pp. 179). SLOC scenario proposes a general vision with two main strategies by which small and local activities may have large-scale impacts:

- Replicating: implies the necessity to experiment, and then consolidate and propagate the best results, by designing a solution locally appropriate;
- Connecting: resulting not only by the accumulation, but by the multiplying effect of a network connection.

In the context of design for social innovation with a post-industrial mindset, designers may contribute to society by applying their expertise in finding solutions that promote sustainability. The collaborative dimension of this process is key to achieve solutions with which people identify themselves and find valuable, based on shared visions associating well-being and sustainability. Designers can also bring their expertise to show how they may be achieved, therefore promoting the motivation needed to act upon such visions (Cross 1981; Manzini 2015, 2016).

4 Design + Craft: An Option to Consider

While in underdeveloped countries the promotion of crafts may represent the impossibility of reaching development (Bonsiepe 1973), in matured markets of industrialized countries is quite different. A report KPMG prepared for Crafts Council (UK) states that "craft skills and knowledge have a strong economic impact and significant potential to drive further growth and innovation in other sectors" (KPMG 2016, p. 3).

Craft has had its share of attention increasing (Jakob and Thomas 2017), has been used to imply quality and authenticity in products of all kinds (Cavalli 2017), and became a market with relevant commercial impact (Luckman 2015). Recent events like London Craft Week,[1] Homo Faber,[2] Revelation,[3] Collect[4] and Doppia Firma[5] promote it by showing actual examples of curated creative practices, like traditional craftsmanship, art-like activities and design and craft collaborations.

In order to assess excellence in craftsmanship, Alberto Cavalli (2017) gathered a list of terms used to described it: authenticity, competence, craftsmanship, creativity, innovation, interpretation, originality, talent, territory, tradition and training . Here we find individual human sensibility, creativity and skill, combined with local and traditional values. Craftsmanship entails both the abstract conceptualization as in design, as well as technical skill and material knowledge. Does so in a dialectic manner, as theory and practice are mutually influenced in the creative process, creating "human world directly out of the raw substance of nature itself. It entails transformation of our direct sensuous experience of nature into a world of culture" (Risatti 2007, pp. 170).

These are features that contribute to create objects with ability to fulfil human drives with emotional engagement, products that can have a meaningful and positive impact in the world, what David Rose calls enchanted objects (Rose 2015). Although the value recognized by people in objects by engaging emotionally in their meaning is something personal and, therefore, different in each individual, this a process in which design can intervene with specific skills—frame creation or sense making—the ability to generate alternative visions that influence people's perception and eventually motivates action (Dorst 2015; Manzini 2015; Norman and Verganti 2014).

A side-by-side view of the fundamental features of industrial and artisanal production (Fig. 1) helps to assess, by means of a scheme, the characteristics of any object, in which the narrative of the object may be based, in terms of its contribution to design, craft and the consumer/market. This will be made by answering the following questions: how is the product made? By which means? With which mindset? With what objective? What type of system is being supported?

[1] https://www.londoncraftweek.com/.

[2] https://www.homofaberevent.com/.

[3] https://www.revelations-grandpalais.com/en/.

[4] https://www.craftscouncil.org.uk/what-we-do/collect/.

[5] http://www.doppiafirma.com/.

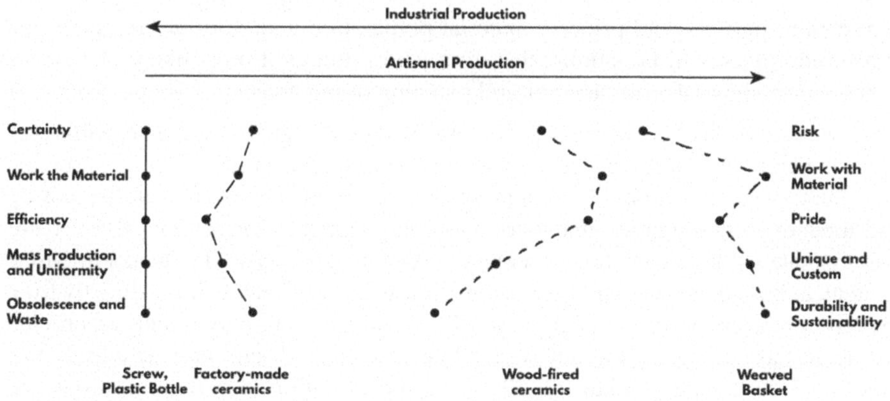

Fig. 1 Side-by-side view of fundamental features of Industrial and Artisanal production, with examples of objects' placement. *Source* Authors (2019)

In industrial production the quality of products is predetermined, materials are developed so machines may work them as needed, in the most efficient way, so they are uniform and mass-produced, resulting in obsolescence and waste. Example of this fully industrialized production are the screw and plastic bottles: made by machines in large series, with industrial material in order to meet predetermined standards (from shape and size to mechanical performance), as cheap as possible.

Artisanal production, on the other hand, always involves some degree of risk in the production, as the quality of the result is constantly dependent on judgment, dexterity and care, so the degree of risk depends on the mastery of the artisan (Pye 1968). Craftspeople have a remarkable sensitivity with regard to material and its possibilities, always trying to take advantage on the natural features it presents (Cavalli 2017), and they have pride in work well done (Sennett 2008). Artisanal products are made one at the time, so adjusting to individual needs is easy, and the result is always a one-of-a-kind object that people will tend to cherish (Risatti 2007; Rose 2015). An example could be a weaved basket, made of unprocessed natural fibers, worked by hand for a particular purpose. Most of the products in our daily life (clothes, furniture, accessories, etc.) will fall somewhere between these antipodes, with characteristics from one and the other end. To assess the variety we may encounter, ceramics are a good example:

- In factory-made ceramics the quality of the resulting object is mostly certain, material is formulated to ensure it, produced in the more efficient way, the use of molds make all products equal, also allows for the productions of virtually unlimited quantity, so the price will tend to be low, promoting the obsolescence;
- In wood-fired ceramics made in traditional way, in which the artisan collects and prepares the material, throws the piece on the wheel and then fires it by burning wood, the degree of risk for the quality of the result is high, great knowledge,

attention, mastery and pride in material preparation and work are required, the resulting piece will be unique, therefore who acquires it most likely recognizes and values the aesthetics and cultural meaning of the object.

New generation of creative people embrace a career in crafts by choice, while previous generations usually became acquainted with their craft as result of economic considerations and availability of apprenticeships, in this perspective craftsmanship and modern technology are understood as complementary (Moritsch 2018). Technology makes it now easy for someone to apprentice in any trade, thanks to online content, although mastery still takes a long time to achieve and learning directly from experts or masters is still very important. Tools and materials may found and ordered online, just as people's work may sold online, also cross-disciplinary activity is now more likely to happen (Adamson 2015). So the evolution in communications and production technology allows the conditions for small-scale fabrication to be possible (Anderson 2012), designers applying their problem-solving and sense-making abilities can contribute to society's resilience by collaborating with local artisans, within a global perspective, what Manzini calls place-making (Manzini 2015).

The way crafts are considered depends on the objectives of those who analyze it, some of the most common approaches presented by Bonsiepe (2011) are:

- Conservative: aims to protect craftspeople from external influences, usually seen in anthropologists trying to keep external influences out;
- Aesthetic: crafts are seen as repository of popular art, sometimes used to collect traditional patterns as basis for design projects;
- Productive: craftspeople are seen as cheap skilled labor, often these initiatives are presented as if it were helping craftspeople, when in deed it perpetuates dependency relations;
- Cultural: crafts are seen as repository of popular culture, often implying the impossibility of evolution;
- Paternalist: sees craftspeople as assistance programs' clients and acts as mediator between craftspeople and retail;
- Innovative: strives for the autonomy of craftspeople, the evolution of their work and the development of their businesses, in this case direct participation is required.

This essay advocates a collaborative creative process between designers and craftspeople, characterized by parity among members, in which the activities are framed within a plan to achieve shared expectations of a better future. It would therefore be included within Bonsiepe's innovative approach, we believe it encompasses a few important aspects to consider in collaborative processes, such as the ability and availability to learn, share and recognize other types of knowledge, as well as the recognition of co-authorship (Azua 2017; Borges 2011).

5 Conclusions

Traditionally, creative people would find inspiration in cultural, scientific and technological novelties, but the end of the millennium created a disturbing situation, in which almost everyone (artists, artisans and designers included) are disconnected from the vanguard of contemporary culture—advanced physics theory and applied technology such as Artificial Intelligence—not by incompetence, but by ignorance, so we feel free to find interest and fun where we can. According to Peter Dormer this is the reason behind crafts' success in the twentieth century (Dormer 1995).

"The question is how can designers change the world. Design cannot. Designers can change the world" (Frog 2013). These words by Harmut Esslinger are still relevant, despite the fact that most designers don't deal directly with the vanguard of technology, as their strategic and frame creation competences are suited to foster new visions of a better world, and the way people "see" the world—our cognitive models—which determines the way people interact with it, maintaining or changing it, in one way or another (Burke 1985). Craft can contribute significantly to a wider range of possibilities within the design realm as it presents opportunities for designers to engage in design for social innovation, through collaborations with craftspeople (Manzini 2015).

Design researchers are unanimous in taking the increase of actionable knowledge within the discipline as the objective of research, which may allow for other professionals to act upon the new knowledge, addressing the challenges they encounter (Branco 2014). The contributions this essay on design and crafts collaboration makes are:

- For design: new knowledge for discipline's culture, by combining designer's way thinking and acting with other creative activity based on practical knowledge;
- For crafts: holistic approach to object ideation and project, focusing on traditional culture and technologies;
- For society: fostering placemaking, contribution to the development and resilience of local economies and culture, availability of objects that promote an emotional relation with its owners.

The authors expect this essay may help designers who want to embrace the creation of meaningful products within a sustainable mindset. The focus on design and craft collaboration is presented here as a viable option within this perspective, which requires practical application for further development, which in turn emphasizes the need for the creation of clear plans and visions of a future that may motivate others to take action.

References

Adamson G (2015) Essay by Glenn Adamson. In: Openshaw J (ed) Postdigital artisans—craftsmanship with a new aesthetics in fashion, art, design and architecture. Frame Publishers, Amsterdam, pp 286–288

Alpay Er H (2015) Padrões de desenvolvimento do design industrial no Terceiro Mundo: um modelo conceitual para países recém-industrualizados. In: Patrocínio G, Nunes JM (eds) Design & Desenvolvimento - 40 anos depois. Blucher, São Paulo, pp 29–54

Anderson C (2012) Makers—the new industrial revolution. The Crown Publishing Group, New York

Azua M (2017) Modelos colaborativos Artesanía/Diseño. In: 7º encontro bid enseñanza y diseño: territorio_diseño. Madrid. Retrieved from https://youtu.be/pq1N_g-6E7Y?list=PLjddHonD1FrjvyBAwMurJY9v7ya6Ki1v9

Bonsiepe G (1973) Development through design. Viena

Bonsiepe G (2011) Design, cultura e sociedade. Blucher, São Paulo

Borges A (2011) Design + Artesanato - o caminho brasileiro. Editora Terceiro Nome, São Paulo

Branco V (2014) Design e investigação em design: alguma reflexões. In: Vilar ET (ed) Design et al. - dez perspectivas contemporâneas. D. Quixotepp, Lisboa, pp 169–185

Bürdek B (2005) Design: the history, theory and practice of product design. Birkhäuser, Basel

Burke J (1985) The legacy of science. In: Burke J, Bergman J, Asimov I (eds) The impact of science on society. NASA, Hampton, pp 3–30. Retrieved from https://history.nasa.gov/sp482.pdf

Casini P (1979) As filosofias da natureza. Editorial Presença, Lisboa

Cavalli A (2017) The master's touch—essential elements of artisanal excellence. Marsilio Editori, Venice

Cross N (1981) The coming of post-industrial design. Des Stud 2(1):3–7

Dormer P (1995) Os significados de design moderno a caminho do séc. XXI. Centro Português de Design, Lisboa

Dorst K (2015) Frame innovation, create new thinking by design. The MIT Press, Cambridge

Frog (2013) Hartmut Esslinger—advice for designers. Retrieved 13 Dec 2017, from https://vimeo.com/59679411

Giedion S (1948) Mechanization takes command: a contribution to anonymous history. Oxford University Press, New York

Jakob D, Thomas NJ (2017) Firing up craft capital: the renaissance of craft and craft policy in the United Kingdom. Int J Cultural Policy 23(4):495–511. https://doi.org/10.1080/10286632.2015.1068765

KPMG (2016) Innovation through craft: Opportunities for growth—a report for the Crafts Council, (July). Retrieved from http://www.craftscouncil.org.uk/content/files/KPMG_CC_Innovation_Report.pdf

Luckman S (2015) Craft and the creative economy. Palgrave Macmillan, Basingstoke

Maciuika JV (2011) The Globalization of the Deutscher Werkbund—Design reform, industrial policy, and German foreign policy, 1907–1914. In: Adamson G, Riello G, Teasley S (eds) Global design history, 1st edn. Routledge, Abingdon, pp 98–106

Manzini E (2015) Design, when everybody designs. The MIT Press, Cambridge

Manzini E (2016) Design culture and dialogic design. Des Issues 32(1):52–59

Margolin V (2015) Design para o desenvolvimento: mapeamento do contexto. In: Patrocínio G, Nunes JM (eds) Design & Desenvolvimento - 40 anos depois. Blucher, São Paulo, pp 13–28

Moritsch HS (ed) (2018) Craft-based design—on practical knowldge and manual creativity. Niggli Verlag, Salenstein

NID (2015) History & background. Retrieved 29 April 2019, from http://www.nid.edu/institute/history-background.html

Nielsen (2015) The sustainability imperative—new insights in consumer expectations. Retrieved 20 Jan 2019, from https://www.nielsen.com/us/en/insights/reports/2015/the-sustainability-imperative.html

Norman DA, Verganti R (2014) Incremental and radical innovation: design research vs technology and meaning change. Design Issues 30(1):78–96. https://doi.org/10.1162/DESI_a_00250
Patrocínio G, Nunes JM (2015) 40 anos depois - entrevista com Gui Bonsiepe. In: Patrocínio G, Nunes JM (eds) Design & Desenvolvimento - 40 anos depois. Blucher, São Paulo, pp 75–82
Paulinyi A (1986) Revolution and technology. In: Porter R, Teich M (eds) Revolution in history. Cambridge University Press, Cambridge, pp 261–289
Pye D (1968) The nature and art of workmanship. Bloomsbury Academic, New York
Risatti H (2007) A theory of craft—function and aesthetic expression. The University of North Carolina Press, Chapel Hill
Robertson J (2008) The sane alternative—a choice of futures, 2nd edn. Retrieved from http://www.jamesrobertson.com/books.htm#sane
Rose D (2015) Essay by David Rose. In: Openshaw J (ed) Postdigital artisans—craftsmanship with a new aesthetics in fashion, art, design and architecture. Frame Publishers, Amsterdam, pp 60–63
Sennett R (2008) The craftsman. Penguin Books, London
UN (n.d.) Sustainable development goals. Retrieved 29 April 2019, from https://www.un.org/sustainabledevelopment/
WCCD (1996) Our creative diversity. Report of the World Commission on Culture and Development. Retrieved from https://unesdoc.unesco.org/ark:/48223/pf0000105586
WCED (1987) Report of the world commission on environment and development: our common future. New York. Retrieved from https://sustainabledevelopment.un.org/content/documents/5987our-common-future.pdf

Portuguese Modernism from *Arquitectura* Magazine (1927–1988). Three Cases of Interior Design in the 1950s

Liliana Neves and Fátima Pombo

Abstract This text intends to contribute to Portuguese Interior Design History based upon the documental study of *Arquitectura* Magazine (1927–1988). This Magazine was, at the time, in Portugal, a way to spread the work and knowledge developed in architecture and interior design, which, despite controversies in social, cultural and political matters, managed to remain outside the dictatorial ideology. The focus of this article is to discuss the decade of 1950s which represents for Portuguese Architecture, namely regarding Interiors, a period of searching for a Portuguese Modernism in critical dialogue with the achievements abroad. The principles of International Style, the influent writings of Le Corbusier and others, international exhibitions (as the 8th Triennial in Milan under the theme The House), numerous architectural projects were presented and discussed in *Arquitectura*. Throughout the interpretation of the 2nd and 3rd series focused in de decade of the 1950s and the analysis of three paradigmatic case studies it is aimed to illustrate the evolution of the architectural interest to design single-family houses interiors as a privileged realm of defining the contours of a modern Portuguese domesticity.

Keywords Interior design · *Arquitectura* magazine · Domestic interiors · 1950s

1 Introduction

The 2nd and 3rd series of the *Arquitectura* Magazine is the documentary base of this research that aims to demonstrate the 1950s as the turning point for the expression of a Modern Movement in Portugal.

Iniciativas Culturais Arte e Técnica (ICAT), the owners of *Arquitectura* during the study period considered in this text and the authors that contributed and organized

L. Neves (✉)
Polytechnic of Castelo Branco, University of Aveiro, Aveiro, Portugal
e-mail: liliana.m.c.neves@gmail.com

F. Pombo
University of Aveiro, Aveiro, Portugal
e-mail: fpombo@ua.pt

D. Raposo et al. (eds.), *Perspective on Design*,
Springer Series in Design and Innovation 1,
https://doi.org/10.1007/978-3-030-32415-5_28

the several issues of the magazine were the key for the development of Portuguese criticism about the modern thinking. Initially in the 2nd series with Keil do Amaral, belonging to the 2nd generation of modern architects, the International Style principals were the inspiration. Then with Nuno Portas in the 3rd series, and following generation of architects the critical thinking was more framed in the Portuguese reality itself.

This text presents three projects that showcase the evolution of the Modern Movement in Portugal, with special attention to domestic spaces. The case studies concern the projects of: (1) House in Vila Nova de Gaia designed by architects Cassiano Barbosa and Arménio Losa in 1948 and published in the magazine in the 2nd series, no. 44 September 1952; (2) Group of Houses in Matosinhos designed by architect Álvaro Siza Vieira in 1954 and published in the magazine in the 3rd series, no. 68 July 1960 and (3) Ofir House designed by architect Fernando Távora in 1957 and published in the magazine in the 3rd series, no. 59 July 1957.

From the point of view of the domestic space, it is noticeable that the projects of the early 1950s were very much influenced by rationalism and functionality. However, through the run of time, the potential needs of the inhabitant to use and transform the space according to his/her exigencies were highly considered in the project, and therefore domesticity gained relevance and the house became the home space to shelter daily life.

2 *Arquitectura* Magazine (1927–1988)[1]

In the 20th Century, in the Portuguese panorama, there were several magazines[2] about architecture but *Arquitectura* Magazine always performed a special role. Authors as Oliveira & Furtado (2009), Ferreira (2001) and Correia (2015) researching specifically about the *Arquitectura* Magazine stressed that, despite being irregular in its publications, it was the one active during a longer time period (1927–1988) resisting to the censorship of dictatorial politics.

As it can be seen in Fig. 1, *Arquitectura* Magazine was published between 1927 and 1988 for a total of 61 years. The 1st series of the magazine begins in 1927 and publishes 43 issues until 1939, coinciding with the Second World War. Between 1939 and 1945 Architect Francisco Costas (the owner and Editor of that time) launched only one issue per year in February in order to keep the name active. These issues (43–49 of the 1st series) were generally very small and with little content depicting almost only the own project of Francisco Costa.

[1]Some articles do not have the indication of the authorship therefore it is used the complete title as reference.

[2]During the 20th Century, in Portugal, were published several magazine related with architecture and in turn also with interior design, like *Construção Moderna* (1900–1919), *Annuario* (1905–1910), *Arquitectura Portugueza* (1908–1958), *Arquitectura* (1927–1988), *Arquitectos* (1938–1942), *Binário* (1958–1977), *Atrium* (1959–1960), *A Propriedade* (1970–1975) and *Jornal Arquitectos* (1981–).

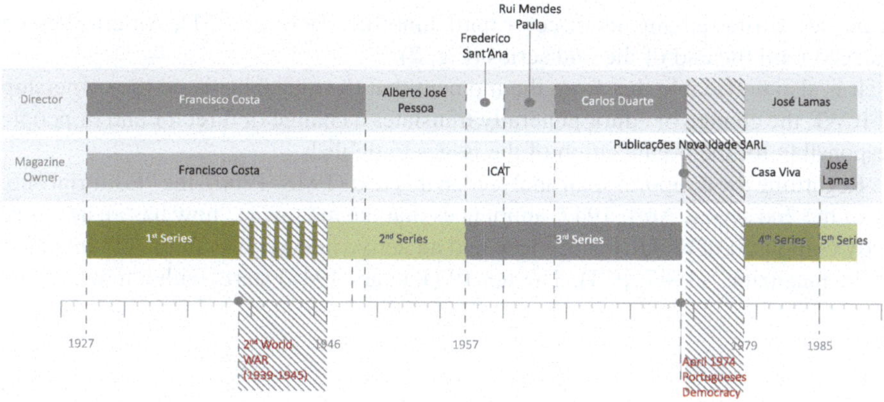

Fig. 1 *Arquitectura* Magazine: analysis of consistency of publication, ownership and director. *Source* Authors

In February of 1946 the 2nd series begins with a new numbering, but still under the ownership of Francisco Costa. In 1948 the magazine was acquired by ICAT and, according to Ana Tostões (1997), when Keil do Amaral began to collaborate with the publishing group, as early as 1947, it was when the magazine performed best.

The 2nd series was published until January/February of 1957, no. 57/58. The no. 59 was the beginning of the 3rd series in July 1957 and although the ownership of the magazine remained with ICAT, this was a milestone in the change of direction and type of publication, since it was led by a new group of architects who "sought a reflection on the Portuguese reality informed about the Modern Movement and revising the International Style" adapting it to the Portuguese reality (Neves and Pombo 2018, p. 7). On April 1974, the magazine was acquired by the *Editora Publicações Nova Idades SARL* and published no. 129, 130 and 131.

The magazine returns in February/March 1979, with no. 132, the first one from the 4th series and on ownership of *Editora Casa Viva* with José Lamas as director.

Over the years the financial problems increased and in September/October 1984 it was published the last issue (no. 153) of this series.

In 1985, Lamas continues to see potential in *Arquitectura* and buying it becomes its owner and decides to start the 5th series with a new face, numbering and name. The *Arquitectura Portuguesa* was born and published the issues 1–12, closing in 1988.

Analysing *Arquitectura* magazine from the point of view of its coordination, it is noticed that between 1927 and 1948 Francisco Costa was not only the owner, but also the editor and director. In March 21, 1948, it was the first publication on the ownership of ICAT and, although Francisco Costa maintained his position as editor until no. 23/24 (May/June 1948), it is in this issue that appears for the first-time other authors associated with the organization of an issue, in this case Vitor Palla, Bento D'Almeida and Manuel Barreira. In no. 25 João Simões becomes editor and

Francisco Costa remains its director until June/July 1949 (no. 31), Alberto Pessoa succeed until the end of the 2nd series (Fig. 2).

It is also noticeable that from the moment that the magazine is on the ownership of ICAT, the change of editor generally translates a change of director and of people responsible for the organization of the issues to publish.

Regarding the temporal span analysed in this text (1948–1960), the "Editorial" no. 14 of the 2nd series, April 1947, announces that the magazine "now has an advisory body composed of architects of merit and an active intervention in the organization of the magazine" (1947, p. 4). The no. 19 (January 1948) starts with a new graphic

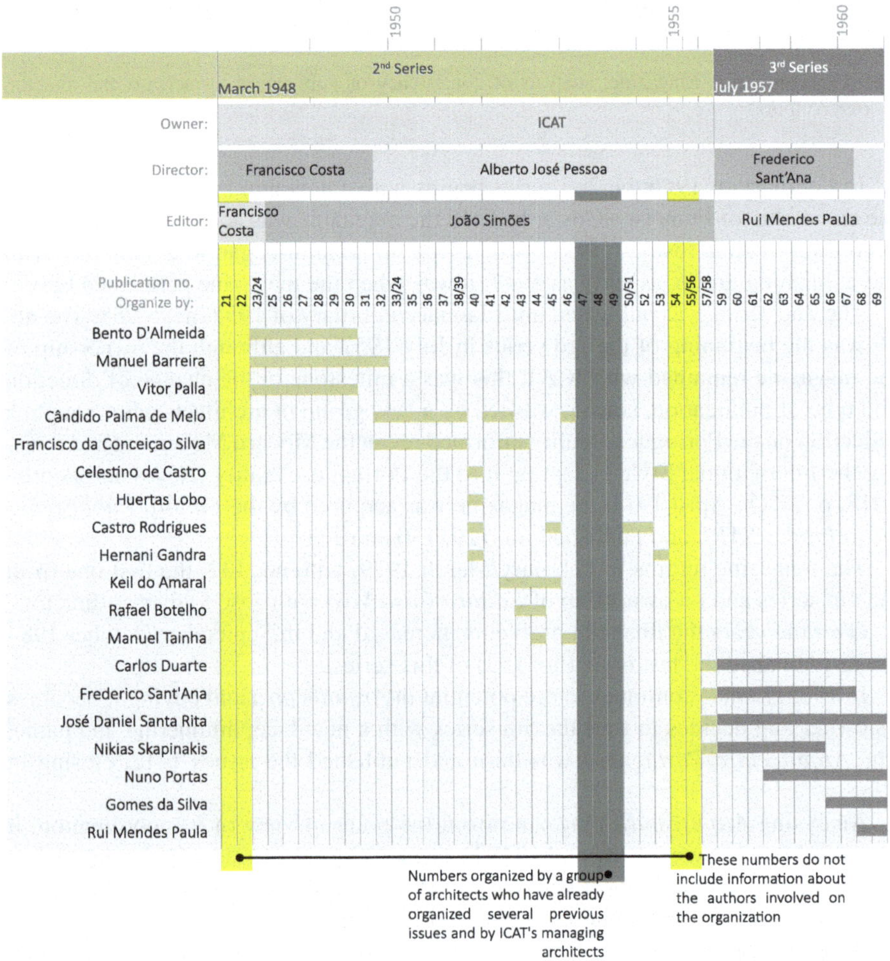

Fig. 2 Analysis of the magazine *Arquitectura* after the ICAT ownership until 1960: evaluation of the envolved names in the organization of the issues, editor, director. *Source* Authors

image in the cover, however, the indication of ownership by ICAT only appears explicitly in no. 21, March 1948.

In the "Editorial" of publication no. 19 became explicit the magazine's concerns in divulging modern Portuguese architecture, in integrating other arts and stimulating the debate by publishing more international projects.

This interest will be notable for the various news regarding CIAM activities and the publication of the translated Athens Charter. From this moment on it was also clear the approach to the architects of Porto, with "more courage to fight against certain prejudices and, above all, with better comprehension about the architectonic solution of our time" (Fernandes 1948, pp. 5–6).

The I National Congress of Architecture in 1948 and later the first Architecture Exhibition held by ODAM[3] in June 1951 were remarkable moments that converged to reinforce the unity of the Portuguese architects in defence of modern architecture inspired by the International Style (Correia 2015, p. 53).

3 Portuguese Modernism Throughout *Arquitectura* Magazine

3.1 International Style Influence in Portugal

"Ecos e Comentários" [Echoes and Comments], no. 19 January 1948 is the first publication of contents referring to conferences and exhibitions that are influential in the propagation of modern international thought - International Style. The first section of this article presents the last CIAM Congress held in Bridgewater and due to the importance of that conference for the development of modern architecture, the readers were informed that it will be published "already in the next issue, […]of the famous «Athens Charter», a historical document where were launched the foundations of the Modern Urbanism" ("Ecos e Comentários", January 1948, p. 23).

Following this comment, the Athens Charter was published in *Arquitectura* between no. 20 and no. 32 (Feb. 1948 and Aug./Sep. 1949), in a total of 11 chapters, a crucial sign that the International Style became an influence for the modern architecture in Portugal.

In the above-mentioned article of "Ecos e Comentários" it is also found references of the evolution of architecture in Brazil through magazines *Architecture d'Aujourd'hui* and *Forum* as well as of Le Corbusier with the goal to awake the interest of Portuguese architects for such development in architecture.

In March 1948, at no. 21, it is found an unsigned 3 pages article dedicated to the 8th Triennial of Milan, where the latest news about architecture and about the fair

[3]ODAM—Organization of Modern Architects. It was a group of Architects from Oporto, Portugal, active between 1947 and 1952 with the purpose of disseminating modern architecture.

Fig. 3 "8ª Trienal de Milão" [8th Triennial of Milan], housing as main theme, images from showrooms. In Arquitectura, 2nd series, no. 21, March 1948, pp. 12 and 13

were reported, highlighting the theme of housing as the main point for debate and for showroom.

Figure 3 depicts the way to create spaces according to the principles of International Style. As it is read:

> The set of this Triennial was resolutely presented under the sign of rationalism and functionalism. All the luxury and all the decoration itself has been eliminated. The 8th Triennial included a section of domestic art studied by great artists who did not disdain to make good drawings for varied objects of daily use from the radio to the cutlery ("8ª Trienal de Milão", March 1948, p. 13)

Almost one year after the I National Congress of Architecture,[4] i.e. in Feb/March 1949 in *Arquitectura* no. 29 were published the conclusions and intentions of the congress, becoming explicit the architects' purpose to follow the Athens Charter contextualized in the Portuguese context:

> Let the study of the urban planning and construction problems consider objectively the principles expressed in the "Athens Charter" [...] never forgetting that its application must be done in close contact with national realities." ("I Congresso Nacional de Arquitectura", February/March 1949, p. 6)

Already in the same issue (no. 29), Formosinho Sanches writes a letter to the readers informing about the exhibition of Brazilian Contemporary Architecture taking place in *Instituto Superior Técnico* in Lisbon in 1948. He also reminds that the students of architecture should look at it as an "evolution" of the International style framed in the Portuguese reality, "solving our problem of Modern Architecture" (Sanches 1949, p. 17). In no. 28 (pp. 4–5), January 1949, Victor Palla, made already a reference to the Brazilian Architecture and its capacities to return to tradition and vernacular language.

[4]The I National Congress of Architecture, took place May 28 and June 4 of 1948 in Lisbon. It was a historical moment for the debate and critical thinking about Portuguese Architecture and the problem of Portuguese dwelling. The several theses presented by the architects tried to approach the international stage and criticized the standard style imposed by the dictatorial politics.

The 3rd series of the magazine approaches mainly the debate and the critical thinking of the international style. At the same time international authors as Carlo Scarpa, Bruno Zevi, Alvar Aalto, Frank Lloyd Wright and their projects became relevant to the construction of the debate in Portugal.

For example, Portas before commenting Carlo Scarpa's project transcribes an excerpt of the text of Bruno Zevi in which he announces the awarding of the prestigious Italian architecture prize 'Olivetti' to Scarpa. Portas' commentary on Scarpa's project refers not only to the structure and architectural space, but also to the dynamic, intentional, and dramatic form of its interior (Portas 1957, p. 26). This commented project reinforces the new editorial orientation ahead from the rationalism and functionalism of the International Style.

3.2 *The Modernist Generation in Portugal*

Table 1 showcases 4 generations of architects in Portugal. The 2nd generation is the one that initiates the ICAT and begins a more evident demand by an approximation of Portugal to the international movement. The 3rd generation seeks a collective architecture more conscious of its roots and organic approach. The 4th generation, seeks to defend a historical continuity and a "cultural unity".

Thus, it can be said that the turning point in the search for a Portuguese modernism happens from the 3rd series of *Arquitectura*.

Table 1 The generation of modern architects in Portugal: most important authors that have seen their work on *Arquitectura* Magazine between the 2nd and 3rd series

Arquitectura series	1st series	2nd series	3rd series	
generation	1st	2nd	3rd	4th
Born in decade	1890	1910	1920	1930
Architects	Carlos Ramos (1897–1969) Cristino da Silva (1896–1976) Pardal Monteiro (1897–1957) Jorge Segurado (1898–1990) Cotinelli Telmo (1897–1948) Rogério de Azevedo (1898–1983)	Keil do Amaral (1910–1975) Viana de Lima (1913–1991) Januário Godinho (1910–1990) Arménio Losa (1908–1988)	Nuno Teotónio Pereira (1922–2016) Fernando Távora (1923–2005) Manuel Taínha (1922–2012)	Nuno Portas (n. 1934) Siza Vieira (n. 1933)

Source Authors based upon Tostões (2015), Correia (2015)

According to Correia (2015, p. 26) in the 1st series of the Magazine, the Modern style was completely absent, privileging the regional style opposing to the modern style. The 2nd series integrate several projects that try to resemble a modern style, initially very marked by the monumentality of the *Estado Novo*[5] expressed mainly in public works. The 3rd series of the magazine begins to be written "for the search of alternative models to the «International Style», and for the attempt to overcome the strictly functional dimensions of the architecture of the Modern Movement" (p. 65).

It can be stated that the moment of transition between the Modern Movement and the Portuguese Modern Movement is demonstrated in *Arquitectura* no. 57/58[6] where it is read in the article *Uma pousada na Nazaré* that «*form follows function*» is today a memory of the past" (January/February 1957, p. 19).

In no. 66, November/December of 1959, Nuno Portas writes that the new generation of the Modern Movement in Portugal was fundamental for the creation of a critical thought in the collective of architects in Portugal (1959, p. 13).

3.3 Three Case Studies of Portuguese Domestic Space in the 1950s

The case studies that are interpreted in this study demonstrate the transformation of the Modern Movement in Portugal. The fact that these projects are single-family houses allows a better illustration of the challenge of designing interior spaces. The 1950s were a turning point from the imposed language and the taste of the Portuguese dictatorial State. Initially when the pioneering projects of modern language emerged in Portugal it is clear the influence of the "International Style", the functionalism of Le Corbusier, the Athens Charter and the CIAM meetings. However, over the time the recognition of Portuguese regional heritage and the criticism of modernist functionalism contributed to the awareness of Portuguese vernacular language and to the direction towards a more organic conception of domestic interiors.

The three chosen projects are representative of such transformation: (1) House in Vila Nova de Gaia from 1948 by architects Cassiano Barbosa and Arménio Losa; (2) Group of houses in Matosinhos from 1954 by architect Álvaro Siza Vieria; and (3) Ofir House from 1957 by architect Fernando Távora.

[5]"Estado Novo" was a dictatorial political regime also known as "Salazarismo" (1933–1974).

[6]No. 57/58 of Arquitectura, in January/February 1957, was a double number issue that mark the end of 2nd series and the beginning of the 3rd series and already organized by architects knowledgeable and critical of the international style and focused on the exhibition of a modernism based on the Portuguese roots. The organizing authors was Carlos Duarte, Frederico Sant'Ana, José Daniel Santa Rita and Nikias Skapinakis.

3.3.1 House in Vila Nova de Gaia by Architects Cassiano Barbosa and Arménio Losa (1948)

This project is part of 3 houses designed for the same client António Oliveira Neves. The house in Vila Nova de Gaia is his official residence. This project was designed during the period that the architects Cassiano Barbosa and Arménio Losa try to promote and to disclose the modern architecture, going against the *Estado Novo* ideology (Neves and Pombo 2018). The project is from 1947, the house was built in 1948 and just appeared in *Arquitectura* in February/March 1952, *2nd series, no. 44.*

In the article "Moradia em Vila Nova de Gaia" the project is presented as "one of those architectural work that clearly reflects the free performance of the architects, understood by a client of the type that can be considered ideal, not only he is a modern architecture lover, but he also surrendered to the architects in total confidence" (September 1952, p. 3).

The house has two mainly volumes (in a "L" shape) with a total of 3 floor levels, as it is possible to see in Fig. 4.

The architects were very strategic about the drawing because of the terrain sloping (Fig. 5). The access to the terrain and into the house is made from the highest quota. The 3rd floor (the one with the highest quota) is not the principal but is the one that gives access to the 2nd floor, the main floor. On the 3rd floor there are the main bedrooms and a small office; on the 2nd floor there is the service wing with kitchen, both bedroom and bathroom for the servant, sewing room and the living space (composed by a dining room, guest bedroom and bathroom and a formal living room connected with the balcony); on the 1st floor there is a laundry (it is a room at the back part of the stairs) and the main living room (it is a multi-task room, just as the modern taste encourages) connected with the outdoor.

This house became known because of the architects' expertise in adapting the tradition with modern elements. Figures 4 and 5 show that tradition is present mainly because of the type of roofing and of the confined definition for space use (Ramos 2004, p. 501). For the interior, the architects were more into following the modern principles on International Style. For example, Fig. 6 demonstrates that the interior space is clearly a modern house plan, where the vertical circulations (stairs) have a big impact in the design. Here the architects elevate the organic drawing, just

Fig. 4 "Moradia em Vila Nova de Gaia", 3 floor levels in "L" shape. In *Arquitectura*, 2nd series, no. 44, 1952. Figures from the magazines, p. 4

Fig. 5 "Moradia em Vila Nova de Gaia", southwest elevation. In *Arquitectura*, 2nd series, no. 44, 1952. Figures from the magazines, p. 3

possible with the use of concrete and finished the walls with a collaboration of an artist (Augusto Tavares) for the creation of a mural painting.

In summary, in this project architects achieved the design by using materials as concrete (in the stairs, in the exterior flooring with concrete tiles), metal banister, false ceilings, big door and window connecting the interior and exterior. The twist of modernism to a more organic, flexible and dynamic perspective about dwelling will appear some years later, namely in the following projects.

3.3.2 Group of Houses in Matosinhos (1954) by Architect Alváro Siza Vieira

The project of the "Group of Houses in Matosinhos" belongs to the architect Álvaro Siza Vieira and was designed in 1954. As it is possible to observe in Fig. 7 it is composed by 4 houses all different from the exterior and interior, inserted in the suburban context (Tostões 2015, p. 466).

The project just became complete in 1957 and it's only presented in *Arquitectura*, 3rd series, no. 68, July 1960 by Nuno Portas, along with two other works from the same architect, an attempt to regain a historical "cultural unity".

In Nuno Portas' commentary accompanying the article, he praises the architect's ability to synthesize architecture and the plastic arts as a way of giving meaning to space (Portas 1960, pp. 13–14). Siza adds color and integrating a sculptural form in

Fig. 6 "Moradia em Vila Nova de Gaia", interior stair with mural painting. In Arquitectura, 2nd series, no. 44, 1952. Figures from the magazines, p. 5

the space, something that had been abandoned by modern language, which assumed the space bare and without ornaments. These experiences are visible in Fig. 8 that exposes the use of a small wall to divide sections, the kind of kitchen's chimneys and by the type and colour of the chosen material for the walls.

Nuno Portas also talks about the connections between the various materials and, to a certain extent, on the influences that the architect sought from *Casabella* and the *Architectural Review*, on the one hand the "revivalist taste" and on the other the "brutalist" experiences, that the architect had in "seeking his own expression" before modern architecture (Portas 1960, p. 17).

According to Ana Tostões, the interior spaces "are conceived with a silent fluidity, forming double-height spaces, with different points of view, accentuated by the wooden stairs completely loose in the space and that makes the connection between floors", where the environments win a "greater value" through the many details that the architect did not leave to chance (2015, p. 467).

In summary, the Modern taste is visible in the exterior through the design of the elevations and the composition between the different type and shapes of the windows

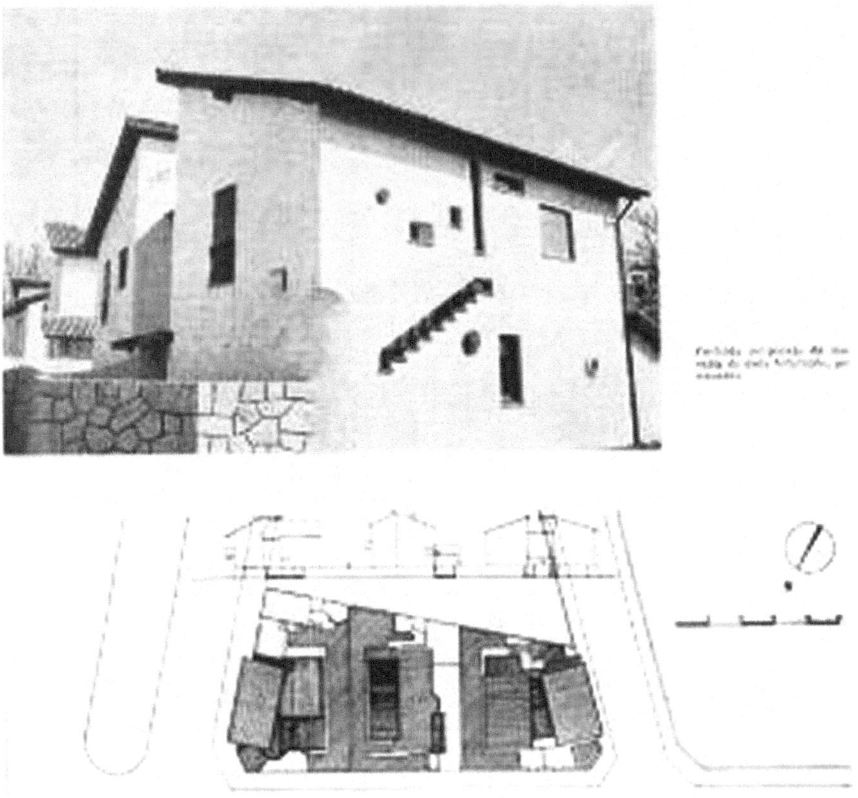

Fig. 7 "Grupo de Moradias em Matosinhos". In Arquitectura, 3rd series, no. 68, 1960. Figures from the magazine, p. 14

and roofing. The interior, however, offers a flow between spaces proposing a plasticity to respond to functions and uses of daily living.

3.3.3 Ofir House (1957) by Architect Fernando Távora

This is a project of a vacation house designed by architect Fernando Távora in Ofir, Esposende, in 1957 for Dr. Ribeiro Silva. This project appears in *Arquitectura*, 3rd series, no. 59 in July 1957. The article is composed by a descriptive memory written by Távora and by a comment by Hestnes Ferreira. Hestnes explains the architect's ability to reconcile traditional building systems and technical progress, either with the use of a wooden roof or an exposed concrete beam, stone wall panels whitewashed walls with large windows, it is an explicit desire to integrate traditional culture into modern culture. Responding to the current problems of architecture with awareness about Portuguese reality (Távora and Ferreira 1957, p. 13).

Fig. 8 Kitchen of one of the houses. *Source* Lucília Monteiro

Távora describes his project as being "a compound of many factors", "of the factors considered, some, [...] are not within the scope of the Architect's responsibilities, others belong to the field of the Architect's training, as well as to his own personality" (1957, p. 12), where each space is the reflection of a certain real use—its program, its meaning, the people to whom it was destined (Trigueiros 1992).

As it can be seen in Fig. 9, the project displays a tripartite plan, differentiating the service area, the bedrooms and the common rooms.

The central living-room is presented to us clearly as an influence of the Modern Movement (Neves and Pombo 2018), but it is the plasticity that the architect gives to the materials and furnishings that brings us back to his more organicist, functionalist, neo-empiricist, cubist and even influences of the spontaneous architectures that he knew in Portugal (Tostões 2015).

Figure 10 shows the central living room inviting to a joyful and comfortable use. It is to note that the architect paid more attention to the details of the central-living room, kitchen and dining area. In addition to detailing all the furniture, he also detailed all the window frames and doors, and defined the most appropriate materials for each element based on the area of Cávado where the house is placed. Távora privileged the use of oak in natural tone, the granite and the apparent concrete. Wall panels that do not go up to the ceiling allow light to penetrate freely the spaces and the fireplace is a sculptural element, to be seen both inside and outside. The architect also stressed the connection of the interior with the exterior mainly in the central-living room.

In summary, this project is known for its vernacular tone within the modern language, emphasized for example by the choice of materials and the connection

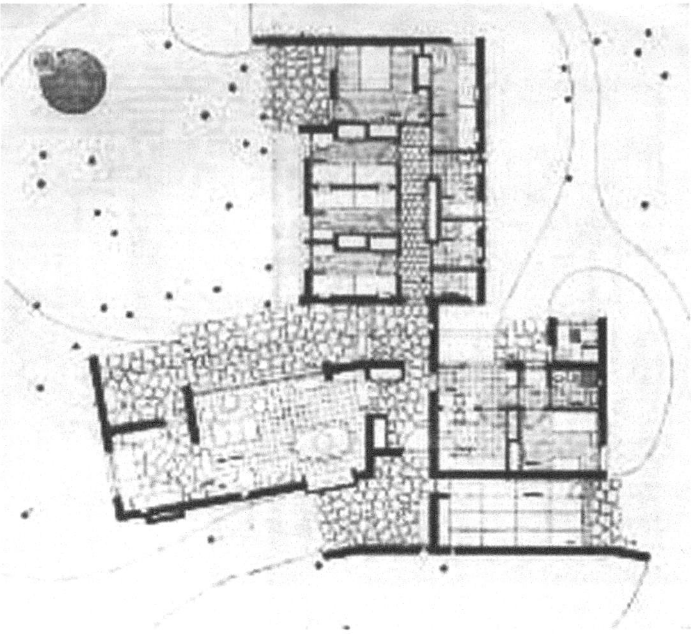

Fig. 9 Ofir House (1957). Floor Plan. In *Arquitectura*, 3rd series, no. 59, 1957. Figures from the magazine, p. 12

Fig. 10 Central-living room in Ofir house. *Source* Trigueiros 1992

between interior and exterior through the strategic window and door placement. Features like the specific design of furniture, flexible to adapt to different uses in the same room or the flow between spaces, accentuate the openness to modern life and modern dwelling.

4 Conclusions

Arquitectura magazine was one of the most prominent in the debate of the Modern Movement in Portugal, especially from the end of the 2nd series until the end of the 3rd series, and its authors and editors were the protagonists of what was best produced in Portugal. The 3rd series of *Arquitectura* marks the beginning of a critical editorial program for the study of contemporary Portuguese architecture.

The three projects of single-family houses voice during the 1950s the evolution of the design understanding for home interiors' quality and character in order to allow domestic life to unfold.

With Cassino Barbosa and Arménio Losa (project from 1948) the intention was to instil the Modern Movement to a house project that was supposed to be traditional. Siza Vieira project (1954) was critical to the construction in series of the same model once he drew four different houses for different families. The project reveals features aligned with the modernism and others representing Portuguese tradition. Fernando Távora project (1957), inset in the spirit of *genius loci*, is the culmination of the organic and vernacular interpretation balanced with an interior a modern frame to accommodate people and their needs at home.

The 1950s' studies acknowledge the translation in architecture of the Modernist ideology changes contributing as well to a deeper under-standing of its impact in the Portuguese architecture, namely regarding the domestic interior space.

References

Correia N (2015) Crítica e debate arquitectónico na 3a série da revista "Arquitectura"—Portugal, 1957/1974. Doctoral thesis in architecture theory and history. Presented to Universitat Politécnica of Catalunya

Ecos e Comentários (1948) In: Arquitectura, 2nd series, vol 19, p 23

Editorial (1947) In: Arquitectura 2nd series, vol 14, p 4

Editorial (1948) In: Arquitectura, 2nd series, vol 19, p 4

Fernandes I (1948) Porto - Intercâmbio profissional. In: Arquitectura, 2nd series, vol 19, p 5

Ferreira P (2001) Teoria da Arquitectura em Portugal, 1935–1945: Debates, Convergências e Dissidências com o Regime Político. Master thesis in art theory. Presented to Fine Arts Faculty of University of Lisbon

I Congresso Nacional de Arquitectura: Conclusões e votos (1949) In: Arquitectura, 2nd series, vol 29, pp 2–7

Moradia em Vila Nova de Gaia (1952) In Arquitectura, 2nd series, vol 44, pp 2–5

Neves L, Pombo F (2018) Interiores domésticos dos anos 50 em Portugal: a construção do fluxo espacial segundo a Revista Arquitectura. *Convergências - Revista de Investigação e Ensino Das Artes* XI(21). Retrieved from http://convergencias.esart.ipcb.pt/?p=article&id=288

Oliveira Â, Furtado G (2009) Revistas Portuguesas de Arquitectura: Evolução nos últimos dois decénios (1988–2008) e revisão dos seus antecedentes. RESDOMUS

Palla V (1949) Lugar da tradição. In Arquitectura, 2nd series, vol 28, pp 4–5

Portas N (1957) Carlo Scarpa. In: Arquitectura, 3rd series, vol 59, p 23

Portas N (1959) A responsabilidade de uma novíssima geração no movimento moderno em Portugal. In: Arquitectura, 3rd series, vol 66, p 13

Portas N (1960) 3 Obras de Alvaro Siza Vieira. I – Grupo de moradias em Matosinhos. In: Arquitectura, 3rd series, vol 68, pp 12–22

Ramos R (2004) A Casa Unifamiliar Burguesa na Arquitectura Portuguesa: mudança e continuidade no espaço doméstico na primeira metade do século XX. Doctoral thesis in architecture. Presented to Architecture Faculty of University of Oporto

Sanches F (1949) Carta aos Leitores. In: Arquitectura 2nd series, vol 29, p 17

Távora F, Ferreira H (1957). Casa de Ofir. In: Arquitectura, 3rd series, vol 59, pp 11–13

Tostões A (1997) Os verdes anos na arquitectura portuguesa dos anos 50. FAUP Publicações, Porto

Tostões A (2015) A Idade Maior. Cultura e tecnologia na arquitectura moderna Portuguesa. FAUP Publicações, Porto

Trigueiros L (1992) Casa de Férias em Ofir - Fernando Távora 1957–1958. Editorial Blau, Lisboa

Uma pousada na Nazaré (1957) In: Arquitectura, 2nd series, vol 57/58, pp 19–21

8ª Trienal de Milão (1948) In: Arquitectura, 2nd series, vol 21, pp 11–13

Challenges to Automotive Interior Design: The Future Is Much More Than Technology—It's About People!

Susana C. F. Fernandes

Abstract Contemporary life is bred of complexity. Technological advancements have revolutionized the automotive sector in general, and driving in particular, to the point where the market has become confusing. When we approach "Interior Automotive Design", it must be borne in mind that design is intended to create pleasant, meaningful and relevant experiences for its users. Today, people discuss level 5 disruptive technology, commonly sharing utopian and dystopian visions about the future of the automobile. This chapter presents a user-centred analysis that assesses driving as a pleasurable activity. Creating personalized and exclusive driving experiences in the same vehicle for a variety of use cases is one of the future challenges of the automotive industry. In this chapter, we review the literature on the main difficulties and expectations of usability of the car interior by many drivers, with a special focus on the display interfaces. We reflect on opportunities for improvement and future developments in the interior of automobiles. Such developments go beyond the expected benefits of new technologies, which in some cases should be considered with caution.

Keywords Automotive interior design · Technological complexity · Display interfaces · Usability · Customization challenges · Driving pleasure · User expectations

S. C. F. Fernandes (✉)
Polytechnic Institute of Maia – IPMAIA, Maia, Portugal
e-mail: sfernandes@ismai.pt

LAETA – Associated Laboratory for Energy, Transports and Aeronautics, Maia, Portugal

N2i – Research Nucleus of the Polytechnic Institute of Maia, Maia, Portugal

D. Raposo et al. (eds.), *Perspective on Design*,
Springer Series in Design and Innovation 1,
https://doi.org/10.1007/978-3-030-32415-5_29

1 The Future of the Car Is Much More Than Technology—It Is About People!

The car of the future must show significant changes in its form and function, when compared to the contemporary car. Although several "commentators" have described the future in terms of the autonomous car, this represents merely a part of the changes that may be coming (Viereckl et al. 2016). The car of the future is already taking different forms, and some trends can be seen, although it is unlikely we will achieve their full fruition on public streets and conventional roads in the next 10 or 20 years, or even longer. Nonetheless, sufficient innovations may emerge, before this step, to transform the car. New levels of connectivity between vehicles are expected, allowing new services inside and outside the car, and new types of cars, many of which dedicated to specific uses (shared transport, collective passenger transport, taxis, etc.). The culture of the car, including conventional wisdom about vehicles and the concepts of "possession" and "driving", is likely to change. The idea of what a contemporary car consists in is perhaps already being radically rethought (Viereckl et al. 2016).

Man has always "dreamed" of aspects for the improvement of mobility. An example of this may be the 1956 advertisement from the magazine Boy's Life (Fig. 1), which displays the image of a family in an autonomous car playing dominoes and a boy making a paper airplane. The same image was presented in the 1957 advertisement of the futuristic family, developed by America's Electric Light & Power Companies, and which, later on, would show a striking resemblance with the family in the episode "Magic Highway, USA", airing on May 14th of 1958, on Disneyland TV. The (now iconic) image is synonymous with automated driving and shows how the "roads will become safe—through electricity!". The text that accompanies the image in the magazine translates the prediction of Isaac Asimov, and says the following:

> «ELECTRICITY MAY BE THE DRIVER. One day your car may speed along an electric super-highway, its speed and steering automatically controlled by electronic devices embedded in the road. Highways will be made safe - by electricity! No traffic jam... no collisions... no driver fatigue.»

Illustration, in the form of an advertisement (1957), on the one hand, satirizes the American road and, on the other, characterizes the history of automobility. This illustration intends to contrast with the reality of driving the first cars, which was a true adventure, due to terrible roads, which, at the time, were nothing more than "bumpy terrain". Furthermore, at the time, there was also a shortage of fuel supplies, spare parts, useful maps and road signals, which rapidly became a matter of everyday life (Kimmel 2015). Although this illustration has no technological similarities with the current prototypes of autonomous cars (of which Fig. 2. is an example), the concept of the interior design reveals the same concern for the satisfaction of users' recreational needs during the trip (Anderson et al. 2016).

Given the various factors that affect the current automotive industry and the many new competitors that seek a share in the market—not to mention the vast advertising,

Fig. 1 Illustration of the autonomous car by H. Miller. Cover image of the 1957 advertisement for "America's Independent Electric Light & Power Companies". *Source image* RAND Corporation. Detailed version of the panorama—Magazine vol. 40, Nr. 5, 30. January 1956, p. 8

which obfuscates what is actually real—there are real opportunities and risks in this industry (Baker et al. 2016). As the revenues and profits change from hardware to software, from products to services, from the industry 3.0 to the new industry 4.0, some brands may be successful while others may not.

Connected cars may become more fun, more efficient and safer than traditional cars, and will free drivers from driving so they can focus on other activities while on their way to work, or on other types of trips. However, this begs the question: at what cost may original equipment manufacturers (OEMs) of the automotive industry actually manufacture a connected car? This is a difficult question to answer. The psychical form that the future car will take remains uncertain, much like the exact nature of the "connected and autonomous packages" that the manufacturers are likely to sell embedded in the vehicles (Bontrager 2018; Baker et al. 2016).

According to Ahlemann and Gerling in "The autonomous frontier" (Ahlemann and Gerling 2016), the totally autonomous vehicle is unlikely to be available to consumers for, at least, another decade. At the moment, there are several criticisms of emerging technology—particularly artificial intelligence (AI), machine learning,

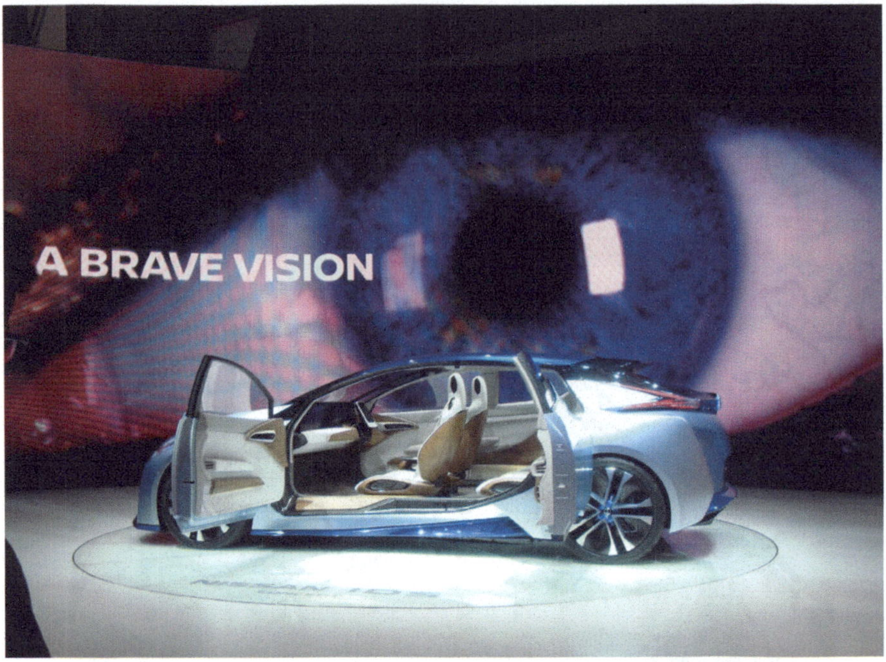

Fig. 2 Autonomous Nissan IDS concept car. Image rights: copyright by the Author

the human-machine interface (HMI), the infrastructure elements underlying the car—and the various resources necessary to achieve a greater level of maturity. At what point are these technologies, currently? Will they achieve such an advanced state that they will become standard resources, even in the least costly vehicle?

At the moment, the technological cost involved in the production of autonomous vehicles is large and, naturally, the manufacturers do not disclose their costs, nor do they estimate prices for the expectable car of the future (Baker et al. 2016; Kimmel 2015).

Thus, this constraint of autonomous cars (among many others identified—Table 1), (Bontrager 2018; Yang and Coughlin 2014; Wood et al. 2012) in the near future, will push a large percentage of consumers of the automotive industry to continue to acquire the conventional car (although it may be supported by numerous technologies for the driver—from level 1 to level 4—Table 2). Out of so many emerging technologies, which are the most advantageous to be incorporated into vehicles? And why? It is important for the automotive industry to enlighten consumers.

Table 1 Main advantages and disadvantages of the levels 5 car

Advantages (if technologies are duly developed – future)	Disadvantages (in the present)
1. Safety—In 2015, 94% of accidents in the USA were due to human error. Computer controlled vehicles may reduce this number	**1. Cost**—Just one of the most expensive components of the hardware puzzle: Waymo Lidar Systems, costs $8000, having previously cost $75,000
2. Time saving—The connection between vehicles (V2V) will allow the calculation of more efficient and less congested transit routes	**2. Privacy and data protection**—Concerns over unauthorized use of information, access to personal data from the computer of the car, among others. Absolute control over routes and destinations, which may compromise the user's privacy
3. Pursuit of free parking spaces—The connection between the vehicle and the infrastructure (V2I) may determine free parking spaces, facilitating the routes	**3. Unemployment and unawareness of driving tasks**—Theoretically it may eliminate jobs in the transport sector (for example: taxis, carriers). Some individuals may fail to learn the rules of the road and the conventional driving of the car
4. Mobility for individuals unable to drive—This may include elderly individuals who live in communities dependent on cars, individuals with motor disabilities or other types of disabilities	**4. Possible malfunction and adaptability**—If other technology malfunctions, such as traffic signals, the vehicle interprets that they do not exist, which may cause accidents. For example, if there is a problem on the road (a natural or human event) will the autonomous car be able to interpret signals and unknown events?
5. Traffic events are the responsibility of the vehicle and not necessarily of the passengers—This is an advantage for the users but not exactly for the manufacturers or political and legislative authorities	**5. Lack of legislation**—There is a need to adjust traffic legislation to integrate autonomous vehicles. This is perhaps the greatest challenge that the governors of future societies will have to face

Model adapted from the source: https://medium.com/swlh/the-race-to-fully-autonomous-cars-8212ff73aad

1.1 Interrupting the Future: What to Do to the Pleasure of Driving?

A vast number of car users express concerns regarding the (announced) speed of technological change, which includes advancements in the connectivity of vehicles, intelligent technology and differentiated motorizing systems (Hirsh et al. 2016; Richtel and Dougherty 2015).

Table 2 Automation levels (SAE)

Levels of Automation	Who does what, when
Level 0	The human driver does all the driving.
Level 1	An advanced driver assistance system (ADAS) on the vehicle can sometimes assist the human driver with either steering or braking/accelerating, but not both simultaneously.
Level 2	An advanced driver assistance system (ADAS) on the vehicle can itself actually control both steering and braking/accelerating simultaneously under some circumstances. The human driver must continue to pay full attention ("monitor the driving environment") at all times and perform the rest of the driving task.
Level 3	An Automated Driving System (ADS) on the vehicle can itself perform all aspects of the driving task under some circumstances. In those circumstances, the human driver must be ready to take back control at any time when the ADS requests the human driver to do so. In all other circumstances, the human driver performs the driving task.
Level 4	An Automated Driving System (ADS) on the vehicle can itself perform all driving tasks and monitor the driving environment—essentially, do all the driving—in certain circumstances. The human need not pay attention in those circumstances.
Level 5	An Automated Driving System (ADS) on the vehicle can do all the driving in all circumstances. The human occupants are just passengers and need never be involved in driving.

Source https://www.nhtsa.gov/technology-innovation/automated-vehicles-safety

Moreover, the experience[1] of driving is appreciated by a high number of car users. It is difficult to describe the attraction of driving a vehicle, but human beings seem to appreciate their dominance over the machine and the "speed" factor. Some people enjoy "aggressive driving", others appreciate "smoother driving", others like the exterior design, the luxury of the interior design and the feeling of "power behind the wheel", the freedom of movement, of defining at what pace they intend to enjoy the views and the "passage of distance over time". Some describe driving as a form of "therapy", such as driving out of the city, on a less crowded road, listening to their favorite music. There are many emotional and psychological reasons for which individuals believe driving to be pleasurable (Chihuri et al. 2016; Haustein and Siren 2014; Helander et al. 2013).

In Business Insider magazine, Yarow (2009) wrote: «Guess What? People Like to Drive, Even If It's Irrational».

The majority of the American and English population (for example) "loves" to drive, even when the cost of trips are higher (when compared to other means of transport) and have several mobility alternatives that allow them to avoid long queues

[1] It is believed that a higher percentage of drivers like to drive (and feel control over the car) instead of being driven/transported. In this context, it is worth noting that the adult population, in general, if they were given a choice, would choose a conventional car rather than an autonomous car (Barkenbus 2018).

Fig. 3 Ford Mustang Sheby GT355, 2016: exterior design. Image rights. Image rights: copyright by the Author

in everyday traffic (Express Jornal 2018; Yarow 2009). In these two countries alone, for example, each family has at least two cars, per household. In the English case, the number of families with two cars actually exceeds the number of families with two children by almost two million—5.1 million compared to 3.2 million (Express Jornal 2018).

Many car brands are oriented towards the profile of the driver who experiences the "pleasure of driving", as is the example of the prestigious "Ford" brand, "Mustang" model. This brand has conducted successive updates of the GT version, over the years (versions from: 2014, 2015, 2016, 2017 and 2018), without detracting from the essence of the manual driving model. Therefore, from the point of view of both exterior design (Fig. 3) and interior design (Fig. 4), there is a concern to maintain the identity of the product. Thus, the "iconic" manual gearbox remains, while the dashboard and central console exhibit small nuances and necessary technological updates (common to other brands in the segment).

Thus, there is to be expected a cohabitation between different technological levels of cars for several years, which will have obvious implications in terms of their interior design.

1.2 Trends: The Display Interfaces and Usage Expectations

Ahlemann and Gerling (2016) state that «currently, even the most intelligent cars are very "dumb", at least when it comes to recognizing the differences between people». These authors argue that intelligent cars are programmed to do the same things for every person – whether they like it or not. For example, they lock the doors, even if one does not want them to be locked; the car does not start unless the driver fastens the seatbelt; they direct the driver (with the GPS) through routes that may not make sense for drivers with local knowledge. They also argue that the automotive industry has created "wonders", but that they are "inflexible wonders", since the intelligent

Fig. 4 Ford Mustang Sheby GT355, 2016: interior design. Image rights: copyright by the Author

car behaves in the same manner «whether it be a young person or a grandmother behind the wheel».

Currently, there is a trend in the automotive sector towards incorporating a high number of devices, applications and features, of a nature often considered complex by the user (see examples in Fig. 5).

Therefore, not only will the informative display interfaces, such the Dashboard and Central Console: Radio, GPS, Telephone, Audio, (among others), incorporate

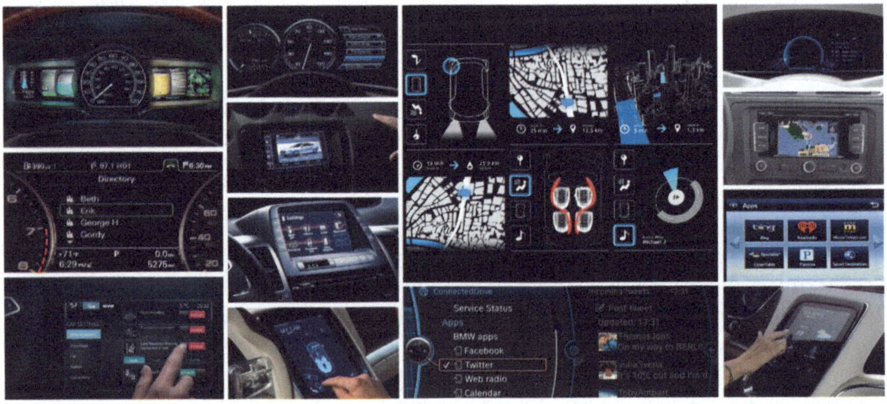

Fig. 5 Different car interfaces (displays: of the dashboard and central console) with interactions considered to be complex by the user. Image rights: https://medium.com/@autoustwo/the-near-future-of-incar-hmi-5b34a76fc7a

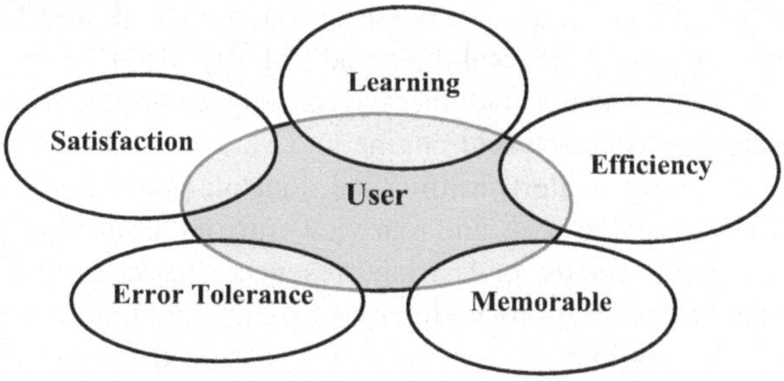

Fig. 6 User expectations regarding the design of the interfaces within the car. Image rights: copyright by the Author

more information, requiring more mental effort and memory, but they may also distract the driver when used (Soares et al. 2014; Wege and Trent 2013; Prem et al. 2011; Gelau and Schindhelm 2010; Pickering et al 2007).

If we consider the differences in the cognitive and perceptual skills between the young and the elderly, which have implications on the tasks within the car, it is expected that older drivers will experience greater difficulties with technological tasks (Caruso et al. 2016; Kim et al. 2012; James 2012).

Indeed, this scenario will increase cognitive requirements, particularly due to the way the devices need to be executed (Montes et al. 2016; Kim et al. 2012), because: (i) they demand an excessive amount of steps in order to conclude; (ii) the steps to reach a certain option are different each time, forcing the drivers to search in their system and; (iii) each solution requires a different action from the driver (for example, adjust the temperature of the interior environment that is controlled by a sliding-type controller, while redirection requires single touches).

The future connectivity between vehicles (V2V) and between vehicles and infrastructures (V2I), of technologically advanced cars, will tend to increase (even more) the complexity of the devices, due to the vast amount of information that will be included, or will be available, for the drivers and passengers of the car. Therefore, it is understood that the design of interfaces, with informative display (as a generalized trend of the automotive industry), for the control of parameters within and outside the vehicle, needs further research. The design of the interfaces with drivers should, thus, be easily understood, and the tasks memorized quickly, as a consequence of the everyday experience of the user (Fig. 6).

1.3 Complexity of Features and Resources

Currently, it has been found that some drivers do not use all features available to them in their vehicles and, in some cases, there is even a lack of knowledge on how to perform certain tasks or how to access certain information, such as adjustments and configurations (for example: elevate/lower or recline a seat (Karali et al. 2017); consult or mark a mileage log on the dashboard; configure the radio, clock, GPS,

Cruise Control, among other situations) (Vink and Lips 2017; Karali et al. 2017; Waller et al. 2015; Rahiman and Zafariq 2013).

If we consider and characterize, for example, a state of the art luxury car, with a differentiated hybrid engine and various driving modes, with adaptive autopilot, driver fatigue alert, traffic lane control, parking assistance that parks on its own, programmable seats and rear view mirrors, blind spot alerts, GPS, collision sensors, voice commands, pedestrian presence sensors, automatic road lighting, and with a panel full of buttons with various innovative features, the following question can be asked: Would we be able to identify all these features inside the car? Would we know how to use all of them intuitively?

This obviously results in the need to consult the vehicle manuals in order to understand certain parameters of use, configuration and even for the interpretation of the possible symbols presented on the dashboard – as an informative display (Andreev 2018; AAA 2017).

In 2015, the US National Security Council and a group of researchers specialized in vehicle safety from the University of Iowa created the website MyCarDoesWhat.org, as a national campaign to help educate drivers about new vehicle safety technologies, designed to help prevent collisions. These technologies range from increasing the stability and control of cars to the systems of information for issuing warnings about collision threats, for subsequent automatic intervention and, thus, avoid or reduce the severity of possible collisions.

The organization MyCarDoesWhat.org has a large number of publications with summarized explanations of various features and technologies present in contemporary cars, as well as in some future developments, with the purpose of clarifying questions and reducing the unknowns among users/consumers. The name chosen for this organization—*«My car does what?»*—in the form of a question, is very elucidative of the problem here identified.

The new generations of vehicles are incorporating countless useful technology to suppress many shortcomings of car driving, but without it being explained, or reflected in the necessary usability of the many driver profiles (Eby et al. 2018). The quantification and technological selection of the inside of cars (currently in commercialization) do not always seem to determine a rigorous, selective and well-founded choice, based on expectations and patterns of use, particularly among the elderly population, and more value-focused assessments for future generations of vehicles (Eby et al. 2016, 2018).

At the same time, a large number of drivers are unaware of how to select a car model that is technologically equipped and adjusted to their driving and usability needs (Stockburger 2018; Eby et al. 2016), thus, there is a proliferation of various applications and consulting services to help select vehicles (Caranddriver.com 2018; Carsguide.com 2018; Vincent 2016). On the other hand, the automotive industry only highlights and divulges new features and technological developments, without the final consumer being informed about the reason for that choice or the respective inclusion in the various models (Heineke et al. 2017; Mohr et al. 2016).

1.4 Conclusions: Future Challenges for the Car – Inclusive Design and Customization

The future of partially automated vehicles and fully automated vehicles "is closer than ever" (MyCarDoesWhat.org 2018; TheHartford.com 2013), although there are still many constraints and questions (economic, legal, of technological feasibility and reliability, consumer/user acceptance, uncertainties about the sector—market and its segments, volatility of social and cultural patterns, costs, environmental issues, safety, maintenance, among others) that will inevitably delay the announced advancements (Viereckl et al. 2016). It is currently important for the designer to analyze this problem in order to point out future solutions for the car and, in particular, for its interior.

Autonomous vehicles may become particularly useful for people with special needs and for elderly drivers. The latter, who are not classified as population groups with "special needs", nonetheless, belong to an important (growing) segment of the world population, whose specificities and characteristics justify segmented research, given that the aging process is related to an increase of illnesses and mutations that produce adverse effects on the physical and cognitive condition in general and, in particular, on the ability to drive vehicles (Quach et al. 2017; Tonetto and Desmet 2016; Karali et al. 2016, 2017). In addition to the adverse effects of the illnesses themselves, which may induce physical limitations analogous to physical limitations due to disabilities in younger groups, there also needs to be consideration of the possible effects of the medication taken to control health problems (Fernandes et al. 2017; Karali et al. 2017). Some types of medication may hinder the ability to drive. If, on the one hand, this argument validates the potential interest of level 5 technology, on the other hand, it cannot be denied that the design process and the conception of this type of vehicles must go beyond the incorporated technology. For example, the design of autonomous vehicles must consider the physical changes that result from the aging of this potential group of users, such as, firstly, the access to the vehicle – the process of entering and exiting. The car may become an extension of the home itself, from which comfort is expected.

Simultaneously, with regard to the mobility of the individual car by special needs drivers and the elderly, in particular, several questions demand answers. Governments, regulatory agencies, car manufacturers, healthcare providers, families and society in general must be interested in obtaining answers related to the issues of mobility, inclusive car design, safety, comfort, autonomy, promotion of social inclusion and active aging (AAA 2017; ACEA 2017; Kaas et al. 2015; Nitsch et al. 2014).

The answers to these questions must be based on solid scientific research. Some data is already becoming available, through searches about "aging and driving", but several questions remain unanswered and without specific solutions in terms of the interior design of the car (Fernandes et al. 2017; Karali et al. 2017).

For vehicles with technological levels between 1 and 4, there needs to be a rigorous assessment of the technology to be incorporated. Consequently, it is imperative to introduce changes to vehicles and incorporate technologically advanced assistance systems, but easy to use for the driver (for example, to compensate for age-related shortcomings) (Trübswetter and Bengler 2013).

Generally speaking, the physical interfaces of cars, that is, all the instruments the driver needs to be able to drive a car, tend to incorporate systems that provide increasingly sophisticated and complex information to the driver, in some cases, possibly requiring a learning process (Reagan and Bliss 2013). Notwithstanding the usefulness of more information, the complex interfaces demand greater attention from the driver in order to handle them, which may negatively affect the attention required for the task of driving. Thus, there should be consideration of the psychology and physiology of the diversity of users in the design process, as well as the possibility of personalization/customization for the benefit of the usability of the vehicle interfaces, to make the systems more effective, efficient and satisfactory. Lastly, the simplification and uniformization of some patterns of use should also be considered.

It is also important to improve the design of the main physical interfaces with the driver, namely the aspects that are considered critical for the driver (Trübswetter and Bengler 2013). For example, among the elderly: the position of the front seat is the main obstacle to entering and exiting the vehicle; the peripheral view and rear view are limited, mainly, by the rear and windshield pillars of the vehicle; other constraints are related to construction aspects in terms of the space of the passenger compartment, the driving assistance system, among others (Nitsch et al. 2014).

Despite the significant evolution and improvements of cars, in general, the interior design can still improve in terms of: the support and guideway of the front seats (Karali et al., 2016, 2017); the instruments and elements of command (Weinberg and Müller 2011); night, peripheral and rear views as well as "blind spots" (Häne et al. 2017; Chun et al. 2013; Brown et al. 2010); entering and exiting the vehicle (Shippen and May 2016); systems to support safe driving (UMTRI 2015; Reimer 2014; de Winter et al. 2014; Strand et al. 2011); emergency braking system (Blower 2014; Wu et al. 2013; Warner et al. 2010); navigation system and the parking assistance system (Douissembekov et al. 2014; Rush et al. 2014; Tachibana 2011), among many others.

It is unquestionable that the improved performance and the autonomy of driving may be achieved through the design. The redesigning of handles, knobs and buttons is important to facilitate their handling, namely by people with physical shortcomings (e.g. arthritis). Other characteristics of the design of the car, such as: the lower limits of the doors; lumbar supports; rearview mirrors; adjustable seats; steering wheel; and the redesign and repositioning of panels and intelligent communication devices, may be particularly useful for drivers (Fernandes et al., 2017; Karali et al. 2017).

Some problems have already been identified (for example: among elderly drivers), such as the difficulty in rotating one's neck and upper body, difficulty in looking to the side and to the rear, thus the interfaces with drivers may incorporate devices to support safe maneuvering. These issues lack in-depth studies for a correct assessment of the problems faced by elderly drivers, which (in many cases) are common to other groups of drivers (Fernandes et al. 2017; Karali et al. 2017).

The vehicle control devices are those that convey information about the functioning cycle of the engine and aid the control and immobilization of the vehicle. These devices are important in that they increase the driver's sensory abilities. For example, by providing visual and sound information, it is possible to control numerous aspects of driving, such as the night vision systems, rearview mirrors, reversing and side view cameras, collision warning sensors and "adaptive cruise control" systems, which adjust the acceleration/deceleration of the vehicle and regulate the distance to other vehicles.

The driving assistance devices provide traffic and navigation information. In turn, the information/communication devices for drivers aid communication and support comfort.

The human body form and posture suffer changes during the aging process (Karthaus and Falkenstein 2016). These changes are pointed out as the main causes of constraints associated with entering and exiting vehicles, the difficulties in accessing the driver's seat caused by loss of lower limb capacity, as well as some weaknesses in muscles, joints, the spine and neck, with these constraints being analogous to physical changes following other causes and pathologies.

Thus, it is not surprising that there are a number of adaptations to cars improvised by the drivers/users themselves, as well as an emerging market growth for the sale of components and accessories to adapt cars so as to increase the standards of comfort.

Therefore, it is justified to design a driver's seat adjusted to these limitations, with the ability to traverse, rotate, as well as laterally and vertically regulate. Furthermore, the position of the seatbelt, steering wheel and airbags should be adapted to the weight, height and stature of the person occupying the seat. These measures may, in the event of collision, reduce the severity of injuries in individuals (Karali et al. 2017; OECD/ITF 2014; Sixsmith 2013).

The interior design of the vehicle (dashboard and user interface devices) should also take into account the changes in vision that occur over time, particularly the loss of sensitivity to contrast, thus the contrast between display elements should be maximized. The displays (digital or analog indicators) should exhibit a different brightness level from the display background. It should also provide appropriate illumination of the information in order to avoid reflectivity. The introduction of digital screens in vehicles (as a general trend in the sector) should allow the adjustment of information display properties, according to the needs of each driver, which, in the case of the elderly, should display the information in a larger and simplified format (Beck et al. 2017; Pschenitza and Unger 2015). The sound systems that convey information regarding navigation, communication, and information devices should also be easily adjustable to the driver's hearing condition (Kang and Momtaz 2018).

Some studies emphasize the application of advanced driver-assistance systems (ADAS) as an important contribution for solving certain types of limitations (Trübswetter and Bengler 2013), when paired with an inclusive design (Paris et al. 2014; Trübswetter and Bengler 2013; Goddard and Nicolle 2012). The expected benefits of the new technologies should, however, be considered with great caution. New technologies must be assessed over time, under conditions as close as possible to the real conditions of use, and with groups that represent potential users.

The sustainable vehicles of the future, with new technologies, should include friendly interfaces, adjusted to the different users, which minimize the factors potentiating accidents and facilitate mobility (Study 2015; Mehler et al. 2014; Neumann et al. 2010; Bernsen and Dybkjaer 2009; Malanowski et al. 2008).

The future generation of cars should be conceived under the premise of more inclusive, interactive and safe mobility, which would answer to the specificities and limitations of the population (for example; seniors, individuals with physical, motor disabilities, etc.) and, simultaneously, be useful for other groups of users (Fernandes et al. 2017; Goddard and Nicolle 2012).

The possibility to customize or personalize a vehicle according to the profile, needs and personal taste of its user is a current added challenge of the future of the car.

References

AAA (2017) Smart features for older drivers. SeniorDrivingAAA.com

ACEA—European Automobile Manufacturers Association (2017) Economic and market report. EU Automotive Industry, Quater 3 2017, December

Ahlemann D, Gerling W (2016) The autonomous frontier. Strategy, pp. 54–60. https://www.strategyand.pwc.com/media/file/Connected-car-report-2016.pdf. Accessed 20 July 2018

Anderson J, Kalra N, Stanley K, Sorensen P, Samaras C, Oluwatola O (2016) Autonomous vehicle technology: a guide for policymakers. RAND Corporation. ISBN 978-0-8330-8398-2

Andreev N (2018). Car dashboard warning lights explained. Cars have an array of dashboard warning lights to let drivers know when something's wrong. But do you know what they mean? [Web log post]. Confused.com. https://www.confused.com/on-the-road/maintenance/dashboard-warning-lights-explained. Accessed 20 July 2018

Baker EH, Crusius D, Fischer M, Gerling W, Gnanaserakan K, Kerstan H, Kuhnert F, Kusber J, Mohs J, Schulte M Seyfferth J (2016) Connected car report 2016: opportunities, risk, and turmoil on the road to autonomous vehicles. Strategy. https://www.strategyand.pwc.com/media/file/Connected-car-report-2016.pdf. Accessed 20 July 2018

Barkenbus J (2018) People aren't ready for self-driving cars [web log post]. Elijah Nouvelage—Reuters. https://www.citylab.com/transportation/2018/01/autonomous-vehicles-consumer-backlash/549650/. Accessed 8 May 2018

Beck D, Lee M, Park W (2017) A comparative evaluation of in-vehicle side view displays layouts in critical lane changing situation. Ergonomics 60(12):1682–1691. https://doi.org/10.1080/00140139.2017.1343958

Bernsen NO, Dybkjaer L (2009) Exploring natural interaction in the car. Philippine Sociol Rev 54(5). https://doi.org/10.3860/psr.v54i0.938

Blower D (2014) Assessment of the effectiveness of advanced collision avoidance technologies. Report No. UMTRI-2014-3. University of Michigan Transportation Research Institute, Ann Arbor, MI. http://hdl.handle.net/2027.42/102534. Accessed 8 May 2018

Bontrager B (2018) The race to fully autonomous cars [web log post]. The startup. https://medium.com/swlh/the-race-to-fully-autonomous-cars-8212ff73aad. Accessed 8 May 2018

Brown T, He Y, Roe C, Schnell T. (2010). Is more better?—Night vision enhancement system's pedestrian warning system modes and older drivers. Ann Assoc Adv Autom Med 54:343–350. PMID: 21050616, PMCID: PMC3242535

Caranddriver.com (2018) What's the right new vehicle for me? https://www.caranddriver.com/features/whats-the-right-new-vehicle-for-me. Accessed 8 May 2018

Carsguide.com (2018). What car should I buy? https://www.carsguide.com.au/simple-search. Accessed 28 Dec 2018

Caruso G, Camere S, Bordegoni M (2016) System based on abstract prototyping and motion capture to support car interior design. Comput-Aided Des Appl 13(2):228–235. https://doi.org/10.1080/16864360.2015.1084194

Chihuri S, Mielenz TJ, DiMaggio CJ, Betz ME, DiGuiseppi C, Jones VC, Li G (2016) Driving cessation and health outcomes in older adults. J Am Geriatr Soc 64(2):332–341. https://doi.org/10.1111/jgs.13931

Chun J, Lee I, Park G, Seo J, Choi S, Han SH (2013) Efficacy of haptic blind spot warnings applied through a steering wheel or seatbelt. Transp Res Part F: Traffic Psychol Behav 21:213–241. https://doi.org/10.1016/j.trf.2013.09.014

de Winter JCF, Happee R, Martens MH, Stanton NA (2014) Effects of adaptive cruise control and highly automated driving on workload and situation awareness: a review of the empirical evidence. Transp Res Part F: Traffic Psychol Behav 27(Part B):196–217. https://doi.org/10.1016/j.trf.2014.06.016

Douissembekov E, Gabaude C, Rogé J, Navarro J, Michael GA (2014) Parking and manoeuvring among older drivers: a survey investigating special needs and difficulties. Transp Res Part F: Traffic Psychol Behav 26(Part A):238–245. https://doi.org/10.1016/j.trf.2014.07.011

Eby DW, Molnar LJ, Kostyniuk LP, St.Louis RM, Zanier N (2017) Characteristics of informal caregivers who provide transportation assistance to older adults. PLoS ONE 12(9):e0184085. https://doi.org/10.1371/journal.pone.0184085

Eby DW, Molnar LJ, Zakrajsek JS, Ryan LH, Zanier N, Louis RMS, Strogatz D (2018) Prevalence, attitudes, and knowledge of in-vehicle technologies and vehicle adaptations among older drivers. Accid Anal Prev 113:54–62. https://doi.org/10.1016/j.aap.2018.01.022

Eby DW, Molnar LJ, Zhang L, St. Louis RM, Zanier N, Ko Styniuk LP, Stanciu S (2016). Use, perceptions, and benefits of automotive technologies among aging drivers. Injury Epidemiology 3(28):1–20. https://doi.org/10.1186/s40621-016-0093-4

Express Jornal (2018) Does your household have two cars? Two-car homes SOAR and this is why. https://www.express.co.uk/life-style/cars/912000/two-cars-household-soar-LV-insurance-competition-holiday-commuting-family-trip. Accessed 20 Sept 2018

Fernandes SCF, Esteves JL, Simoes R (2017) Characteristics and human factors of older drivers: improvement opportunities in automotive interior design. Int J Veh Des 74(3):167–203. https://doi.org/10.1504/IJVD.2017.086418

Gelau C, Schindhelm R (2010) Enhancing the occlusion technique as an assessment tool for driver visual distraction. IET Intel Transp Syst 4(4):346–355. https://doi.org/10.1049/iet-its.2009.0115

Goddard N, Nicolle CA (2012) What is good design in the eyes of older users? Designing inclusive systems: designing inclusion for real-world applications. Springer, London, pp 175–184. ISBN 9781447128663

Häne C, Heng L, Lee GH, Fraundorfer F, Furgale P, Sattler T, Pollefeys M (2017) 3D visual perception for self-driving cars using a multi-camera system: calibration, mapping, localization, and obstacle detection. Image Vis Comput 68:14–27. https://doi.org/10.1016/j.imavis.2017.07.003

Haustein S, Siren A (2014) Seniors' unmet mobility needs—how important is a driving licence? J Transp Geigrraphy 41:45–52. https://doi.org/10.1016/j.jtrangeo.2014.08.001

Helander MG, Khalid HM, Lim TY, Peng H, Yang X (2013) Emotional needs of car buyers and emotional intent of car designers. Theor Issues Ergonomics Sci 14(5):455–474. https://doi.org/10.1080/1463922X.2012.656152

Heineke K, Möller T, Padhi A, Tschiesner A (2017) How mobility players can compete as the automotive revolution accelerates. In: McKinsey & Company McKinsey, automotive & assembly insights. https://www.mckinsey.com/industries/automotive-and-assembly/our-insights/how-mobility-players-can-compete-as-the-automotive-revolution-accelerates. Accessed 8 May 2018

Hirsh E, Jullens J, Wilk R, Singh A. (2016) 2016 Auto industry trends, automakers and suppliers can no longer sit out the industry's transformation. Strategy. https://www.strategyand.pwc.com/media/file/2016-Auto-Trends.pdf. Accessed 20 July 2018

James T (2012) Vehicle technology and the older driver. E&T—Engineering and Technology Magazine. http://eandt.theiet.org/magazine/2012/10/car-design-for-all-ages.cfm?origin=EtOtherStories. Accessed 20 July 2018

Kaas H, Tschiesner A, Wee D, Kässer M (2015) How carmakers can compete for the connected consumer. In: McKinsey & Company, automotive & assembly insights. https://www.mckinsey.com/industries/automotive-and-assembly/our-insights/how-carmakers-can-compete-for-the-connected-consumer. Accessed 8 May 2018

Kang MW, Momtaz SU (2018) Assessment of driver compliance on roadside safety signs with auditory warning sounds generated from pavement surface—a driving simulator study. J Traffic Transp Eng (English Edition) 5(1):1–13. https://doi.org/10.1016/j.jtte.2017.09.001

Karthaus M, Falkenstein M (2016) Functional changes and driving performance in older drivers: assessment and interventions. Geriatrics 1(2):12. https://doi.org/10.3390/geriatrics1020012

Karali S, Gyi DE, Mansfield NJ (2016) Driving a better driving experience: a questionnaire survey of older compared with younger drivers. Ergonomics 60(4):1–8. https://doi.org/10.1080/00140139.2016.1182648

Karali S, Mansfield NJ, Gyi DE (2017) An approach to vehicle design: in-depth audit to understand the needs of older drivers. Appl Ergonomics 58:461–470. https://doi.org/10.1016/j.apergo.2016.07.017

Kim S, Hong JH, Li KA, Forlizzi J, Dey AK (2012) Route guidance modality for elder driver navigation. In: lecture notes in computer science (including subseries lecture notes in artificial intelligence and lecture notes in bioinformatics), vol 7319 LNCS, pp 179–196. https://doi.org/10.1007/978-3-642-31205-2_12

Kimmel S (2015) Driving innovation: a recap of the 2015 transportation research board annual meeting and connected-autonomated vehicle technology [web log post]. IEEE USA Insight. https://insight.ieeeusa.org/articles/driving-innovation-a-recap-of-the-2015-transportation-research-board-annual-meeting-and-connected-autonomated-vehicle-technology/. Accessed 8 May 2018

Malanowski N, Ozcivelek R, Cabrera M (2008) Active ageing and independent living services: the role of information and communication technology. Eur Commun 39. http://www.vditz.de/fileadmin/media/publications/pdf/Active_Ageing.pdf. Accessed 20 July 20

Mehler B, Reimer B, Lavallière M, Dobres J, Coughlin JF (2014) Evaluating technologies relevant to the enhancement of driver safety. AAA Foundation for Traffic Safety, Washington, DC

Mohr D, Kaas H, Gao P, Wee D, Möller T (2016) Automotive revolution—perspective towards 2030: How the convergence of disruptive technology-driven trends could transform the auto industry. In: McKinsey & Company, advanced industries. https://www.mckinsey.com/. Accessed 8 May 2018

Montes SA, Introzzi IM, Ledesma RD, López SS (2016) Atención selectiva y propensión al error en la conducción: estudio mediante una tarea de búsqueda visual conjunta. Avances en Psicología Latinoamericana 34(2):195–203. https://doi.org/10.12804/apl34.2.2016.01

Neumann I, Cocron P, Franke T, Krems JF (2010) Electric vehicles as a solution for green driving in the future? A field study examining the user acceptance of electric vehicles. In Krems JF, Petzoldt T, Henning (Hrsg). Proceedings of the European conference on human interface design for intelligent transport systems, Berlim, Germany, April 29–30, Humanist Publications, Lyon, pp S445–S453

Nitsch M, Lambacher, O. & Howe, J (2014) Cars for elderly drivers. mopact, mobilising the potential of active ageing in Europe. Braunschweiger Gerontological Series, 2. https://doi.org/10.15491/JKH297182

OECD/ITF—International Transport Forum (2014) Road safety annual report 2014. https://doi.org/10.1787/23124571. https://www.oecd-ilibrary.org/transport/road-safety-annual-report-2014_irtad-2014-en. Accessed 10 Sep 2015

Paris J-C, Bellet T, Cour M, Marin-Lamellet C, Deleurence P, Moreau F, Boverie S, Andre J-M, Claverie B (2014) Driving assistances for senior drivers: a human centered design approach. In: Paper No. 18291. Transport research arena—5th conference, April 2014, France

Pickering, CA, Burnham KJ, Richardson MJ (2007) A review of automotive human machine interface technologies and techniques to reduce driver distraction. In 2nd IET international conference on system safety, 22–24 Oct 2007, London, UK, pp 223–228. https://doi.org/10.1049/cp:20070468

Prem B, Arora K, Manager P, Hmi A, Verma M, Biz A, Manager D (2011) Designing Next-generation automotive center consoles. In: EE times design. http://eetimes.com/design. Accessed 8 May 2018

Pschenitza M, Unger T (2015) Field of vision of modern cars—a study to improve the evaluation of car geometries based on real world accident scenarios documented in the ADAC Accident research. Berichte der Bundesanstalt fuer Strassenwesen. Unterreihe Fahrzeugtechnik 102:14. http://bast.opus.hbz-nrw.de/volltexte/2015/1348/pdf/F102_ESAR_komplett_CD.pdf%5Cn, https://trid.trb.org/view/1359582. Accessed 8 May 2018

Quach A, Levine ME, Tanaka T, Lu AT, Chen BH, Ferrucci L, Ritz B, Bandinelli S, Neuhouser ML, Beasley JM, Snetselaar L, Wallace RB, Tsao PS, Absher D, Assimes TL, Stewart JD, Li Y, Hou L, Baccarelli AA, Whitsel EA, Horvath S (2017) Epigenetic clock analysis of diet, exercise, education, and lifestyle factors. Aging (Albany NY) 9(2):419–446. https://doi.org/10.18632/aging.101168

Rahiman W, Zafariq Z (2013) An overview of development GPS navigation for autonomous car. In: 8th IEEE conference on industrial electronics and applications (ICIEA), 2013. IEEE Publisher, Melbourne, VIC, Australia. https://doi.org/10.1109/iciea.2013.6566533

Rakotonirainy A, Steinhardt D (2009) In-vehicle technology functional requirements for older drivers. In: Schmidt A, Dey A, Seder T, Kern O (eds) Proceedings of the first international conference on automotive user interfaces and interactive vehicular applications, automotive UI 2009, 21–22 Sept 2009, Essen, Germany, pp 27–33

Reagan IJ, Bliss JP (2013) Perceived mental workload, trust, and acceptance resulting from exposure to advisory and incentive based intelligent speed adaptation systems. Transp Res Part F 21:14–29. https://doi.org/10.1016/j.trf.2013.07.005

Reimer B (2014) Driver assistance systems and the transition to automated vehicles: a path to increase older adult safety and mobility? Public Policy Aging Rep 24:27–31. https://doi.org/10.1093/ppar/prt006

Richtel M, Dougherty C (2015) What we learned designing a car user experience [web log post]. The New York Times. https://www.nytimes.com/2015/09/02/technology/personaltech/google-says-its-not-the-driverless-cars-fault-its-other-drivers.html. Accessed 8 May 2018

Rusch ML, Schall MC Jr, Lee JD, Dason JD, Rizzo M (2014) Augmented reality cues to assist older drivers with gap estimation for left-turns. Accid Anal Prev 71:210–221. https://doi.org/10.1016/j.aap.2014.05.020

Shippen J, May B (2016) Constitutive kinematic modes and shapes during vehicle ingress/egress. Appl Ergonomics 56:127–135. https://doi.org/10.1016/j.apergo.2016.03.017

Sixsmith A (2013) Technology and the challenge of aging. In: Sixsmith A, Gutman G (eds) Technologies for active aging. international perspectives on aging, vol 9. Springer, Boston, MA. https://doi.org/10.1007/978-1-4419-8348-0_2

Soares F, Simões P, Simoes R (2014). Automotive central console interface design. In: Proceedings of the 5th international conference on applied human factors and ergonomics, 19–23 July 2014, Kraków, Poland, pp 2047–2058. ISBN 9781495121036

Stockburger J (2018) Top 25 new cars for senior drivers [web log post]. Consumerreposrts.org. https://www.consumerreports.org/elderly-driving/top-25-new-cars-for-senior-drivers/. Accessed 8 May 2018

Strand N, Nilsson J, Karlsson ICM, Nilsson L (2011) Interaction with and use of driver assistance systems: a study of end-user experiences. In: Proceedings of the 18th ITS world congress. ITS America, Washington, DC

Study AL (2015) Keeping older adults driving safely: A research synthesis of advanced in-vehicle technologies. The National Academies of Sciences, Engineering, and Medicine. https://www.aaafoundation.org/sites/default/files/SeniorsAndTechnologyReport.pdf. Accessed 8 May 2018

Tachibana H (2011) The development of a parking support system considering driving characteristics. In: Proceedings of the SAE 2011 world congress and exhibition. SAE International, Warrendale, PA

TheHartford.com (2013) Top technologies for mature drivers: consumer insights. http://www.thehartford.com/mature-market-excellence/consumer-technology-insights. Accessed 8 May 2018

Tonetto LM, Desmet PMA (2016) Why we love or hate our cars: a qualitative approach to the development of a quantitative user experience survey. Appl Ergonomics 56:68–74. https://doi.org/10.1016/j.apergo.2016.03.008

Trübswetter N, Bengler K (2013) Why should I use ADAS? Advanced driver assistance systems and the elderly: knowledge, experience, and usage barriers. In: Proceedings of the seventh international driving symposium on human factors in driver assessment, training, and vehicle design. University of Iowa, Ames, IA

UMTRI—University of Michigan Transportation Research Institute (2015) Curve-speed warning systems. University of Michigan Transportation Research Institute, Ann Arbor, MI. http://www.umtri.umich.edu/our-focus/curve-speed-warning-systems. Accessed 8 May 2018

Viereckl R, Ahlemann D, Koster A, Hirsh E, Kuhnert E, Mohs J, Fischer M, Gerling W, Gnanasekaran K, Kusber J, Stephan J (2016). Connected car report 2016: opportunities, risk, and turmoil on the road to autonomous vehicles. Strategy. https://www.strategyand.pwc.com/reports/connected-car-2016-study. Accessed 8 May 2018

Vincent JM (2016) What kind of car do you need? [web log post]. Cars US News, Car Rankings & Advice. https://cars.usnews.com/cars-trucks/how-to-choose-the-right-car-for-you#need. Accessed 8 May 2018

Vink P, Lips D (2017) Sensitivity of the human back and buttocks: the missing link in comfort seat design. Appl Ergonomics 58:287–292. https://doi.org/10.1016/j.apergo.2016.07.004

Waller S, Bradley M, Hosking I, Clarkeson PJ (2015) Making the case for inclusive design. Appl Ergonomics 46(Part B):297–303. https://doi.org/10.1016/j.apergo.2013.03.012

Warner HW, Özkan T, Lajunen (2010) Drivers' propensity to have different types of intelligent speed adaptation installed in their cars. Transp Res Part F 13(3):206–214. https://doi.org/10.1016/j.trf.2010.04.005

Wege CA, Trent WV (2013) Distraction and inattention prevention by combining behaviour-based safety with advanced driver assistance systems. IET Digital Library, pp 215–237. https://doi.org/10.1049/pbsp009e_ch11

Weinberg G, Müller C (2011) Multimodal input in the car, today and tomorrow. IEEE Multimedia 18(1):98–103. https://doi.org/10.1109/MMUL.2011.14

Wood SP, Chang J, Healy T, Wood J (2012) The potential regulatory challenges of increasingly autonomous motor vehicles. Santa Clara Law Rev 52(4):1423–1502. ISSN 0146-0315

Wu J, Subramanian R, Craig M, Starnes M, Longthorne A (2013) The effect of earlier or automatic collision notification on traffic mortality by survival analysis. Traffic Injury Prevent 14:50–57. https://doi.org/10.1080/15389588.2013.799279

Yang J, Coughlin JF (2014) In-vehicle technology for self-driving cars: Advantages and challenges for aging drivers. Int J Autom Techn 15(2):333−340. https://doi.org/10.1007/s12239-014-0034-6

Yarow J (2009) Guess what? People like to drive, even if it's irrational [web log post]. Bus Insider. http://www.businessinsider.com/guess-what-people-like-to-drive-2009-9. Accessed 8 May 2018

Conclusions

The chapters present in this book a broad scope where Designers assume leading roles in Design and non-design contexts. In this context, design schools play a crucial role and can play a leading role in design education.

Adapting to new curriculum contexts presupposes changes but not only changes in the structure and content of educational programs because the model behind these programs and the didactic approaches to transfer them require applied research.

The disciplinary and design-focused curriculum should focus on peer collaboration, paving the way for multidisciplinary perspectives and changing a landscape dominated by purely personal and disciplinary arguments.

The issues of interdisciplinarity and multidisciplinarity are unavoidable both in practices and in information technology support systems.

Geographic referencing systems and augmented reality are intelligent and interactive systems that allow users to interact with space through technological interfaces. They will enable the individualization of spatial experiences, and foster new modalities of practices in space, recreating symbolic representations generating new identity and memory proposals.

The communication of experiences is central in the design communication project, even in the non-formal communication process in the transmission of commercial brand perception and where the stakeholders have an active relational component. The study of new markets is a guarantee that the trademark structure can be adapted and the strategies created in the Design project, allow the brands to present stronger identities.

The intersection of systemic practices, concepts, and technologies allows the Designer a tactical adaptation to contemporary needs. We can conclude that new technologies, technically and conceptually, offer the designer new avenues, saving him time and facilitating many of the daily tasks, becoming indispensable in many phases of the work.

D. Raposo et al. (eds.), *Perspective on Design*,
Springer Series in Design and Innovation 1,
https://doi.org/10.1007/978-3-030-32415-5

However, should not overlook the tangible experiences of reflection through design, as the Design experience can provide attentive attention and add reflective value while also playing a critical role within the creative process of Design.

This mediated capacity for reflection is capable of creating adaptive structures and evolutionary relationships as a way of creating creative challenges, interests, and bridges in speculative scenarios of man and nature. The "tangible" reflection performed by sketches of solutions via drawing or physical reflection systems allows for a tangibility in the graphic reasoning that accompanies the search for new solutions.

To propose new solutions with different innovations, is crucial to have an analytical, critical, curious, creative, and collaborative attitude. Designers should identify and understand the various existing collaborative processes associated with Design in its multiple strands.

In summary, the Project and the disciplinary knowledge of Design is a living rhizome, with multiple relationships, multidisciplinary and interdisciplinary, but "organized" according to the project. The forms of information organization raise awareness about the focus of the project, providing conceptual and practical framework for the embodiment of the Design Project practices.

The manufacturer's authorised representative in the EU is Springer Nature Customer Service Centre GmbH, Europaplatz 3, 69115 Heidelberg, Germany. If you have any concerns regarding our products, please contact ProductSafety@springernature.com

Printed and bound by CPI Group (UK) Ltd, Croydon, CR0 4YY
05/10/2026
02228145-0012